R. Mahrwald (Ed.)

Modern Aldol Reactions

Vol. 1: Enolates, Organocatalysis, Biocatalysis and Natural Product Synthesis

Also of Interest

Grubbs, R. H. (Ed.)

Handbook of Metathesis

3 Volumes

2003

ISBN 3-527-30616-1

Nicolaou, K. C., Snyder, S. A.

Classics in Total Synthesis II
More Targets, Strategies, Methods

2003

ISBN 3-527-30685-4 (Hardcover with CD-Rom)
ISBN 3-527-30684-6 (Softcover)

De Meijere, A., Diederich, F. (Eds.)

Metal-Catalyzed Cross-Coupling Reactions

Second, Completely Revised and Extended Edition

2 Volumes

2004

ISBN 3-527-30518-1

Krause, N., Hashmi, A. S. K. (Eds.)

Modern Allene Chemistry

2 Volumes

2004

ISBN 3-527-30671-4

Rainer Mahrwald (Ed.)

Modern Aldol Reactions

Vol. 1: Enolates, Organocatalysis, Biocatalysis and Natural Product Synthesis

WILEY-VCH Verlag GmbH & Co. KGaA

PD Dr. Rainer Mahrwald
Department of Organic Chemistry
Humboldt University
Brook-Taylor-Str. 2
12489 Berlin
Germany

Library of Congress Card No.: Applied for
British Library Cataloguing-in-Publication Data: A catalogue record for this book is available from the British Library.
Bibliographic information published by Die Deutsche Bibliothek
Die Deutsche Bibliothek lists this publication in the Deutsche Nationalbibliografie; detailed bibliographic data is available in the Internet at http://dnb.ddb.de

Printed in the Federal Republic of Germany.
Printed on acid-free paper.

Typesetting Asco Typesetters, Hong Kong
Printing Strauss Gmbh, Mörlenbach
Bookbinding Litges & Dopf Buchbinderei GmbH, Heppenheim

ISBN 3-527-30714-1

Foreword

Historically, the stimulus for the development of a particular reaction has been interconnected with a class of natural products whose synthesis would be greatly facilitated by the use of that particular bond construction. For example, the steroid synthesis challenges proved instrumental in the development of the Diels–Alder reaction. So too the synthesis challenges associated with the macrolide antibiotics have provided the motivation for the development of the full potential of the aldol addition reaction. R. B. Woodward's 1956 quote on the "hopelessly complex" architecture of the erythromycins was probably stimulated, in part, by the fact that the aldol reaction existed in a completely underdeveloped state five decades ago.

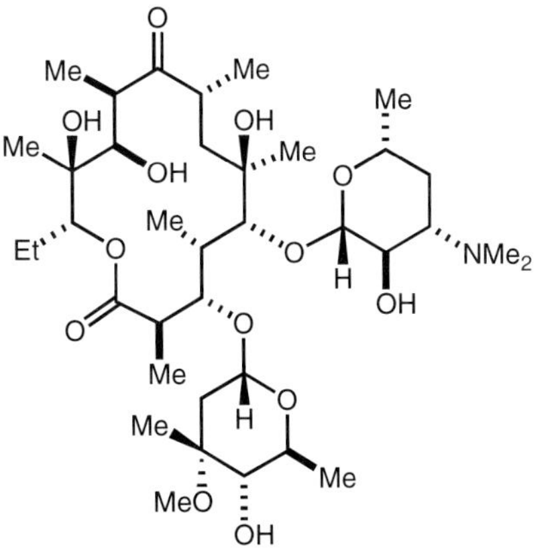

The erythromycin-A structure, as viewed by Woodward in the '50s

"Erythromycin, with all of our advantages, looks at present quite hopelessly complex, particularly in view of its plethora of asymmetric centers."

R. B. Woodward in *Perspectives in Organic Chemistry*; Todd, A. Ed.; Wiley-Interscience, New York, 1956, page 160.

The challenges associated with the development of this reaction are also embodied in the more general goals of acyclic stereocontrol that have been under active investigation for nearly twenty-five years. In these studies, the goal of understanding pi-face selectivity at trigonal carbon centers for a multitude of organic transformations has been the ultimate objective. From these research activities, a host of stereochemical models have evolved, such as the Felkin–Anh model for carbonyl addition and the Zimmermann–Traxler aldol stereochemical model for aldol diastereoselection.

The development of modern aldol reaction methods has evolved through a succession of pivotal discoveries that have advanced the whole field of stereoselective synthesis:

A. Development of enolization strategies for the formation of (*E*) and (*Z*) enolates.
B. Development of kinetic diastereoselective aldol addition variants through the discovery of optimal metal architectures [B(III), Ti(IV), Sn(II)].
C. Discovery of aldol reaction variants such as the Lewis acid catalyzed addition of enolsilanes to aldehydes (Mukaiyama aldol variant).
D. Development of chiral enolates exhibiting exceptional pi-face selectivities.
E. Development of chiral metal complexes as Lewis acid aldol catalysts.

This two-volume series on aldol addition reaction methodology brings together an up-to-date discussion of all aspects of this versatile process. The reader will gain an appreciation for the role of metal enolate architecture in aldol diastereoselectivities (Vol. I; Chapters 1–3) and for the utility of chiral metal complexes in the catalysis of the Mukaiyama aldol reaction (Vol. II; Chapters 1–3, 5). In Vol. II; Chapter 6, enantioselective catalytic processes incorporating both enolization and addition are surveyed, as is the exciting progress being made in the use of chiral amines as aldol catalysts (Vol. I; Chapter 4). This highly active area of research will continue to develop ever more versatile chiral catalysts and stereochemical control concepts.

Students and researchers in the field of asymmetric synthesis will greatly profit from the contributions of this distinguished group of authors who have so insightfully reviewed this topic.

May 2004

David A. Evans
Harvard University

Contents

Volume 1

Volume 2

Preface

The aldol reaction was first described by Kane in 1848. Thus it is high time to provide a comprehensive overview of the different developments in aldol chemistry, especially those of the past few decades. Demands for this important method of C–C–bond formation came and continue to come from every field of synthetic chemistry, particularly from natural product synthesis. Here, challenging problems in regioselectivity, chemoselectivity, diastereoselectivity and enantioselectivity frequently arise, many of which are still awaiting a solution. Symptomatically the word "selectivity" in its various connotations occurs no fewer than 1,100 times in both volumes, i.e. an average of twice a page.

This book examines the enormous variety of aldol chemistry from the view of both organic as well as inorganic and bioorganic chemistry. It presents a wide range of potent syntheses based on the discoveries from enolate chemistry or the catalysis of Lewis acids and Lewis bases, for instance. The important role of metal catalysis, organocatalysis and direct aldol addition is described, along with enzymatic methods. However, it was not our intention to simply list all existing publications about aldol chemistry. Instead, we wanted to point out fundamental and at the same time efficient ways leading to defined configured aldol products. Two of these are depicted on the cover: the metal catalysis and the enzymatic method.

It is now my pleasure to express my profound gratitude to the 22 authors and co-authors, all belonging to the elite of aldol chemistry, for their outstanding contributions and their professional cooperation. Special thanks are due to Wiley-VCH, especially Elke Maase and Rainer Münz, for their fine work in turning the manuscript into the finished book. Finally, I am indebted to my wife and my son for countless hours of assistance.

Last but not least, this book is also a tribute to the works of Teruaki Mukaiyama, who has done tremendous work in the field of aldol reaction and now celebrates his 77th birthday.

Berlin, Germany — Rainer Mahrwald
May 2004

List of Contributors

Editor

PD Dr. Rainer Mahrwald
Institut für Organische und Bioorganische Chemie
der Humboldt-Universität zu Berlin
Brook-Taylor-Str. 2
12489 Berlin
Germany

Authors

Prof. Dr. Carlos F. Barbas, III
The Skaggs Institute for Chemical Biology
and the Department of Molecular Biology
The Scripps Research Institute
10550 North Torrey Pines Road
La Jolla, CA 92037
USA

Prof. Dr. Manfred Braun
Institut für Organische Chemie und
Makromolekulare Chemie I
Heinrich-Heine-Universität Düsseldorf
Universitätsstr. 1
40225 Düsseldorf
Germany

Prof. Dr. Scott E. Denmark
245 Roger Adams Laboratory, Box 18
Department of Chemistry
University of Illinois
600 S. Mathews Avenue
Urbana, IL 61801
USA

Prof. Dr. Wolf-Dieter Fessner
TU Darmstadt
Department of Organic Chemistry and
Biochemistry
Petersenstr. 22
64287 Darmstadt
Germany

Shinji Fujimori
236 Roger Adams Laboratory, Box 91-5
Department of Chemistry
University of Illinois
600 S. Mathews Avenue
Urbana, IL 61801
USA

Prof. Dr. Arun K. Ghosh
Department of Chemistry
University of Illinois at Chicago
845 West Taylor Street
Chicago, IL 60607
USA

Prof. Dr. Kazuaki Ishihara
Graduate School of Engineering
Nagoya University
Chikusa
Nagoya, 464-8603
Japan

Prof. Dr. Jeffrey S. Johnson
Department of Chemistry
University of North Carolina at Chapel
Hill
Chapel Hill, NC 27599-3290
USA

Prof. Dr. Shū Kobayashi
Graduate School of Pharmaceutical
Sciences
The University of Tokyo
Hongo, Bunkyo-ku
Tokyo 113-0033
Japan

Naoya Kumagai
Graduate School of Pharmaceutical Sciences
The University of Tokyo
Hongo 7-3-1, Bunkyo-ku
Tokyo, 113-0033
Japan

Prof. Dr. Benjamin List
Max-Planck-Institut für Kohlenforschung
Kaiser-Wilhelm-Platz 1
45470 Mülheim an der Ruhr
Germany

PD Dr. Rainer Mahrwald
Institut für Organische und Bioorganische Chemie
der Humboldt-Universität zu Berlin
Brook-Taylor-Str. 2
12489 Berlin
Germany

Prof. Dr. Shigeki Matsunaga
Graduate School of Pharmaceutical Sciences
The University of Tokyo
Hongo 7-3-1, Bunkyo-ku
Tokyo, 113-0033
Japan

Dr. Jun-ichi Matsuo
The Kitasato Institute
Center for Basic Research
(TCI) 6-15-5 Toshima
Kita-ku, Tokyo 114-003
Japan

Prof. Dr. Teruaki Mukaiyama
The Kitasato Institute
Center for Basic Research
(TCI) 6-15-5 Toshima
Kita-ku, Tokyo 114-003
Japan

David A. Nicewicz
Department of Chemistry
University of North Carolina at Chapel Hill
Chapel Hill, NC 27599-3290
USA

Prof. Dr. Dieter Schinzer
Otto-von-Guericke-Universität Magdeburg
Chemisches Institut
Universitätsplatz 2
39106 Magdeburg
Germany

Michael Shevlin
Department of Chemistry
University of Illinois at Chicago
845 West Taylor Street
Chicago, IL 60607
USA

Prof. Dr. Masakatsu Shibasaki
Graduate School of Pharmaceutical Sciences
The University of Tokyo
Hongo 7-3-1, Bunkyo-ku
Tokyo, 113-0033
Japan

Prof. Dr. Isamu Shiina
Department of Applied Chemistry
Faculty of Science
Tokyo University of Science
Kagurazaka, Shinjuku-ku
Tokyo 162-8601
Japan

Prof. Dr. Fujie Tanaka
Department of Molecular Biology
The Scripps Research Institute
10550 North Torrey Pines Road
La Jolla, CA 92037
USA

Prof. Dr. Hisashi Yamamoto
Department of Chemistry
The University of Chicago
5735 S. Ellis Avenue
Chicago, IL 60637
USA

Dr. Yasuhiro Yamashita
Graduate School of Pharmaceutical Sciences
The University of Tokyo
Hongo, Bunkyo-ku
Tokyo 113-0033
Japan

Prof. Dr. Akira Yanagisawa
Department of Chemistry
Faculty of Science
Chiba University
Inage, Chiba 263-8522
Japan

1
Fundamentals and Transition-state Models. Aldol Additions of Group 1 and 2 Enolates*

Manfred Braun

1.1
Introduction

In an aldol reaction, an enolizable carbonyl compound reacts with another carbonyl compound that is either an aldehyde or a ketone. The enolizable carbonyl compound, which must have at least one acidic proton in its α-position, acts as a nucleophile, whereas the carbonyl active component has electrophilic reactivity. In its classical meaning the aldol reaction is restricted to aldehydes and ketones and can occur between identical or non-identical carbonyl compounds. The term "aldol reaction", in a more advanced sense, is applied to any enolizable carbonyl compounds, for example carboxylic esters, amides, and carboxylates, that add to aldehydes or ketones. The primary products are always β-hydroxycarbonyl compounds, which can undergo an elimination of water to form α,β-unsaturated carbonyl compounds. The reaction that ends with the β-hydroxycarbonyl compound is usually termed "aldol addition" whereas the reaction that includes the elimination process is denoted "aldol condensation". The "traditional" aldol reaction [1] proceeds under thermodynamic control, as a reversible reaction, mediated either by acids or bases.

In contrast, modern aldol methods rely on the irreversible formation of "preformed enolates" which are added to aldehydes or ketones. In any case, the aldol reaction has proven itself by a plethora of applications to be one of the most reliable methods for carbon–carbon bond-formation yielding either carbon chains, with oxygen functionality in 1,3-positions, or alkenes, by a carbonyl olefination process [2, 3].

The first example of this reaction, the acid-catalyzed self-condensation of acetone to give mesityl oxide, was reported more than one and a half centuries ago by Kane [4]. The condensation of an aromatic aldehyde with an aliphatic aldehyde or ketone, obviously the first example of an aldol condensation under basic conditions, was reported by Schmidt [5] and by

* This chapter is dedicated to the memory of Ulrike Mahler (deceased 1995) and Ralf Devant (deceased 2002).

Modern Aldol Reactions. Vol. 1: Enolates, Organocatalysis, Biocatalysis and Natural Product Synthesis.
Edited by Rainer Mahrwald

ISBN: 3-527-30714-1

Claisen and Claparède [6], and named after the inventors ("Claisen–Schmidt condensation") [2]. Obviously, Wurtz first recognized [7] the simultaneous presence of aldehyde and alcohol moieties in the "aldol" **1** resulting from the acid-induced reaction of acetaldehyde, and the reaction was named after the product it leads to later on (Eq. (1)).

H_3C–CHO —(HCl, H_2O)→ H_3C–CH(OH)–CH_2–CHO **1** (1)

1.2 The Acid or Base-mediated "Traditional" Aldol Reaction

Several reaction conditions feature in this "traditional" transformation. First, the reaction is run in protic solvents and can be mediated either by acid or by base. Second, the reaction is reversible, particularly under these conditions. Finally, the enol or the enolate, which acts as a nucleophile, is inevitably generated in the presence of the aldehyde or ketone that functions as an electrophile. The aldol reaction performed under these conditions (Scheme 1.1) was "the state of the art" until the early nineteen-seventies, when the chemistry of "preformed enolates" emerged.

The "traditional" aldol reaction has been the subject of several reviews, among which the summary by Nielsen and Houlihan in "*Organic Reactions*" in 1968 is a classical contribution and a very valuable survey [1]. The subject has also been treated in House's monograph [2] and, more recently, by Heathcock [3]. Thus, only general features and few representative examples of "traditional" aldol addition will be given here; the reader is referred to the above-mentioned surveys for more details.

Synthetically the reversibility of the aldol addition can cause substantial problems. Investigations performed in order to determine the relative energies of an enolate and an aldehyde on the one hand and the aldolate on the other revealed the outcome of the aldol reaction to be slightly exergonic [8]. The aldol formed by either acid- or base-catalyzed reaction is significantly stabilized by a strong OH bond in the aldol **2** which arises either directly from acid-mediated addition or on protonation of the aldolate **3** in the base-catalyzed variant, as shown in Scheme 1.1. Alternatively, chelation of the counter-ion in aldolates resulting from preformed enolates in non-protic media serves as the driving force [9]. As a general rule, applicable for protic solvents, the equilibrium in an aldol addition is located on the product side when aldehydes react with each other (Eq. (2)), but on the side of the starting materials for ketones (Eq. (3)).

2 R–CH_2–CHO ⇌ R–CH_2–CH(OH)–CH(R)–CHO (2)

acid catalysis

basic catalysis

Scheme 1.1
Aldol addition: general acidic and basic catalysis.

(3)

As a consequence, self-addition of enolizable aldehydes is usually readily accomplished in aqueous basic media if sufficient solubility is not prevented by the extended length of the carbon chain. The aldol addition (Eq. (4)) and condensation (Eq. (5)) of butanal giving 2-ethyl-3-hydroxyhexanal [10] and 2-ethylhexenal [11], respectively, can serve as illustrative examples. The self-addition of enolizable ketones, on the other hand, does not, per se, lead to substantial amounts of the aldol product, because of the unfavorable equilibrium mentioned above. This equilibrium can, however, often be shifted by a subsequent elimination step, so the aldol condensation of ketones is more frequently applied than addition. There are also special procedures that enable shifting of the equilibrium in the aldol reaction of ketones towards the products. Among these the formation of "diacetone" **4** by heating acetone under reflux in a Soxhlet apparatus filled with calcium or barium hydroxide is a well-known procedure which avoids contact of the base with the non-volatile hydroxy ketone **4** thus preventing a retro aldol reaction from occurring (Eq. (6)) [12, 13].

KOH, H_2O 75% (4)

NaOH, H_2O 80°C 86% (5)

$Ba(OH)_2$ Soxhlet, reflux 75% **4** (6)

Mixed aldol reactions between different aldehydes or ketones are usually plagued by formation of a mixture of products, because each component can function as a CH-acidic and carbonyl-active compound. Whereas the "directed aldol reaction" [14–16] is a rather general solution to this problem, the traditional aldol addition of non-identical carbonyl compounds is only successful when applied within the framework of a limited substitution pattern. Thus, a fruitful combination in mixed aldol reactions is that of an aldehyde with an enolizable ketone. Obviously, the aldehyde, having higher carbonyl reactivity, reacts as the electrophilic component, whereas the ketone, with comparatively lower carbonyl reactivity, serves as the CH-acidic counterpart. Because the self-aldolization of ketones is endothermic, this type of side reaction does not occur to a significant extent, so the product of the mixed aldol condensation is obtained in fair yield, as illustrated by the formation of ketone **6** from citral **5** and acetone, a key step in the synthesis of β-ionone (Eq. (7)) [17].

5 + NaOEt 45-49% **6** (7)

The most efficient variant of this combination is based on reaction of an enolizable ketone with a non-enolizable aldehyde, so that self-condensation of the latter cannot occur. Several examples of this type of combination in aldol reactions are given in Scheme 1.2. Usually in situ elimination occurs, so α,β-unsaturated ketones result, in particular when aromatic aldehydes are condensed with ketones ("Claisen–Schmidt reaction") [18–21].

The intramolecular aldol condensation of dialdehydes, ketoaldehydes, and

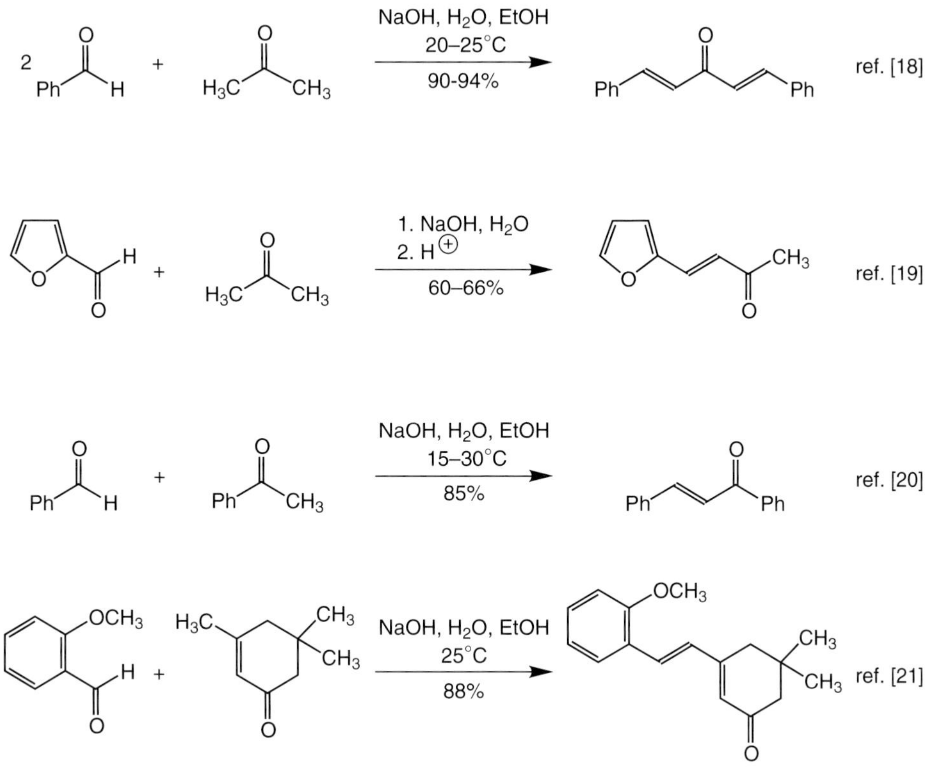

Scheme 1.2
Illustrative examples of the condensation of aromatic aldehydes with enolizable ketones ("Claisen–Schmidt reaction").

diketones is one of the most efficient means of synthesizing five, six, and seven-membered rings. There are numerous applications of this variant of the aldol reaction, in particular in the context of the Robinson annelation reaction, described in the literature for decades. Because this topic has been reviewed comprehensively [22, 23], a few illustrative examples only will be given here. An early and rather prominent example of the intramolecular aldol condensation is found in Woodward's synthesis of cholesterol [24]. Because the precursor **7** is an unsymmetrical dialdehyde, the problem of regioselectivity arises. Nevertheless, the α,β-unsaturated aldehyde **8** is formed in excess and only minor amounts of the regioisomer **9** are obtained (Eq. (8)). Complete regioselectivity was observed, however, when dialdehyde **10** was submitted to an intramolecular aldol condensation. Thus, the enal **11** was obtained exclusively in the first synthesis of genipin described by Büchi and coworkers (Eq. (9)) [25].

7 → 8 + 9 (piperidine, AcOH, benzene, 60°C; 66%) (8)

10 → 11 (piperidine, AcOH; 68%) (9)

When ketoaldehydes, compounds that are unsymmetrical per se, are submitted to intramolecular aldol condensation the ketone usually acts as the CH-acidic component whereas the aldehyde plays the role of the carbonyl active counterpart. This regiochemical outcome is also favored when the conditions of a thermodynamic control are used. Again, this type of aldol condensation has been used in a variety of natural products synthesis. A steroid synthesis, the aldolization step of which is given in Eq. (10), is an illustrative example [26].

KOH, H_2O, 50–80°C; 73% (10)

When this type of stereochemical outcome is prevented by steric hindrance, the two carbonyl groups can play opposite roles in the sense that the aldehyde, deprotonated in its α-position, functions as nucleophile whereas the ketone acts as the carbonyl-active compound, as shown in the example given in Eq. (11) [27].

Na_2CO_3, EtOH, H_2O; 46% (11)

Although problems of regiochemistry are inherent, the aldol condensation of diketones has found wide application. Typical examples are syntheses of cyclopentenones and cyclohexenones from 1,4- and 1,5-diketones, respectively. The concept is illustrated by a synthesis of jasmone **12** (Eq. (12)) [28] and of the homosteroid derivative **13**, the latter arising under thermodynamic control in a Robinson annelation reaction (Eq. (13)) [29].

12 (12)

$MeO^{\ominus}$

13 (13)

Intramolecular aldol condensations also serve as the key step in "biomimetic" syntheses of polyketides, synthetic strategies that try to imitate in vitro a proven or an assumed biosynthetic pathway [30]. Although the first attempts in this direction go back to the early 20th century [31], practical and efficient syntheses based on this concept were elaborated much later.

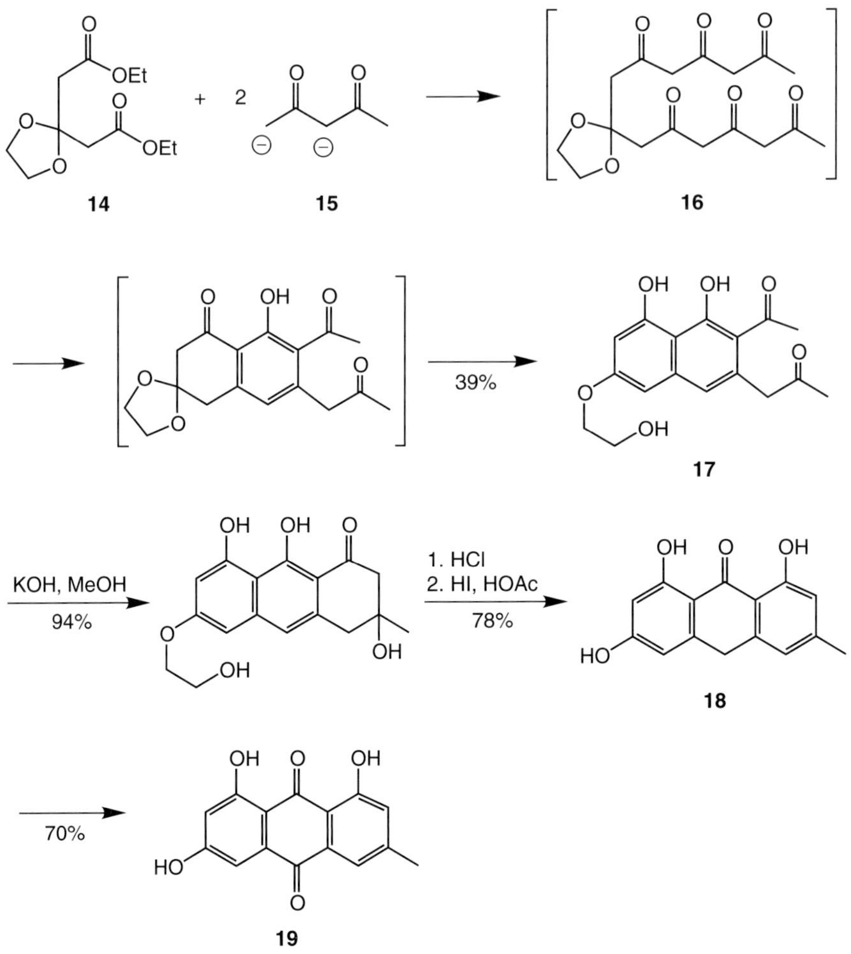

Scheme 1.3
Biomimetic synthesis of emocline **19** involving intramolecular aldol reactions.

The concept is illustrated in Scheme 1.3 – when intermediate **16** containing a carbon chain with six free keto groups and one protected keto group has been generated by twofold Claisen condensation of the diester **14** with the highly reactive dianion **15**, it undergoes spontaneous aldol condensation followed by aromatization to give the naphthalene derivative **17**. A further aldol addition, which leads to the formation of a third six-membered ring, needs treatment with potassium hydroxide. Finally, dehydration and deprotection lead to the anthrone **18**, which is readily oxidized to the natural product emodine **19** [32]. Similar approaches based on intramolecular aldol reactions have been applied to the synthesis of naturally occurring anthracyclinones [33] and isoquinolines [34]; the biosynthesis of these is known to involve polyketone intermediates.

The problem of stereochemistry has very rarely been addressed by traditional aldol addition. The question of relative stereochemistry has been studied occasionally in the context of intramolecular, in particular transannular, aldolizations. There were, however, few diastereoselective variants. Control of enantioselectivity was achieved to a remarkable extent in the cyclization of 1,3-cyclopentanediones. Independent work by research groups at Hoffmann LaRoche and Schering AG in the early nineteen seventies revealed that highly enantioselective cyclization of triketones **20a**, **b** can be accomplished by treatment with catalytic amounts of L-proline, as shown in Eq. (14) [35, 36]. Thus, the products **21a** and **21b** are obtained from the Hajos–Parrish–Eder–Sauer–Wiechert reaction, the intramolecular aldol condensation, in 93 and 99% ee, respectively. The method has been applied successfully to enantioselective steroid syntheses, and provided a route to the skeleton of several other natural products [37]. The source of stereoselectivity in this cyclization has been investigated carefully [38]. Very recently, proline catalysis has also been applied to enantioselective intermolecular aldol additions [39]. This promising approach will be discussed in detail in Chapter 4 of Part I.

20a: R = Me
20b: R = Et

21a: 93.4% *ee*
21b: 99.5% *ee*

(14)

Although the number of applications of the "traditional" aldol reaction "is legion", and despite its undoubted versatility, the reaction suffers from general lack of control of stereochemistry and from the difficulty of reliable determination of the carbonyl-active and CH-acidic components. Both problems have been solved by the technique of directed aldol addition based on preformed enolates.

1.3 The Aldol Addition of Preformed Enolates – Stereoselectivity and Transition-state Models

The chemistry of preformed enolates emerged in temporal and causal coherence with the "LDA area". Although lithium and magnesium salts of diisopropylamine were first developed in the nineteen-fifties [40], lithium diisopropylamide (LDA) has been a widely used reagent since 1970, because of its behavior as a soluble, strong, and non-nucleophilic base [14]. LDA and related bases, for example lithium hexamethyldisilazane (LIHMDS) [41], lithium *N*-isopropylcyclohexylamide (LICA) [42], and lithium 2,2,6,6-

tetramethylpiperidide (LITMP) [43] turned out to be the reagents of choice for conversion of a variety of carbonyl compounds into their enolates in an irreversible reaction which also enabled control of regiochemistry. This is illustrated in the kinetically controlled deprotonation of 2-methylcyclohexanone, **22**, which leads to the formation of the enolate **23** with remarkable regioselectivity (Eq. (15)) [44]. Complementary routes that lead to the formation of the regioisomeric enolate **24** with a more substituted double bond have also been elaborated; they are based on a deprotonation under thermodynamic control or use of enol acetates, silyl enol ethers, or α,β-unsaturated carbonyl compounds as precursors. In addition, procedures for formation of (*E*) and (*Z*) enolates were elaborated. The formation of preformed enolates has been reviewed comprehensively [45, 46]. In addition, the determination of enolate structures by crystal structure analyses, pioneered by the research groups of Seebach, Boche, and Williard, and NMR spectroscopic investigations and theoretical calculations led to insight into their reactivity [47].

$$\mathbf{22} \xrightarrow[\text{MeOCH}_2\text{CH}_2\text{OMe, -78°C}]{\text{LiN}(i\text{Pr}_2),} \mathbf{23} + \mathbf{24} \qquad (99 : 1) \tag{15}$$

Preformed enolates can be obtained not only from aldehydes and ketones, but also from carboxylic esters, amides, and the acids themselves. The corresponding carbonyl compound always acts irreversibly as the CH-acidic component. Thus, the term aldol reaction is no longer restricted to aldehydes and ketones but extended to all additions of preformed enolates to an aldehyde or a ketone. In contrast with the "traditional" aldol reaction, this novel approach is based on a three-step procedure (usually, however, performed as a one-pot reaction). First, the metal enolate **25** is generated irreversibly, with proton sources excluded, and, second, the compound serving as the carbonyl active, electrophilic component is added. The metal aldolate **26** thus formed is finally protonated, usually by addition of water or dilute acidic solutions, to give the aldol **27** (Scheme 1.4) [45, 46].

The principal aim in the development of the "modern" aldol reaction was stereochemical control, a field that has been treated in a series of review articles [46, 48–60]. In stereochemical terminology, the topic is discussed in the terms of "simple diastereoselectivity" and "induced stereoselectivity" [61]. Except for relatively rare examples when R^1 is identical with R^2 and R^3 is identical with R^4, all aldol additions are stereogenic. If the carbonyl-active compound is either an aldehyde (except formaldehyde – $R^4 = H$) or a prochiral ketone ($R^3 \neq R^4$), addition of the enolate leads to formation of either

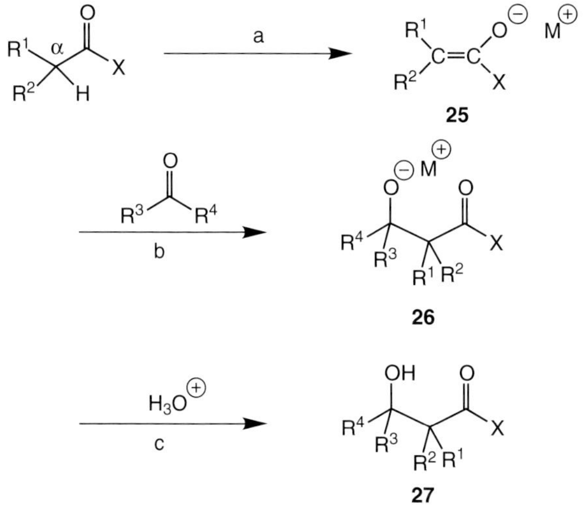

R^1, R^2: H, alkyl, aryl, OR, NR_2
R^3, R^4: H, alkyl
X: H, alkyl, aryl, OR, NR_2, OM

Scheme 1.4
Aldol reaction of preformed enolates. (a) irreversible enolate formation; (b) addition of the preformed enolate to aldehydes or ketones; (c) protonation.

one or two stereogenic centers. This depends on whether an enolate with identical α-substituents (mostly $R^1 = R^2 = H$) or an enolate with different α-substituents is used. Under the latter conditions one of the substituents R^1 and R^2 is usually a hydrogen atom (Scheme 1.4).

A general stereochemical pattern of the aldol addition is shown in Scheme 1.5. When a carbonyl compound **28** with an α-substituent R^2 (which can be an alkyl or an aryl group, or a hetero substituent, for example alkoxy, or a protected amino group, but which is not identical with hydrogen) is converted into the "preformed" enolate **29** and added to an aldehyde, four stereoisomeric products **30a**, **30b**, **31a**, **31b** can result.

When neither the enolate **29** nor the aldehyde contains stereogenic units, both reactants have enantiotopic faces and **30a** and **30b** are enantiomers. The same is true for the pair **31a** and **31b**. However, **30** and **31** form a pair of diastereomers. When an aldol addition leads to an excess of one of these diastereomers **30** or **31**, it is said to exhibit simple diastereoselectivity. Several notations that assign descriptors to diastereomeric aldols are found in the literature. The classical *erythro*/*threo* nomenclature, which is based on Fischer projection formulas [62], will not be used in this chapter, because it can cause considerable confusion with branched carbon chains. Among the

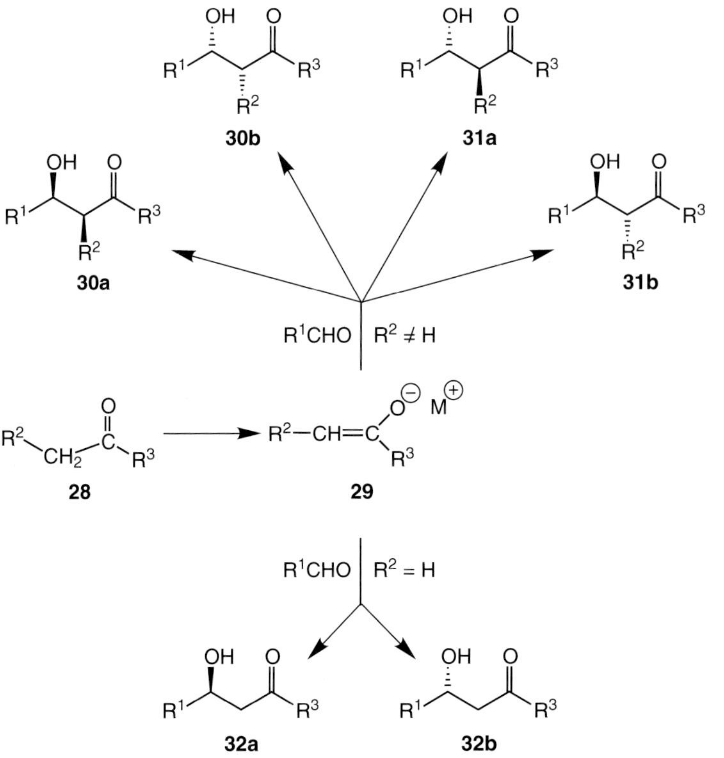

Scheme 1.5
Stereochemical pattern of aldol addition.

different alternatives proposed in the literature, the Prelog–Seebach notation deserves mention: their *l* (like) and *u* (unlike) descriptors are systematic and unambiguous, because they correlate strictly with Cahn–Ingold–Prelog nomenclature [63]. For practical reasons, however, the *syn/anti* notation, introduced by Masamune [64], will be used in this chapter. Thus, the carbon chain that contains the two stereogenic centers is drawn in a zigzag fashion. In the *syn* diastereomers both substituents at the stereocenters are directed either toward or away from the viewer. In the *anti* isomers one of the substituents is directed toward the viewer, the other one away from the viewer, or vice versa. According to this notation the stereoisomers **30a** and **30b** are termed *syn* whereas **31a** and **31b** are defined as the *anti* isomers. When the topicity of an aldol addition is of interest the *lk* (like) and *ul* (unlike) notation [63] will be used in this chapter to describe the pathway by which the enolate approaches the carbonyl compound (aldehyde or ketone).

The problem of simple diastereoselectivity does not arise in aldol additions when an α-unsubstituted enolate **29** ($R^2 = H$) or an enolate with two identical α-substituents reacts with an aldehyde or a prochiral ketone. The products **32a** and **32b** obtained from this combination are enantiomers, if neither the aldehyde nor the enolate is a chiral molecule.

If an aldol addition is performed using either an enolate with stereogenic units, which can be located in the α-substituent R^2 or in the ipso substituent R^3, or if a chiral aldehyde is used as the electrophilic component, the aldol products **30a**, **31a**, and **32a** are diastereomers of **30b**, **31b**, and **32b**. In these combinations not only is simple diasteroselectivity has to be achieved (except for the α-unsubstituted case leading to **32a/b**) but also induced stereoselectivity. In detail, induced stereoselectivity is postulated for the following combinations of reactants:

1. reaction of chiral enolates with achiral or chiral aldehydes;
2. reaction of achiral enolates with chiral aldehydes; and
3. reaction of achiral enolates with achiral aldehydes, if they are mediated by use of a chiral catalyst (including enzymes or antibodies), chiral ligands at the metal M, or all kinds of chiral solvents.

When, in reactions 1–3, the enolate does not have an α-substituent, e.g. **29** ($R^2 = H$), induced stereoselectivity is highly desirable. In addition, the corresponding reaction of α-substituted enolates **29** ($R^2 \neq H$) should not only provide induced stereoselectivity but also simple diastereoselectivity. In this chapter we will not use "induced diastereoselectivity" as a subdivision applicable to reactions 1 and 2 and "induced enantioselectivity" for reaction 3 (leading to enantiomeric aldol products), because, irrespective of the stereochemical relationship of the products (diastereomers or enantiomers), the transition states leading to the different stereoisomeric compounds are always diastereotopic, even if the products are enantiomers (reaction 3). The term "induced stereoselectivity", which includes the different variants 1–3, is therefore used here [57, 61].

Most aldol additions of preformed enolates are run under kinetic control. In some such kinetically controlled aldol reactions simple diastereoselectivity is related to the configuration of the enolate. The seminal investigations of Dubois [65], then intensive studies by the research groups of

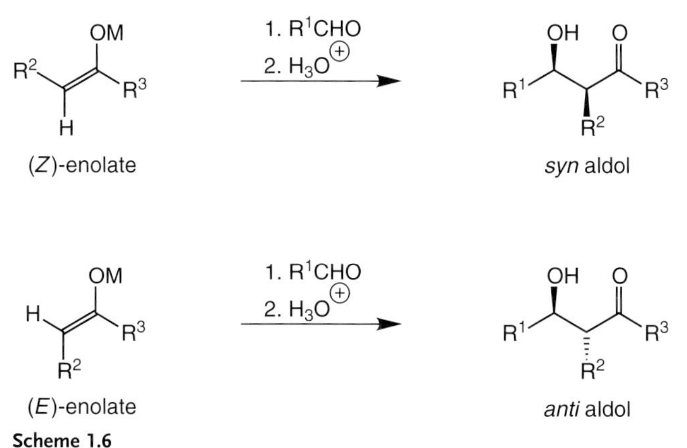

Scheme 1.6
Correlation between enolate geometry and aldol configuration.

House, Heathcock, and Ireland [48–51], revealed that (*Z*)-configured enolates furnish mainly *syn*-aldols whereas *anti*-*β*-hydrocarbonyl compounds arise predominantly from (*E*) enolates. In this context, the descriptors *E* and *Z* refer to the relative position of the *α*-substituent R^2 and the oxygen–metal bond (Scheme 1.6).

Because procedures for the selective generation of (*Z*) and (*E*) enolates have been elaborated for a variety of carbonyl compounds [45], the kinetically controlled aldol addition offers a solution to the problem of simple diastereoselectivity. Representative examples of the *Z*/*syn* and *E*/*anti* correlation are given in Eqs. (16) and (17) [66, 67].

(*Z*)-enolate → 1. PhCHO, THF, -78°C; 2. $H_3O^{\oplus}$ → *syn* + *anti*

R	*syn* : *anti*
CMe_3	98.7 : 1.3
Et	90 : 10

(16)

(*E*)-enolate → 1. PhCHO, THF, -78°C; 2. $H_3O^{\oplus}$ → *syn* + *anti*

X	*syn* : *anti*
OMe	62 : 38
$OCMe_3$	51 : 49
O–aryl (Me, Me)	12 : 88
O–aryl (Me_3C, Me_3C, Me)	2 : 98

(17)

As shown by Eq. (16), (*Z*) lithium ketone enolates yield *syn* aldols, if the substituent R at the carbonyl group is sterically demanding. Because carboxylic amides and thioamides, like ketones, form (*Z*) enolates predominantly, their aldol addition also leads to the predominant formation of *syn*-β-hydroxycarboxylic acids (or their corresponding derivatives). In general, *Z*-configured boron and titanium enolates result in higher simple diastereoselectivity in favor of *syn* aldols than the corresponding lithium or magnesium enolates. Also, with regard to the induced stereoselectivity boron enolates are usually more selective than lithium enolates [49, 52]. This might be because the boron–oxygen bond in enolates is shorter than the lithium–oxygen bond, so cyclic transition states involving boron as the metal are tighter, and steric repulsion is more effective and chiral information is transferred more efficiently (for chiral auxiliary groups or ligands). On the other hand, carboxylic esters and thioesters, which form predominantly (*E*) enolates, react with aldehydes with substantial *anti* selectivity. As shown in Eq. (18), this selectivity is, however, restricted to reactions in which bulky aromatic substituents form the alcoholic moiety of the ester. The *anti* diastereoselectivity of (*E*) enolates is usually lower then the *syn* selectivity of comparable (*Z*) enolates. One must also take into account the effect of the α-substituent. As shown in Eq. (18) the *Z–syn* correlation can be completely reversed for α-substituents that are bulky, sterically demanding alkyl groups [68].

OLi, R, H, CMe_3 — 1. Me_3CCHO, Et_2O, 20°C; 2. $H_3O^{\oplus}$ → OH, O, Me_3C, CMe_3, R (*syn*) + OH, O, Me_3C, CMe_3, R (*anti*)

R	*syn* : *anti*
Me	100 : 0
Et	100 : 0
n-Pr	98 : 2
$CHMe_2$	29 : 71
CMe_3	0 : 100

(18)

In the Mukaiyama addition of the aldol reaction [16], silyl ketene acetals or silyl enol ethers are added to aldehydes in a reaction mediated by Lewis acids or fluoride. Here again the *Z–syn* correlation is sometimes not observed [69, 70]. Thus, the *Z–syn*, *E–anti* correlation seems to be a rule with several exceptions [71].

The stereochemical outcome of the different aldol additions of preformed enolates calls for plausible transition state models. Two kinds of explanation seem suitable for rationalizing the different stereochemical results, which even seem to contradict each other: Thus, in a single type of transition state model a different substitution pattern might cause the reactants to have different orientations to each other so that formation of different stereoisomers results. On the other hand, different reaction conditions and reactants can be used, even if the way the aldol reaction is run is restricted to the use of preformed enolates. Thus, the latter might have very different counter-ions, and the solvents might also be different. It seems plausible that this can cause the reaction to occur via completely different types of transition state.

The most widely accepted transition state hypothesis for aldol additions is the Zimmerman–Traxler model. This was originally developed to explain the stereochemical outcome of the Ivanoff reaction – addition of the dianion of carboxylic acids with magnesium counter-ions to aldehydes and ketones [72]. On the basis of investigation of the stereochemical outcome of the reaction of doubly deprotonated phenyl acetic acid to benzaldehyde (Eq. (19)), Zimmerman and Traxler proposed in a seminal paper a transition state model that involves a six-membered chair-like assembly of the reactants [73].

i-PrMgBr, Et_2O

OMgBr

OMgBr

1. PhCHO

2. $H_3O^{\oplus}$

69%

22%

(19)

This model offers a plausible explanation of the (*Z*)–*syn*, (*E*)–*anti* correlation, as shown in Scheme 1.7. The diasteromeric transition states **33a** and **33b**, which emerge from addition of a (*Z*) enolate to an aldehyde, differ in the position of the substituent R^1, which is equatorial in **33a** and axial in **33b**. By analogy with conformational analysis of the cyclohexane system [62], the transition state **33a** is expected to have a lower energy than the diastereomeric alternative **33b**. As a consequence the predominant formation of *syn* aldolates results from this kinetically controlled reaction.

When the (*E*) enolate is chosen as the starting material, the analogous argument indicates the transition state **34a** with R^1 in an equatorial position to be favored compared with the alternative **34b**, in which the substituent R^1 occupies an axial position. Accordingly, the *anti* aldolate is expected to be the predominant product.

Although first developed for a magnesium enolate, the Zimmerman–Traxler model could be used very successfully to explain the stereochemical

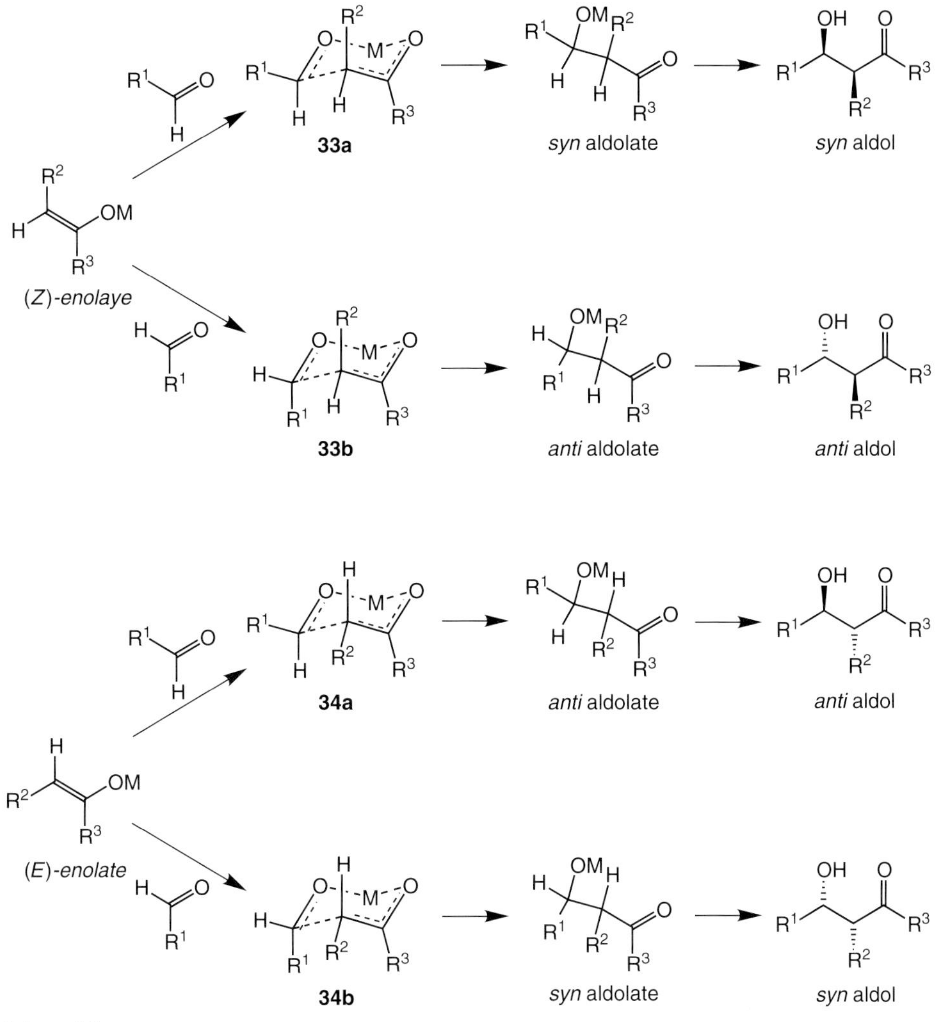

Scheme 1.7
Zimmerman–Traxler transition state models in the aldol additions of (*Z*) and (*E*) enolates.

outcome of aldol additions of boron and titanium enolates. This might be because they are monomeric, in contrast with the enolates of lithium and magnesium, known to form aggregates. Not only is (*Z*)–*syn* [49, 52] and (*E*)–*anti* [74, 75] correlation better for boron enolates, they also usually result in greater induced stereoselectivity. This also is easily explained by the Zimmerman–Traxler model if it is assumed that for boron and titanium enolates, stronger Lewis acids than lithium and magnesium, the six-membered transition state is tighter, so steric effects are maximized.

The validity of the Zimmerman–Traxler model for alkali metal or magnesium enolates could be questioned, because these strongly electropositive metals might form ionic rather than covalent bonds to the enolate/aldolate oxygen atom. Even if there was a contact ion-pair of metal cation and oxygen anion, however, the geometry of a six-membered chair in the transition state would be very similar to that shown in Scheme 1.7. Even the aggregation of lithium enolates, well recognized today [47], does not severely contradict the Zimmerman–Traxler model – indeed, a six-membered transition state, postulated to occur at a tetrameric lithium pinacolone enolate [76], is very compatible with the "closed" model proposed by Zimmerman and Traxler.

The Zimmerman–Traxler model has, however, been challenged by the frequent observation that (*Z*) enolates result in higher simple diastereoselectivity, giving *syn* aldols, compared with the lower *anti* selectivity of (*E*) enolates [49–51, 68]. Assuming the classical chair transition state, it has been remarked that for (*E*) enolates models **34a** and **34b** might both be plagued by unfavorable steric repulsion. Thus, the equatorial orientation of R^1 in **34a** avoids repulsion by R^3, but at the expense of a steric hindrance between R^1 and R^2, which is enhanced because the torsional angle at the forming carbon–carbon bond is less than 60°. This has been postulated in skewed transition state models, proposed by Dubois [68] and Heathcock [51], in which the dihedral angle between the enolate double bond and the carbonyl group approaches 90°, as shown in Scheme 1.8 for the *E–syn* and *E–anti* correlation. The alternative, **34b**, takes advantage of minimizing the R^1–R^2 repulsion, but on the other hand, is disfavored by the axial R^1–R^3 hindrance. As a result, both transition states **34a** and **34b** become similar in energy, so stereoselectivity is reduced.

An important modification of the classical Zimmerman–Traxler model, which still relies on the idea of a "pericyclic-like" transition state, considers

H O---M R1 R2 O H R3 (*E*)-enolate → OH O R1 R3 R2 *anti*

H O---M H R2 O R1 R3 (*E*)-enolate → OH O R1 R3 R2 *syn*

Scheme 1.8
Skewed transition state model for (*E*) enolates.

(Z)-enolate → syn

(Z)-enolate → anti

Scheme 1.9
Boat transition state models for (*Z*) enolates.

boat conformations as alternatives. Thus, Evans [49, 77] has suggested boat transition states when there is substantial steric hindrance between groups R^1 and R^2 on the forming carbon–carbon bond. The model plausibly explains the results given in Eq. (18), which show that an increase of the α-substituent R^2 in a (*Z*) enolate leads to a higher proportion of the *anti* aldol. The alternative orientations outlined in Scheme 1.9 for the *Z–syn* and *Z–anti* correlation show the latter to be a reasonable alternative for large substituents R^2.

Hoffmann, Cremer and co-workers have proposed a transition state model for addition of enol borates to aldehydes [78]. The authors pointed out that a twist-boat **36b** could easily be formed from the U-conformation of the boron enolate (*E*)-**35b** whereas the intermediate with the W-orientation, (*Z*)-**35a**, is a suitable precursor of the chair-type transition state **36a**. The consequence of the assumption that (*Z*) enolates react via a chair transition state whereas (*E*) enolates react via a boat conformation is that *syn* aldols are formed irrespective of enolate geometry, as shown in Scheme 1.10 [79]. This type of stereochemical outcome has been observed in various examples of the Mukaiyama-type aldol addition [80] and in aldolizations of enol stannanes and zirconium enolates [81, 82]. In contrast, a clear (*Z*)–*syn*, (*E*)–*anti* correlation has been observed in the addition of allylboronates to aldehydes [83], a reaction that is closely related to the aldol addition. The stereochemical hypothesis shown in Scheme 1.10 has been underscored by the semi-empirical calculations of Gennari and coworkers [84] which show that, starting from (*Z*) enolates, the half-chair transition state leading to *syn* aldols is preferred. For (*E*) enolates, a preference for either *syn* or *anti* aldols is predicted, depending on whether or not the metal carries a bulky substituent. The diminished simple diastereoselectivity in aldol additions of (*E*) enolates can be interpreted as a result of competition between the chair

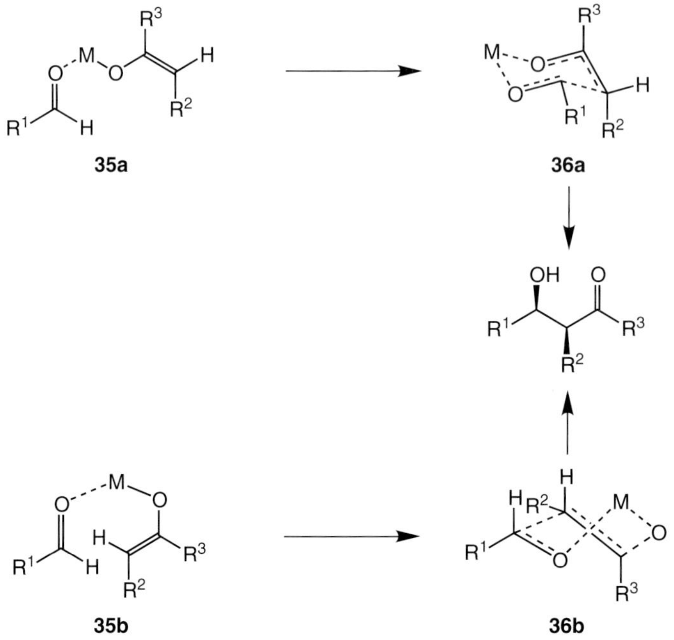

Scheme 1.10
Different transition state models (chair and twist-boat) for (*Z*) and (*E*) enolates ($M = BL_2$).

and twist-boat transition states [85]. More recently, six-membered transition state hypotheses have also been proposed for "direct" aldol additions [86] and for the phosphoramide-catalyzed addition of trichlorosilyl enolates to aldehydes [87].

A different cyclic transition state model which does, however, not incorporate the metal, has been proposed by Mulzer and coworkers. It was developed to explain the observation that in the addition of doubly deprotonated phenyl acetic acid to pivaldehyde the highest *anti* selectivity is obtained with the most "naked" enolate anions (e.g. K/18-crown-6). The hypothesis, which might explain this stereochemical result, assumes that the approach of the enolate to the aldehyde is dominated by the interaction of the enolate HOMO and the π^* orbital of the aldehyde that functions as the LUMO. The favored approach of the reactants occurs when the substituents of the enolate (phenyl) and the aldehyde (*t*-butyl) are oriented in a *trans* orientation at the forming carbon bond, so that their mutual steric repulsion is minimized (Scheme 1.11). The expected transition state **37** has some similarity to that of a 1,3-dipolar cycloaddition, although the corresponding cycloadduct **38** does not form, because of the weakness of the oxygen–oxygen bond. Instead, the doubly metalated aldol adduct **39** results [88]. In a similar

M	anti : syn
½ Mg	58 : 42
Li	70 : 30
K/18-crown-6	97 : 3

$R^1 = C(CH_3)_3$; $R^2 = C_6H_5$

Scheme 1.11
Anti selectivity in the aldol addition of doubly deprotonated phenylacetic acid and 1,3-dipolar cycloaddition transition state model.

way, Anh and Thanh emphasized that frontier orbital interactions played an essential role in determining the stereochemical outcome of the aldol reaction [89].

A completely different rationale for the stereochemical outcome of aldol additions relies on open-transition-state models. These involve antiperiplanar orientation of enolate and carbonyl group, in contrast with their *syn*-clinal conformation assumed in the six-membered cyclic transition states. Open-transition-state structures have been proposed to offer a rationale for those aldol additions that give predominantly *syn* products, irrespective of enolate geometry [90]. This outcome has been observed in aldol reactions of tin and zirconium enolates and of "naked" enolates generated from enolsilanes by treatment with tris(diethylamino)sulfonium difluoromethylsiliconate [70]. As shown in Scheme 1.12, the driving force for the

enolate (*Z*) : (*E*)	product ratio *syn* : *anti*
99 : 1	95 : 5
9 : 91	94 : 6

$R^1 = R^3 = Ph$; $R^2 = Me$

Scheme 1.12
Formation of *syn* aldols irrespective of enolate geometry. Open-transition-state models.

open-transition-state model is the tendency of the negatively charged oxygen atoms to be as far apart from each other as possible. It is assumed that both transition states that lead to the formation of the *anti* aldol, i.e. **40a** formed from the (*Z*) enolate and **40b** from the (*E*) enolate, are disfavored, because of the steric repulsion of substituents R^1 and R^2, which are oriented in a *gauche* conformation. This type of steric hindrance is avoided in the transition state structures **41a** and **41b**, so both (*Z*) and (*E*) enolates give the *syn* aldol predominantly.

The basic assumption of open-chain transition-state models is the antiperiplanar orientation of the enolate and the carbonyl double bond. This

Scheme 1.13
Calculated reaction pathway for addition of acetaldehyde lithium enolate to formaldehyde.

type of model has also been used to explain the stereochemical outcome observed in aldol additions of thioester silylketene acetals [84]. An open transition has also been proposed in additions of silyl ketene acetals to aldehydes, mediated by chiral copper complexes [91].

Even if a particular enolate with a distinct geometry is reacted with an aldehyde, the question whether the transition state is "closed" or "open" cannot be answered by simple "either–or". More recent discussions have, instead, led to an "as well as", because the role of the counter-ion becomes more evident. Thus, ab-initio calculations of Houk and coworkers [92] predict an open-transition-state structure for metal-free, "naked" enolates and closed transition states for lithium enolates. For addition of acetaldehyde lithium enolate to formaldehyde, the lowest-energy reaction pathway (shown in Scheme 1.13) has been studied on the basis of on ab-initio (3–21 G) calculations [93].

The reactants first reach the coordination complex **40**, a local minimum on the energy hypersurface [94]. In this complex the O–Li–O angle can vary from 145° to 180°. The transition state of carbon–carbon bond formation is calculated to have the half-chair conformation **41**. The angle of nucleophilic attack on the carbonyl group is 106.9°, consistent with the Bürgi–Dunitz trajectory [95] and in accordance with calculations of Houk and coworkers [92]. The transition state structure **41** finally collapses to the aldolate **42** with the lithium atom coordinating the two oxygen atoms. The activation barrier of the reaction is calculated to be 1.9 kcal mol^{-1} and the overall exothermicity is 40.2 kcal mol^{-1}.

Concerning the question of the conformation of the six-membered cyclic transition state, the different possibilities, for example chair, half-chair, or twist boat, seem to be quite close in relative energy so the particular substitution pattern is assumed to determine the favored conformation. The key role of the counter-ion has been confirmed by experimental results obtained from the intramolecular aldol addition of bicyclic keto aldehydes. Here again, enolates with a strongly coordinating metal counter-ion, for example Mg^{2+}, have strong preference for a reaction via a closed transition state in which the metal counter-ion is coordinated both to the enolate and to the carbonyl oxygen atom. "Naked" enolates, on the other hand, have a pronounced tendency to react through an open transition structure with an *anti*-periplanar conformation of the enolate and the carbonyl moiety [96].

It is self-evident that the transition state hypotheses discussed above are exclusively relevant to kinetically controlled aldol additions. Although this type of reaction control is the rule when preformed enolates are used, one should be aware that the reversibility of aldol additions cannot be excluded a priori and in any instance. In aldol reactions of preformed enolates, reversibility becomes noticeable in equilibration of *syn* aldolates with *anti* aldolates rather than in an overall low yield as found in the traditional aldol reaction. Considering the chair conformations of the *syn* and the *anti* aldolates, the former seem to be thermodynamically less stable, because of the axial position of the α-substituent R^2. This situation is avoided in the *anti* adduct (Eq. (20)). Indeed, the *anti* diastereoisomer is favored in most aldol additions run under thermodynamic control. This has been observed, for example, in aldolates *syn*-**43** and *anti*-**43**, which arise from addition of doubly lithiated phenylacetic acid to pivalaldehyde. Whereas the kinetically controlled reaction gives a *syn*/*anti* ratio of 1.9:1, equilibration occurring after several hours in tetrahydrofuran at 25 °C leads to a 1:49 in favor of the *anti* products (Eq. (21)) [88].

(20)

(21)

The enolate counter-ion has an important effect on the rate of the reverse aldol reaction. Boron enolates usually undergo completely irreversible addition to aldehydes. The more "ionic" of the alkali metals, for example sodium and potassium, have a greater tendency to undergo retro aldol reactions than lithium. Thus the potassium aldolate *anti*-**44** formed from deprotonated ethyl mesityl ketone and benzaldehyde undergoes equilibration to *syn*-**44** even at −78 °C, whereas the corresponding lithium aldolate isomerizes at 0 °C (Eq. (22)) [66]. That, in this reaction, the *syn*-aldolate is thermodynamically favored is possibly explained by the steric hindrance between the α substituent (methyl) and the bulky *ipso* substituent (mesityl) in the *anti* isomer.

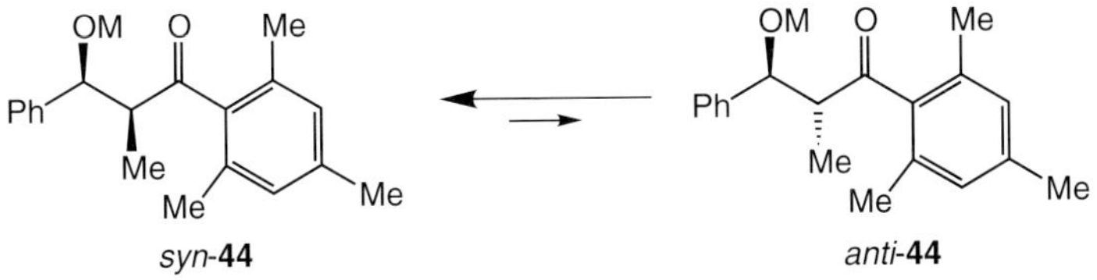

syn-**44** ⇌ *anti*-**44** (22)

The influence of further counter-ions like ammonium, magnesium and zinc on the reversibility has been studied [65, 71]. Another influence comes from the stability of the enolate. As a rule, the rate of the retroaldol reaction correlates with the stability of the enolate. In stereoselective aldol addition, the reversibility is, in general, rather considered as a complication than a tool to obtain high selectivity. In particular, thermodynamically controlled aldol additions are usually not suitable to obtain non-racemic aldols.

1.4 Stereoselective Aldol Addition of Lithium, Magnesium and Sodium Enolates

Modern synthetic methods in organic chemistry are aimed at obtaining chiral products in a non-racemic, if at all possible, enantiomerically pure form. In addition, the products should be accessible as pure diastereomers. This section therefore focuses on aldol additions that provide pure stereoisomers. Amongst the enolate counter-ions of groups 1 and 2, only magnesium, sodium, and, particularly, lithium are important for this synthetic purpose. Although it should be remarked that most aldol additions leading to enantiomerically pure products rely on boron, tin, titanium, and zirconium enolates, topics that will be discussed in Chapters 1 and 3 of Part I of this book and Chapters 3–5 of Part II, there are also advantages of the more polar lithium and magnesium enolates. They are, in particular, highly reactive and can be added to aldehydes under mild conditions at low temperatures. Furthermore, their handling is easy, and they can be used on a large scale. As a consequence, a variety of useful aldol additions that rely on lithium and magnesium enolates have been developed and are applied fairly frequently.

1.4.1
Addition of Chiral Enolates to Achiral Carbonyl Compounds

1.4.1.1 α-Substituted Enolates

A variety of carbohydrate-derived ketones have been converted into their corresponding lithium enolates and used as chiral nucleophiles in additions to aldehydes. Induced diastereoselectivity was, however, found to be moderate only. Chiral ketones, oxazolidinones, amides, and esters, on the other hand, performed amazingly well when used as boron, titanium, or tin enolates in aldol additions [49, 52, 55, 57]. The corresponding lithium enolates, however, resulted in substantially lower stereoselectivity. Interestingly, the lithium enolate generated from α-siloxy ketone **45** leads, stereoselectively, to the *syn* aldol **46a**; the diastereomeric ratio (*dr*) exceeds 95:5, defined as the ratio of the major isomer to the sum of all other isomers. In contrast, the corresponding boron enolate furnishes the diastereomeric *syn* aldol **46b**, which results from the opposite induced stereoselectivity (Scheme 1.14). The different behavior of the enolates is explained by a transition state model **48a** with a chelated lithium counter-ion and a non-chelating boron atom in **48b**. When ketone **45** is deprotonated by treatment with bromomagnesium tetramethylpiperidide, the *anti* diastereomer **47a** is obtained in substantial excess relative to the minor product, the *anti* stereoisomer **47b**. It turns out that the enolates with the different counter-ions lithium, boron, and magnesium are, in a sense, complementary. The *anti* selectivity of the magnesium enolate is rationalized by assuming that it has the (*E*) configuration (Scheme 1.14) [45, 97–100].

Procedure: 6-Hydroxy-3-trimethylsilyloxy-4-alkanones 47a by Magnesium-mediated Addition of (*S*)-5,5-Dimethyl-4-trimethylsiloxy-3-hexanone to Aldehydes [100]. An oven-dried 5-mL Wheaton vial is flushed with nitrogen and 2,2,6,6-tetramethylpiperidine (0.26 mL, 1.5 mmol), dry THF (0.5 mL) and ethylmagnesium bromide (1.2 M in THF, 1.16 mL, 1.4 mmol) are added. The vial is capped securely and heated with stirring at 70 °C for 24 h. The resulting solution is cooled to 0 °C, and (*S*)-5,5-dimethyl-4-trimethylsiloxy-3-hexanone (0.10 mL, 0.08 g, 0.50 mmol) in THF (0.25 mL) is added over 30–60 min with a syringe pump. The solution is stirred for 1.5 h at 0 °C after addition is complete. The enolate solution is then cooled to −78 °C and the aldehyde (2 mmol) is added dropwise. After 30 min the reaction is quenched by pouring the mixture into satd aq. $NaHCO_3$ (5 mL). The layers are separated, and the aqueous phase is extracted with diethyl ether (5 × 10 mL). The combined organic layers are washed with cold HCl (1%, 10 mL) and satd aq. $NaHCO_3$ (10 mL). After drying, the solution is concentrated to yield the product as a clear oil. Ratios of diastereomers are determined by integration of the relevant peaks in the ^{1}H NMR spectra of the crude products. Purification is accomplished by flash chromatography (diethyl

1. $LiN(i\text{-}Pr)_2$/THF -78°C
2. TMEDA
3. RCHO

d.r. > 95 : 5

Me_3SiO **46a**

1. $Bu_2BOTf/(i\text{-}Pr)_2NEt$
2. RCHO
3. H_2O_2, OH

d.r. > 95 : 5

Me_3SiO **46b**

Me_3SiO **45**

1. NMgBr
2. RCHO

d.r. > 92 : 8

Me_3SiO **47a**

1. NMgBr
2. HMPA/$ClTi(Oi\text{-}Pr)_3$
3. RCHO

d.r. > 95 : 5

Me_3SiO **47b**

Me_3C, $SiMe_3$, H, Li, R, Me — **48a** → **46a**

Me_3SiO, CMe_3, H, BBu_2, R, Me — **48b** → **46b**

Scheme 1.14
Stereodivergent aldol addition of (*S*)-5,5-dimethyl-4-(trimethylsilyloxy)-3-hexanone (**45**).

ether:hexanes, 5:95). The products **47a** are thus obtained: R = *i*-Pr, 70%; R = *t*-Bu, 80%; R = Ph, 80%.

Because the titanium enolate of the ketone **45** affords the stereoisomeric product **47b** this completes a method of stereodivergent aldol addition. Starting from the identical chiral ketone **45** they lead to all of the different stereoisomeric products in a controlled manner [45].

The chiral α-benzoyloxyketone **49**, accessible from mandelic acid, also reacts stereoselectively with aldehydes to give the *syn* aldols **50a**/**b**. Both the lithium [101] and the titanium [102] enolates lead to the predominant formation of the diastereomer **50a**. However, the stereoselectivity obtained by use of the titanium enolate surpasses that of the lithium analog (Eq. (23)).

1. $LiN(i\text{-}Pr)_2$/THF
2. RCHO
or
1. $LiN(i\text{-}Pr)_2$
2. $ClTi(Oi\text{-}Pr)_3$
3. RCHO

49 → **50a** + **50b**

Enolate counterion	R	ratio **50a : 50b**
Li	Et	96 : 4
Li	Me_2CH	96 : 4
Ti	Et	97 : 3
Ti	Me_2CH	97.7 : 2.3

(23)

Remarkably high stereoselectivity is obtained by means of the sodium enolate of α-*N*,*N*-dibenzylamino-substituted ketone **51**, a counter-ion not very frequently used in stereoselective aldol additions. In this instance, however, the sodium enolate turned out to be more efficient than the lithium analog. The predominant formation of the main diastereomeric product **52a** rather than **52b** is explained by an open transition state, assumed to be strongly favored over the cyclic transition state, when the more "ionic" sodium enolate is used rather than the corresponding lithium reagent (Eq. (24)) [103].

A large variety of propionic acid esters and higher homologs having a chiral alcohol moiety have been used in additions to aldehydes [56, 57]. It turned out, however, that the lithium enolates result in only moderate simple diastereoselectivity and induced stereoselectivity, in contrast with the corresponding boron, titanium, tin, or zirconium enolates and silyl ketene acetals, with which stereoselectivity is excellent. The same feature has been observed in enolates derived from chiral amides and oxazolidinones, as

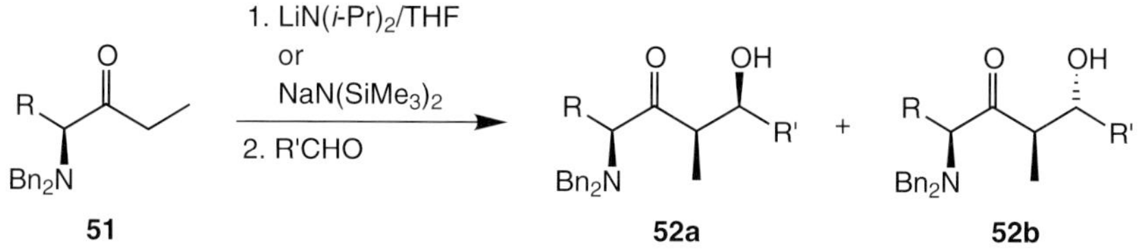

enolate counter ion	R	R'	ratio **52a** : **52b**
Li	Me	Ph	89 : 11
Na	Me	Ph	94 : 6
Li	$CHMe_2$	Ph	63 : 37
Na	$CHMe_2$	Ph	>95 : 5

(24)

outlined in Chapter 2 and in Chapter 3 of Part I. The aldol additions of the chiral propanoate **53**, generated from triphenylglycol, serve as an illustrative example. Whereas the lithium enolate **54** (M = Li) gives the diastereomers **55a**, **b** and **56a**, **b** in a more or less stereo-random manner, acceptable simple diastereoselectivity in favor of the *anti* product combined with high induced stereoselectivity was obtained only after transmetalation of the lithium enolate into the zirconium species (Eq. (25)) [104, 105].

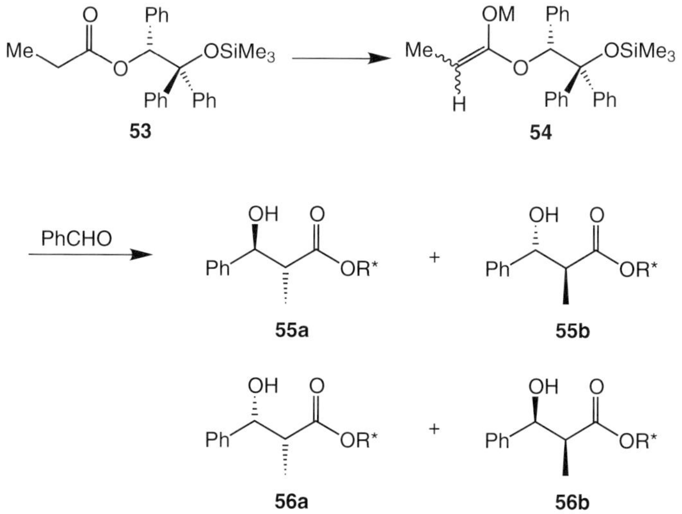

(25)

enolate counterion	ratio *anti*-**55** : *syn*-**56**	ratio **55a** : **55b**
Li	70 : 30	68 : 32
Cp_2ZrCl	90 : 10	>97 : 3

The aldol reaction of chiral lactones, developed in the context of "self-reproduction of chirality" [106], takes advantage of the easy generation and high reactivity of lithium enolates. When the chiral *cis* dioxolanone **57**, readily available from (*S*)-lactic acid and pivalaldehyde, is treated with LDA, the lactone enolate **58** is generated. Although the stereogenic center originating from lactic acid has vanished, because of enolate formation, the acetal carbon atom maintains the chiral information and the *t*-butyl residue directs the topicity in the addition to aldehydes and unsymmetrical ketones, so high stereoselectivity is achieved (Eq. (26)) [107].

57 →[$LiN(i\text{-}Pr)_2$, THF; -78°C] **58** →[PhC(O)Me]

(26)

By analogy, a series of heterocyclic compounds **59** have been deprotonated to give the corresponding α-hetero-substituted cyclic enolates **60**. On addition to aldehydes they are found to react with both high simple diastereoselectivity and induced stereoselectivity (Eq. (27)) [106–108].

59 → **60**

(27)

X = O, NR', S; Y = O, NR'; R = H, Alkyl

The concept has also been extended to a six-membered homolog, the dioxanone **61**, which furnishes the enolate **62** by deprotonation. Here again, subsequent aldol addition proceeds with high stereoselectivity and the diastereomer **63a** results predominantly or almost exclusively (Eq. (28)) [109, 110]. With all heterocyclic enolates **58**, **60**, and **62** stereocontrol is relatively easily accomplished, because of the rigid structure of the heterocycle.

Procedure: Aldol Addition of 2-*tert*-Butyl-6-methyl-one-1,3-dioxan-4-one to Propanal [110]. An ice-cold solution of $(i\text{-}Pr)_2NH$ (11.13 mL, 79.4 mmol, 1.14 equiv.) in THF (160 mL) is treated with a solution of *n*-butyllithium in hexane (1.4 M, 53 mL, 79.4 mmol, 1.14 equiv.), kept at 0 °C for 15 min, then cooled to −78 °C. To this solution of LDA is added the dioxanone **61** (12.0 g, 69.7 mmol) in THF (80 mL) at such a rate that the temperature never ex-

CMe3 LiN(i-Pr)2 THF; -78°C CMe3 RCHO
61 62 OLi

(28)

63a 63b

ceeds −70 °C; the mixture is then maintained at −78 °C for 45 min. To the resulting enolate solution are added (7.15 mL, 99 mmol, 1.42 equiv.) propanal in THF (80 mL), the temperature never being allowed to rise above −70 °C. The reaction mixture is stirred at −78 °C for 3 h then quenched at −78 °C by the addition of satd aq. NH_4Cl (200 mL) then diethyl ether (200 mL). The two phases are separated, and the aqueous phase is extracted with diethyl ether (2 × 200 mL). The combined organic extracts are dried ($MgSO_4$) and the volatile compounds removed by rotary evaporation and then with a high-vacuum pump. The crude product obtained, a 7:1 ratio of epimers at C(1′), is (1′*S*,2*R*,5*R*,6*R*)-2-*t*-butyl-5-(1-hydroxypropyl)-6-methyl-1,3-dioxan-4-one **63a** (R = Et), yield 9.6 g (60%).

Efficient stereochemical control is also provided by the chiral lithium ketone enolate **64**, addition of which to a variety of aldehydes leads to the formation of the corresponding *β*-hydroxy ketones **65**, usually as single products (Eq. (29)) [111].

LiN(i-Pr)2 THF
64

RCHO

(29)

65

R = $CHMe_2$, $(CH_2)_3Me$, $CH{=}C(Me){-}CO_2Et$

The deprotonation of chiral iron acyl complexes, which can be obtained as enantiomerically pure compounds, leads to the corresponding enolates, as shown by the research groups of Davies and Liebeskind [112–115]. The lithium enolate **67a**, however, which originates from propanoate **66a**, reacts stereoselectively with aldehydes or ketones only if it has been transmetalated into the corresponding copper or aluminum enolate (Eq. (30)) [116].

66a: R = Me
66b: R = OBn

n-BuLi, THF, -78°C, R = Me → **67a**

n-BuLi, THF, -78°C, R = OBn → **67b** → (acetone) 40% → **68**

(30)

Stereoselective aldol addition to the lithium enolates themselves has been achieved by reaction of the deprotonated benzyloxy-substituted iron complex **66b** and subsequent reaction with symmetrical ketones. The enolate involved in this procedure is assumed to exist as a chelated species **67b**. The aldol **68** is obtained in a diastereomeric ratio higher than 99:1. The reaction is, nevertheless, plagued by low chemical yield, because of deprotonation of acetone. Because a symmetric ketone is used as an electrophile, the reaction leads to the formation of just one new stereogenic center [117, 118].

1.4.1.2 *α*-Unsubstituted Enolates

When, instead of an α-substituted enolate **29** ($R^2 \neq H$), the α-unsubstituted enolate **29** ($R^2 = H$) is used in an aldol addition, the stereochemistry is, at first glance, simplified, as outlined in Scheme 1.5. Formation of the aldol products **32a** and **32b** shows that now only one new stereogenic center is formed. Nevertheless, aldol addition of α-unsubstituted chiral enolates (or achiral enolates, mediated by chiral ligands or additives) has been a problem, because of insufficient induced stereoselectivity [53]. In fact, using the same chiral auxiliary group R^3 transition from an α-substituted to an α-unsubstituted enolate is often accompanied by complete loss of the ability to discriminate between the enantiotopic faces of an aldehyde, so that more or less equal amounts of the stereoisomers **32a** and **32b** result (cf. Scheme

1.5). The problem occurs in aldol additions of methyl ketones and all kind of acetic acid derivative, esters, amides, and thioamides [53, 54].

In a pioneering investigation, the lithium enolate derived from 3-methyl-2-pentanone was added to aldehydes. Only moderate diastereoselectivity was obtained, however [119]. Exceptionally high induced stereoselectivity was observed when camphor-derived ketone **69** was converted into the lithium enolathe and subsequently added to aldehydes. α-Cleavage at the carbonyl group enabled the formation of β-hydroxy aldehydes and acids in high enantiomeric excess (Eq. (31)) [60]. The work on the aldol addition of methyl ketones led to a variety of stereoselective variants which rely mainly on boron enolates and will be discussed in Chapter 3 of Part I of this book. On the other hand, the transition from lithium to boron enolates did not significantly improve the induced stereoselectivity of chiral acetamides [53]. Several highly stereoselective procedures based on tin and titanium enolates have been developed, however.

OSiMe$_3$ **69** → 1. LiN(i-Pr)$_2$, THF, -78°C; 2. RCHO; $d.r.$ > 95 : 5 → OSiMe$_3$ OH R → X = H, OH

(31)

Many attempts have been made to add chiral acetates to aldehydes or prochiral ketones, to obtain non-racemic β-hydroxycarboxylic esters. Here again, several variants based on boron and titanium enolates and on Mukaiyama aldol additions of silyl ketene acetals have been developed, and will be described in Chapter 2 (titanium enolates), Chapter 3 (boron enolates) and in Part II (Mukaiyama reaction). For enolates of group 1 and 2 elements the following fruitful approaches were elaborated.

First, sulfinyl acetates **70** [120, 121], carrying their chiral information in the sulfoxide moiety were efficiently deprotonated with t-butylmagnesium bromide and added to aldehydes to give the aldol adducts **71**. Removal of the sulfinyl residue is accomplished – in an immolative manner – by reduction with aluminum amalgam to furnish t-butyl β-hydroxycarboxylic esters **72**. Remarkably, the method developed by Solladié and Mioskowski is not restricted to aldehydes as electrophilic components, but has been extended to prochiral ketones also (Scheme 1.15) [122]. The transition-state model **73**, in which the magnesium atom is chelated by the enolate, sulfoxide, and carbonyl oxygen atoms, serves to explain the stereochemical outcome of the reaction. It is plausible that the aldehyde approaches the enolate from the side of the non-bonding electron pair of the sulfoxide (opposite to the aryl residue) and that the larger group R occupies a position *anti* to the sulfinyl substituent.

1. t-BuMgBr THF, -78°C
2. R^1COR^2

Al/Hg

Starting Material		t-Butyl 3-Hydroxy Ester **72**	
R^1	R^2	*e.e.* (%)	Yield (%)
H	C_7H_{15}	86	80
	$C{\equiv}CC_3H_7$	80	73
	$C{\equiv}CC_6H_{13}$	70	53
	C_6H_5	91	85
CH_3	C_7H_{15}	95	88
	$C{\equiv}CC_3H_7$	48	60
	$C{\equiv}CC_6H_{13}$	36	72
	$COOC_2H_5$	8	80
	$C_2H_4OCOCH_3$	40	90
	C_6H_5	68	75
C_6H_5	CF_3	20	75

Scheme 1.15
Aldol addition of (R)-tolylsulfinylacetate **70** to aldehydes and prochiral ketones. Transition state model **73** (reaction of the magnesium enolate of **70** with RCHO).

Procedure: 3-Hydroxy carboxylic esters by addition of *tert*-butyl (+)-(R)-2-(4-methylphenylsulfinyl)acetate to carbonyl compounds [123]. A solution of *t*-BuMgBr (40 mL; prepared from 3 g Mg, 20 g *t*-BuBr, and 50 mL Et_2O) is added to a solution of sulfinyl ester **70** (1.5 g, 5.9 mmol) in THF (400 mL), at −78 °C, over a period of 20 min, under argon. The mixture is then stirred for 30 min and the carbonyl compound (2 g) in THF (30 mL) is added. After 12 h at −78 °C the mixture is hydrolyzed by addition of satd aq. NH_4Cl (50 mL) and extracted with $CHCl_3$ (2 × 50 mL). The extract is dried with Na_2SO_4 and concentrated. The residue, the β-hydroxy-α-sulfinyl ester, is diluted with THF (400 mL) and water (40 mL) and then treated with aluminum amalgam (4 × 5-g portions) while maintaining the temperature at 15–20 °C. The solvent is evaporated and the residue, the β-hydroxy ester **72**, is purified by column chromatography (silica gel; Et_2O–petroleum ether 20:80).

The Solladié procedure has been successfully applied in a synthesis of maytansin [124]. In an analogous manner, chiral sulfinyl acetamides **70** (NMe_2 instead of $OCMe_3$) can be added to aldehydes and deliver β-hydroxy amides in high enantiomeric excess, again after reductive removal of the chiral auxiliary group [125]. More recently, the lithium enolate derived from the acetate **74** with axial chirality has been found to react with aldehydes in a highly stereoselective manner to give β-hydroxy esters **75** that can be converted into the corresponding carboxylic acids by alkaline hydrolysis (Eq. (32)) [126].

1. LiN(*i*-Pr)$_2$, THF, −78°C
2. RCHO

d.r. > 97 : 3

74 → **75** (32)

Chiral acetyl iron complexes **76** also seem predestined to serve as reagents that enable introduction of a chiral acetate unit into aldehydes. In contrast with the benzyloxy-substituted derivative **66b** (Eq. (30)), however, only marginal induced stereoselectivity is achieved when the lithium enolate of **76** is added to aldehydes, and the diastereomer **77a** is formed in low preference compared with **77b** (Eq. (33)). High diastereoselectivity is obtained only after transmetalation [112, 114, 115].

A significant improvement – as far as chiral lithium enolates of acetyl iron complexes are concerned – came from the complex **78**, which carries a (pentafluorophenyl)diphenylphosphane ligand instead of the usual triphenylphosphane. Thus, the enolate **79**, generated by treatment with LDA, gives the diastereomeric adducts **80a** and **80b** in a diastereomeric ratio of 98.5:1.5 on treatment with benzaldehyde. A donor–acceptor interaction between the enolate oxygen atom and the fluorinated aromatic ring, supported

1. n-BuLi, THF, -78°C; 2. PhCHO

76 → 77a + 77b

d.r. 57 : 43

(33)

by spectroscopic studies, in the boat like transition state model **81** is assumed to be responsible for the observed *lk*-topicity (i.e. the (*S*)-enolate **79** approaches the aldehyde from its *Si* face) of the reaction (Scheme 1.16) [127]. Despite the elegance of the concept of Davies–Liebeskind enolates, one should be aware that preparation of the acyl iron complexes definitely needs resolution. It is somewhat typical that a series of procedures has been elaborated by using the racemic iron complexes [115].

It is highly desirable that chiral α-unsubstituted enolates should be available by simple methods from enantiomerically pure starting materials that are inexpensive and readily accessible in both enantiomeric forms. This postulate seems to be fulfilled to a reasonable extent by (*R*)- and (*S*)-2-hydroxy-1,1,2-triphenylethyl acetate **83** ("HYTRA") [53, 128, 129]. It is readily prepared from methyl mandelate which is first converted into triphenylglycol **82** and subsequently converted into the acetic ester **83** by treatment with acetyl chloride (Eq. (34)). Both enantiomers of the reagent are readily accessible, because both (*R*)- and (*S*)-mandelic acid are industrial products [130]. Diol **82** and acetate **83** are commercially available.

(*R*)-methyl mandelate → PhMgBr, Et_2O, reflux, 77% → **82**

→ acetyl chloride, pyridine, 92% → (*R*)-**83**

(34)

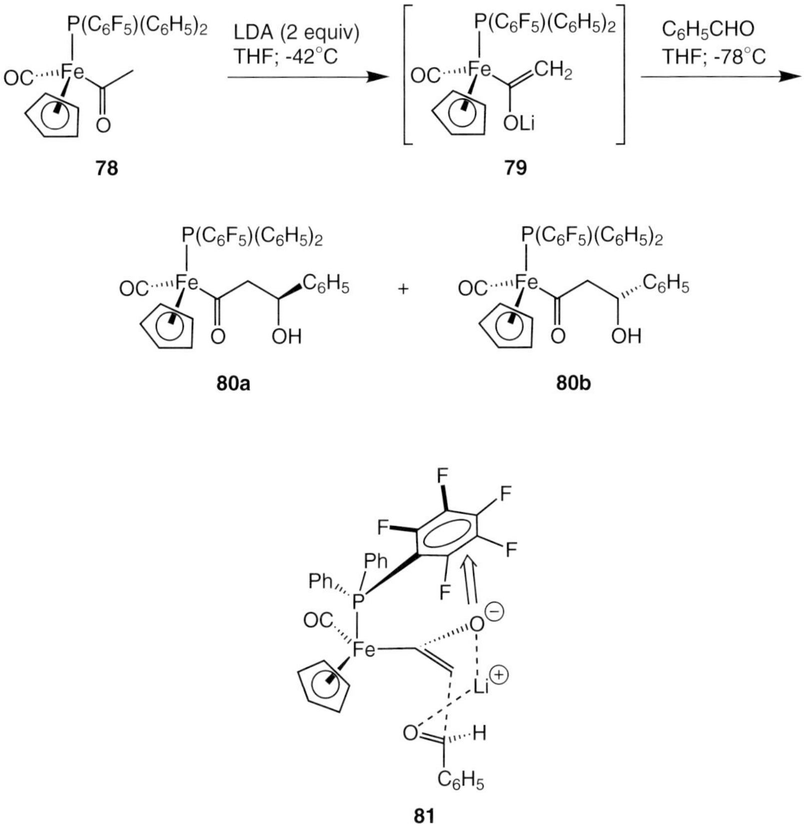

Scheme 1.16
Diastereoselective aldol addition of lithiated (pentafluorophenyl)diphenyl-substituted acetyl iron complexes **78**. Transition state model **81**.

Double deprotonation of the chiral acetate (*R*)-**83** by treatment with 2 equiv. LDA enables generation of the enolate **84**. Remarkably, the dilithiated reagent **84** dissolves in THF whereas the ester **83** is fairly insoluble. Bridging of the oxygen anions by the lithium cations in the enolate **84** might explain the enhanced solubility. When the lithium enolate is added to aldehydes at −78 °C, the diastereomeric aldol adducts **85a** are formed predominantly, the ratio of diastereomers **85a**:**85b** ranging between 10:1 and 12:1. Enhancement of the induced stereoselectivity can be accomplished by transmetalation to the corresponding magnesium enolate and by performing the addition to the aldehyde at lower temperatures (−110 °C to −135 °C, using 2-methylbutane as co-solvent). Thus diastereomeric ratios up to 50:1 can be achieved. A single recrystallization of the crude product mixture

Scheme 1.17
Stereoselective aldol additions of the chiral acetate **83**.

usually gives the major diastereomer **85a** in the pure form. Mild alkaline hydrolysis of the aldol adducts **85** furnishes the β-hydroxycarboxylic acids **86** and triphenylglycol **82**, which is easily separated and can be reused (Scheme 1.17) [131].

When the lithium enolate **84** is added to propenal (without transmetalation), the diastereomeric esters **87a** and **87b** are formed in the ratio 92:8. In this reaction the crude mixture **87a**/**87b** was hydrolyzed to give the carboxylic acid (*R*)-**88** in 83.5% ee. To obtain the enantiomerically pure 3-hydroxy-4-pentenoic acid, enrichment was performed by single recrystallization of the ammonium salt, formed from (*S*)-**1**-phenylethylamine. When the amine has been liberated from the salt the carboxylic acid (*R*)-**88** is obtained in $>$99.8% ee and 41% overall yield (Scheme 1.18) [132]. The (*S*) enantiomer, but not the (*R*) enantiomer, of 3-hydroxy-4-pentenoic acid **88** (both prepared according to this procedure) has been shown to be a substrate for the enzyme 3-hydroxybutanoate dehydrogenase – another example of the different biological activity of enantiomeric compounds [133].

Procedure: (*R*)-3-Hydroxy-4-pentenoic acid (88) by aldol addition of doubly deprotonated (*R*)-HYTRA (83) [132]. A 250-mL, two-necked, round-bottomed flask is equipped with a magnetic stirrer, a septum, and a connection to a combined vacuum and nitrogen line. The air in the flask is replaced by nitrogen and dry THF (100 mL) and diisopropylamine (37.7 mL, 0.264 mol)

1. 2 LiN(*i*-Pr)$_2$
2. acrolein

83

87a + 87b

KOH
MeOH, H_2O

– HO… 82

88
83.5% *e.e.*

1. (Ph)(Me)CH–NH_2
2. recrystallization
3. NaOH

88
99.8% *e.e.*

Scheme 1.18
Synthesis of (*R*)-3-hydroxy-4-pentenoic acid **88** by stereoselective aldol addition of the chiral acetate (*R*)-**83**.

are injected via syringes via the septum. The mixture is cooled to –78 °C and treated, while stirring, with a solution of *n*-butyllithium (15%, 168 mL, 0.269 mol) in hexane. The dry ice–acetone bath is replaced with an ice bath and stirring is continued for 30 min.

A 2-L, three-necked, round-bottomed flask equipped with a mechanical stirrer, a septum, and a connection to a combined vacuum and nitrogen line is charged with (*R*)-**83** (40.0 g, 0.120 mol). The air in the flask is replaced by nitrogen and dry THF (400 mL) is added through a cannula of 2 mm i.d., during which the flask is slightly evacuated. The suspension is stirred at –78 °C in a dry ice–acetone bath. The ice-cold solution of LDA is added via a cannula, with vigorous stirring, during which the 2-L flask is slightly evacuated. The mixture is stirred at 0 °C for 30 min to complete double deprotonation. A clear, orange solution forms. This is subsequently cooled to below −70 °C (dry ice/acetone bath). The septum is removed cautiously (overpressure) and dry 2-methylbutane (900 mL) is poured into the flask (alternatively, a low-boiling (30–37 °C) fraction of petroleum ether can be used) during which a vigorous stream of nitrogen is maintained. The flask is immediately closed with a septum and a thermocouple, connected to a resistance thermometer, is introduced via the septum. The reaction flask is plunged into a liquid-nitrogen bath, the depth of immersion being 2–3 cm. When the temperature of the suspension has reached −125 °C, a solution of propenal (acrolein; 19.2 mL, 0.285 mol) in dry THF (30 mL) is added dropwise via a syringe through the septum at such a rate that the temperature does not exceed −120 °C. Stirring is continued for 30 min at −120° to −125 °C, during which time the yellowish color turns to pale blue. The mixture is treated with satd aq NH_4Cl (250 mL) and left to warm to r.t. The organic solvents are removed in vacuo. The precipitate in the aqueous suspension is separated by suction filtration, washed with several portions of water (total 400 mL), and transferred to a 4-L, round-bottomed flask equipped with a magnetic stirrer and a condenser. Methanol (2.4 L), water (1 L), and potassium hydroxide (72 g) are added, and the mixture is heated under reflux for 3 h. After cooling to r.t. the organic solvent is removed in vacuo. The residual aqueous alkaline suspension is shaken with 200-mL portions of CH_2Cl_2. The aqueous solution is transferred to a 2-L, round-bottomed flask, immersed in an ice bath, and acidified to pH 3 by cautious addition of hydrochloric acid (6 mol L^{-1}); the mixture is stirred vigorously with a magnetic stirrer and the pH is controlled carefully in order to avoid over-acidification. The clear solution is saturated with NaCl and extracted with ethyl acetate (8 × 200 mL), during which the pH of the aqueous layer is monitored and, if necessary, readjusted to pH 3 by the addition of hydrochloric acid (6 mol L^{-1}). The combined organic layers are dried with $MgSO_4$, concentrated, and the oily residue is distilled under reduced pressure in a short-path distillation apparatus to afford the colorless acid **88**; yield 8.32 g (59.6%); b.p. 69 °C/0.04 Torr (5.3 Pa); $[\alpha]_D^{17}$ −21.7 ($c = 1.95\%$ aq. ethanol); 83.5% ee.

A solution of the acid (8.32 g, 0.072 mol; 83.5% ee) in dry diethyl ether (300 mL) at r.t. is placed in a 500-mL, round-bottomed flask equipped with a magnetic stirrer. (−)-(*S*)-1-Phenylethylamine (20 mL, 0.157 mol) is added in one portion with vigorous stirring. A white precipitate forms immediately and the flask is closed with a drying tube filled with calcium chloride. After stirring for 30 min at 25 °C the precipitate is separated by suction filtration and washed with ice-cold diethyl ether (2 × 50-mL). Recrystallization from dry THF (240 mL) affords 13.80 g of a colorless salt, m.p. 127 °C; $[\alpha]_D^{20}$ −9.7 ($c = 1.998$, deionized water). The salt is treated with aq. sodium hydroxide (2%, 400 mL) and the mixture is washed with $CHCl_3$ (3 × 70 mL). The acidified aqueous solution is extracted with ethyl acetate (as described above) and the combined organic extract is dried with $MgSO_4$ and concentrated. To remove remaining traces of solvent the flask containing the product is connected via a short, curved glass tube to a two-necked, liquid-nitrogen-cooled flask, which is connected to an efficient oil pump; yield 5.77 g (69.4%, 41.4% relative to (*R*)-2-hydroxy-1,2,2-triphenylethyl acetate); $[\alpha]_D^{17}$ −26 ($c = 0.996$, 95% aqueous ethanol); >99.8% ee (determined by ^{1}H NMR measurement of the methyl ester in the presence of $Eu(hfc)_3$, no signals of the (*S*) enantiomer are detected).

Since the first report of stereoselective aldol additions of the chiral acetate **83**, the reagent has been applied frequently in syntheses of natural products and biologically active compounds. Among these are γ-amino-β-hydroxybutanoic acid ("GABOB") [134], the enantiomeric naphthoquinones shikonin and alkannin [135], D- and L-digitoxose [136], desoxy and aminodesoxy furanosides [132], detoxinine [137], tetrahydrolipstatin and related pancreatic lipase inhibitors [138], statin [139] and statin analogs [140], compactin and mevinolin [141], fluoroolefin peptide isoesters [142], the HMG-CoA synthase inhibitor F-(244) [143], epothilone A [144], the A-ring building block of 1α,25-dihydroxyvitamin D_3 [145], intermediates for (23*S*)-hydroxyvitamin D3 derivatives [146], building blocks of lankacidin C [147], the synthetic statin NK-104 [148], the C1–C9 segment of bryostatin [149], the C20–C34 segment of the immunosuppressant FK-506 [150], and pyranoyl steroids having hypocholesterolemic properties [151]. Selected examples are shown in Scheme 1.19. Furthermore, a large variety of synthetic inhibitors of HMG-CoA reductase have been synthesized by including the HYTRA aldol procedure as one of the key steps [152–158]. The enzyme is responsible for reduction of hydroxymethylglutaryl CoA to mevalonic acid, a key step in cholesterol biosynthesis. HMG CoA reductase inhibitors contain a β,δ-dihydroxycarboxylic ester or carboxylate moiety or the corresponding δ-lactone. Several of these compounds, which can be regarded as synthetic analogs of compactin and mevinolin, have found their way to the marketplace, because of their hypocholesterolemic activity. Selected examples of the structures of these drugs are given in Scheme 1.20. As in the previous scheme, the stereogenic center generated by aldol addition of the chiral acetate **83** is marked with an asterisk.

The (*R*)-configured reagent **83** always attacks the aldehyde predominantly from the *Re* side, the (*S*) acetate **83** correspondingly from the *Si* side. Thus, there is a predictable *lk* topicity in HYTRA aldol additions (Scheme 1.21).

It has been reported that the induced stereoselectivity in aldol additions of the chiral lactate **83** can be improved by using an excess of base (LDA or lithium hexamethyldisilazane) in the deprotonation step [159].

1.4.2
Addition of Achiral Enolates to Chiral Carbonyl Compounds

In a chiral aldehyde or ketone the two carbonyl faces are diastereotopic and the products resulting from either *Re* or *Si* face attack of the enolate are

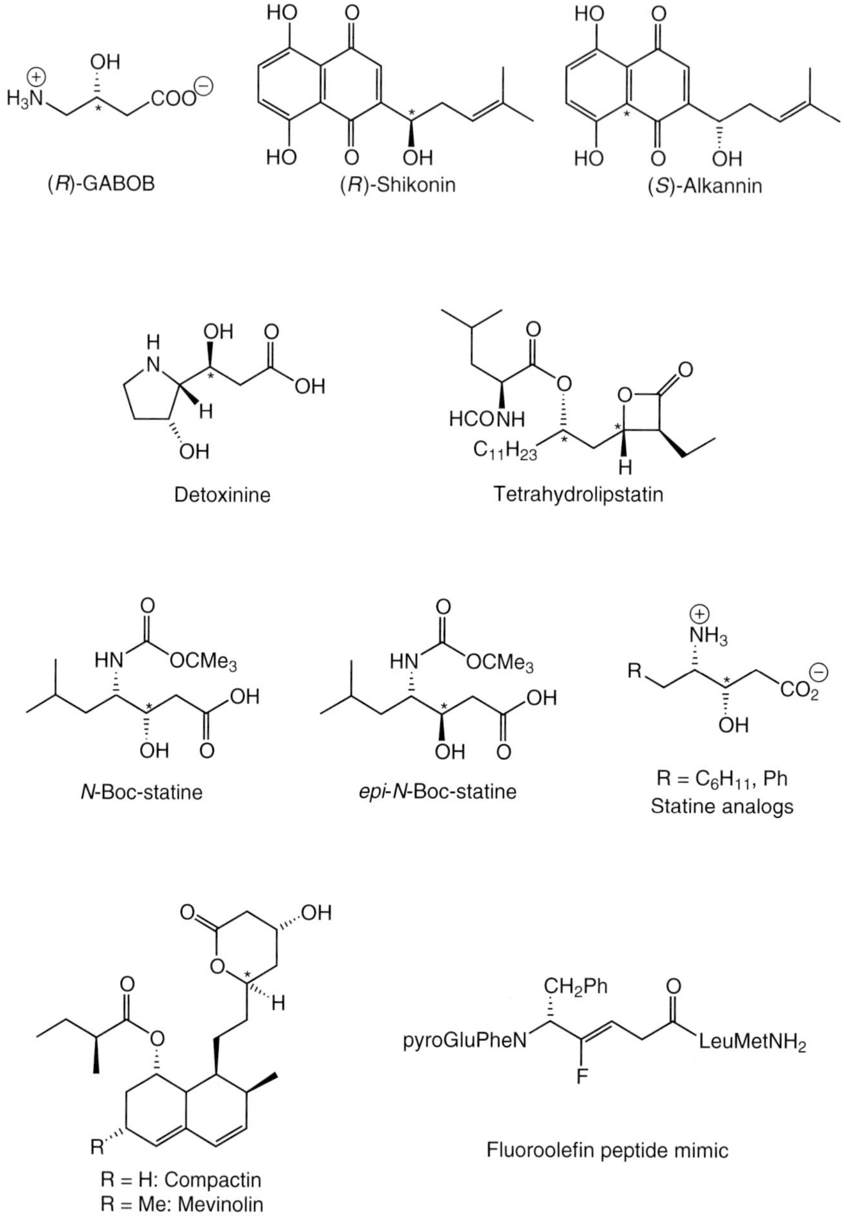

(*R*)-GABOB

(*R*)-Shikonin

(*S*)-Alkannin

Detoxinine

Tetrahydrolipstatin

N-Boc-statine

epi-N-Boc-statine

R = C_6H_{11}, Ph
Statine analogs

R = H: Compactin
R = Me: Mevinolin

Fluoroolefin peptide mimic

Scheme 1.19
Natural products and biologically active compounds synthesized by use of the HYTRA aldol method.

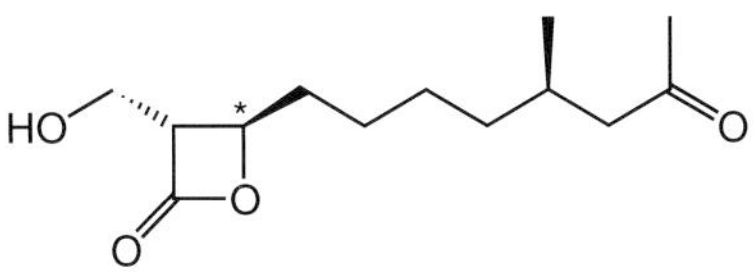

HMG-CoA synthase inhibitor

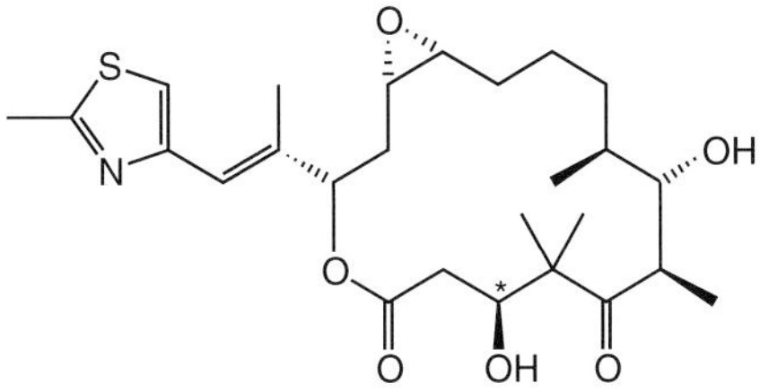

Epothilone A

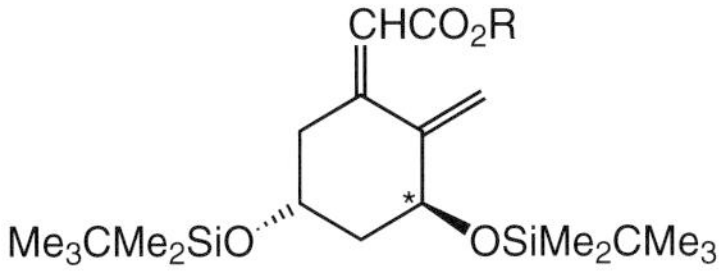

Vitamin D_3 A-ring building block

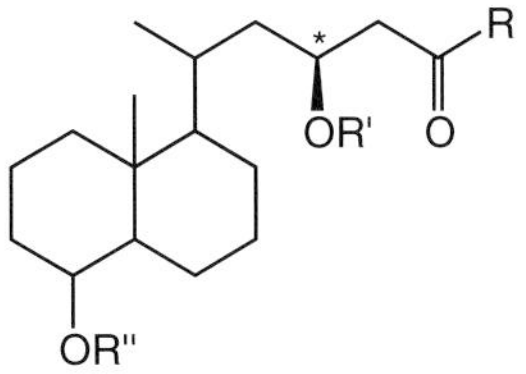

23-Hydroxy vitamin D_3 building blocks

Scheme 1.19 *(continued)*

diastereomers. The question of the stereochemistry of nucleophilic addition to chiral aldehydes or ketones has been addressed by the seminal studies of Cram and co-workers [160–162]. Indeed, the stereochemical outcome of this type of reaction is best rationalized by either the "Cram–Felkin–Anh model" [163] or "Cram's cyclic model" when the carbonyl group is substituted by ether or amino residues. The latter type of stereochemical result has also been termed "chelation control", the former type of reaction accordingly as "non chelation control" [164]. Most of these investigations have been performed on racemic substrates. In view of the tendency towards enantiomerically pure products, emphasis is given here to those procedures that start from non-racemic aldehydes or ketones. In the aldol reaction "chelation control" is usually provided by those variants that use strong Lewis acidic enolate metals, for example boron, titanium, and tin. In particular, "chelation control" results very frequently in Mukaiyama aldol additions to α-oxygen- and α-nitrogen-substituted aldehydes [165].

Lithium enolates, in contrast, either give predominantly the product predicted by the "Cram–Felkin–Anh model" or react more or less non-stereoselectively. Thus, the favored formation of the *syn*-aldol product in the reaction of 2-phenylpropanal with the lithium enolates of acetone, pinacolone, methyl acetate, or *N*,*N*-dimethylacetamide is in accordance with Cram's rule or the Felkin–Anh model (Eq. (35)). However, a rather moderate *syn*:*anti* ratio of 3:1 is typical of this type of reaction [51, 67].

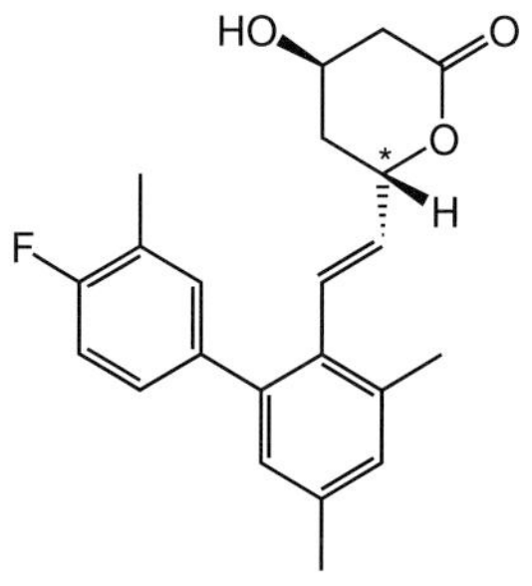

ref. [141] 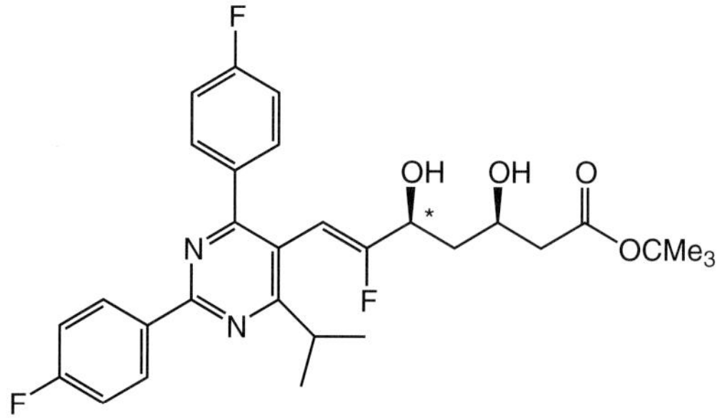 ref. [152]

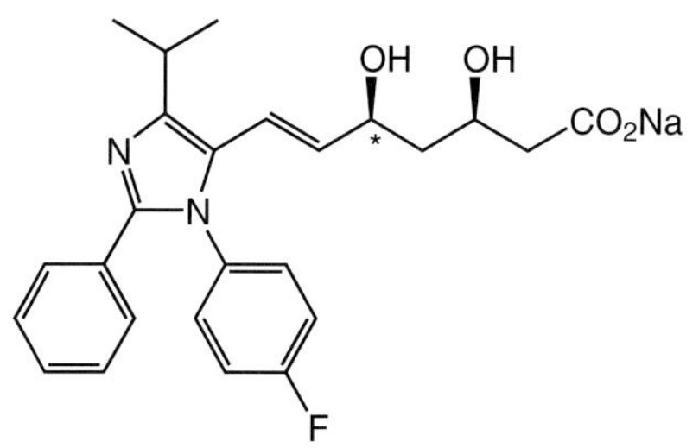

ref. [153] 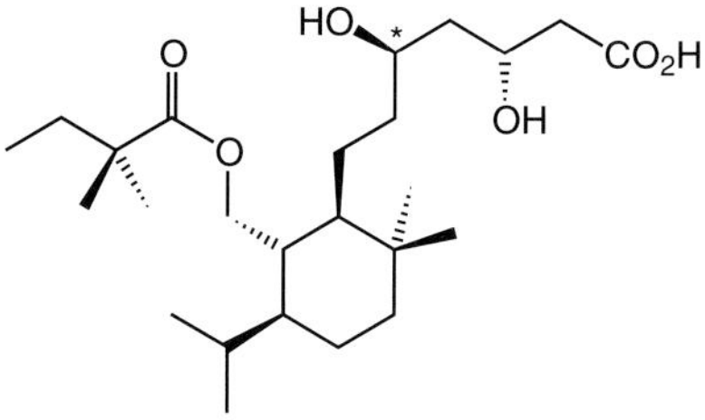 ref. [154]

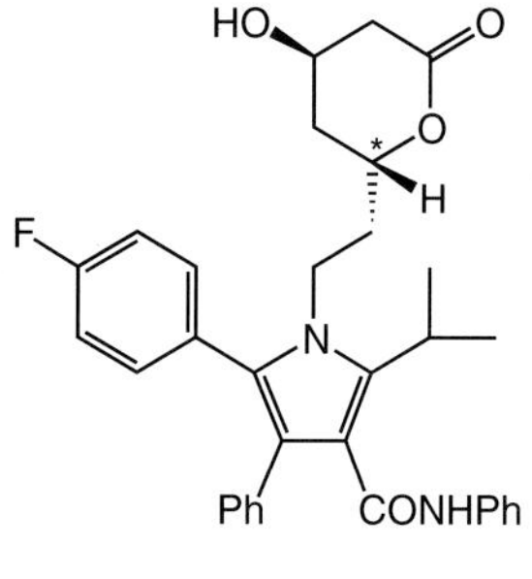

ref. [155] 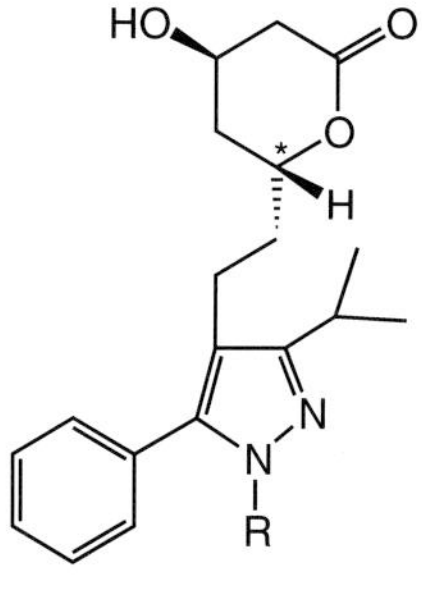 ref. [156] 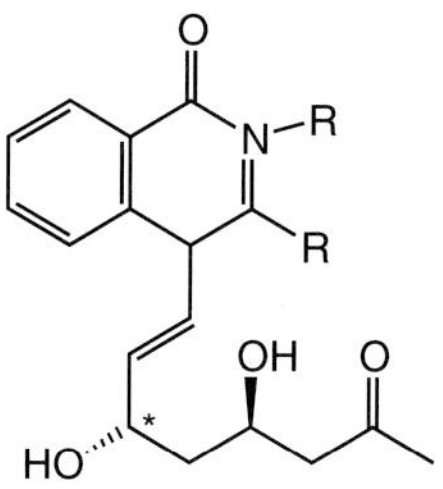 ref. [157]

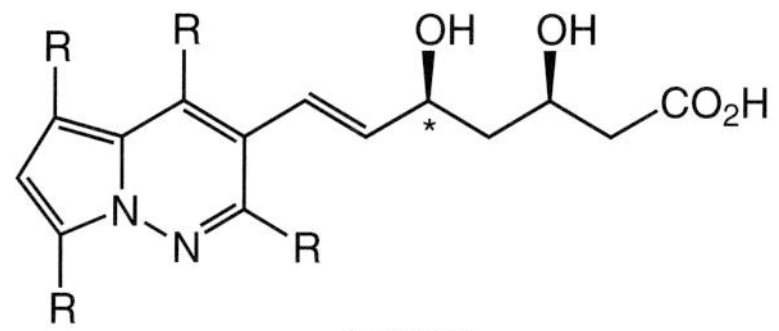

ref. [158]

Scheme 1.20
Hypocholesterolemic-active drugs (HMG-CoA reductase inhibitors) prepared by use of the HYTRA aldol method (selected examples).

(*R*)-**83** (*S*)-**83**

(*R*)-**84** *Re* *Si* (*S*)-**84**

Scheme 1.21
Lk-topicity in aldol additions of (*R*) and (*S*) acetate **83**.

syn 75 : *anti* 25

(35)

When the α-substituted (*E*) enolate **89** is added to 2-phenylpropanal, control of simple diastereoselectivity is provided in as far as the products with the 2,3-*anti* configuration result exclusively. Induced stereoselectivity is lower, however, as indicated by the 80:20 ratio of diastereomeric β-hydroxy esters **90a** and **90b** (Eq. (36)) [51].

89

90a 80 : **90b** 20

(36)

A single diastereomer **93**, however, results from addition of the lithium enolate **92** derived of *t*-butyl thiopropanoate to the chiral, enantiomerically pure aldehyde **91**. The transformation is a key carbon-chain-elongation step in Woodward's synthesis of erythromycin A (Eq. (37)) [166]. Somewhat lower diastereoselectivity is observed in the aldol reaction between the lithium enolate **95** and the chiral aldehyde **94**, a transformation used in a synthesis of maytansin (Eq. (38)). The diastereomeric adducts **96a** and **96b** result in a ratio of 90:10 [167].

MeO O O O OAc O O H O + SCMe$_3$ OLi

91 **92**

→ MeO O O O OAc O O OH O SCMe$_3$

93

(37)

Me_3SiO O H O H + SLi SEt →

94 **95**

Me_3SiO O H OH S SEt + Me_3SiO O H OH S SEt

96a : **96b**

(38)

A variety of α-alkoxy-substituted aldehydes have been submitted to aldol addition of lithium enolates. "Cram–Felkin–Anh" selectivity is usually observed, although often with rather low stereoselectivity. Exceptionally high diastereoselectivity results from the aldol reaction between the lithium enolate of pinacolone and isopropylidene glyceraldehyde. Thus, the β-hydroxy ketone **97** is obtained as a single product (Eq. (39)). Distinctly lower selectivity is observed when the same aldehyde is submitted to aldol additions of ester enolates, however [168].

97

(39)

The *N*-dibenzyl protecting group has been developed as a tool to provide non-chelate-controlled additions to α-amino aldehydes. Thus, *anti*-configured aldol adducts are obtained predominantly when α-*N*-dibenzyl-protected aldehydes **98** are submitted to aldol additions of lithium enolates as shown in Eqs. (40) and (41) [169, 170].

98

anti *syn*

(40)

R^1	R^2	diastereomeric ratio *anti* : *syn*	yield
Me	H	95 : 5	82
$PhCH_2$	H	90 : 10	83
Me	Me	97 : 3	84

82%

d. r.: >96 : <4

(41)

Further selected examples of diastereoselective aldol reactions between lithium enolates and chiral aldehydes are given in Eqs. (42) [171], (43) [172], and (44) [173]. In the last example, the salt-free generation of the lithium enolate was occasionally found to be crucial to stereoselectivity [174].

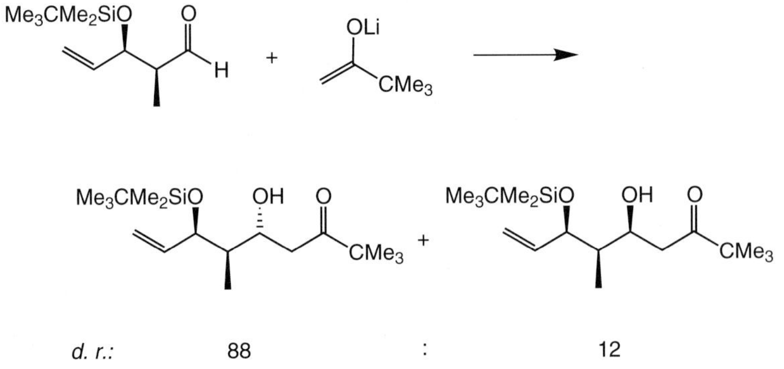

d. r.: 88 : 12

(42)

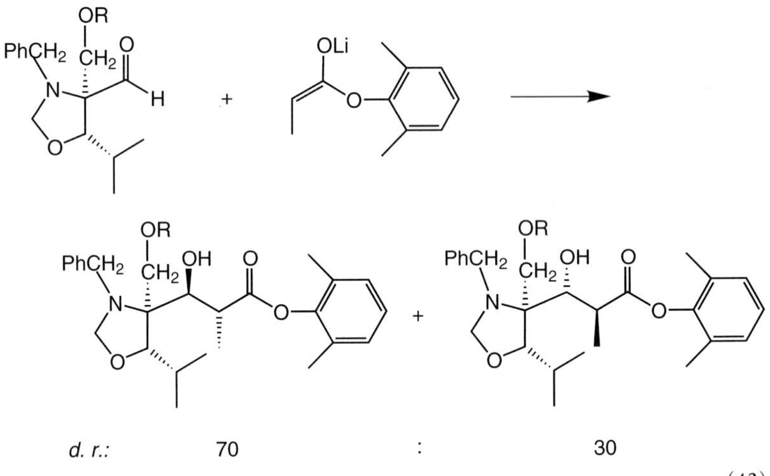

d. r.: 70 : 30

(43)

87%

d. r.: 82 : 18

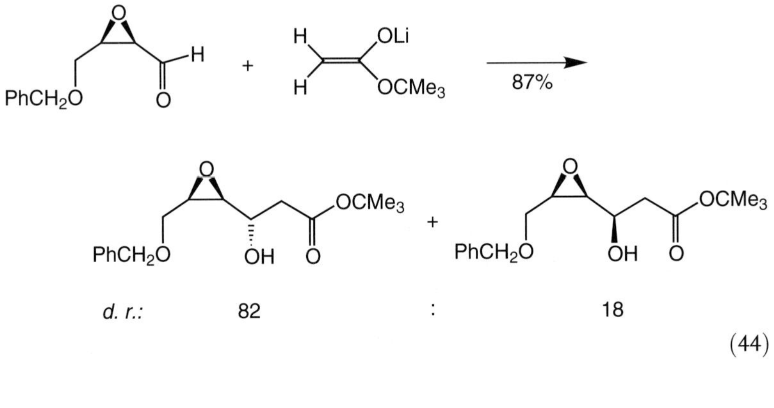

(44)

1.4.3 Addition of Chiral Enolates to Chiral Carbonyl Compounds

If a chiral aldehyde reacts with an achiral enolate the induced stereoselectivity is determined by the "inherent" preference of the aldehyde to be attacked from its *Re* or *Si* face. If, however, a chiral aldehyde is combined with a chiral enolate one must consider whether the inherent selectivities of the two reagents will be consonant in one of the combinations ("matched pair"), but dissonant in the other combination ("mismatched pair"). Thus, different diastereoselectivity results from each combinations. The problem of insufficient stereoselectivity in the "mismatched" combination can be solved by means of highly efficient chiral enolates which can "outplay" the inherent selectivity of the aldehyde. The concept has been applied extensively in the context of boron enolates, a topic that has been reviewed comprehensively [52] and is discussed in detail in Chapter 3 of Part I of this book.

The principle of "enolate controlled" stereochemistry can be demonstrated by use of the chiral acetate **83**. When doubly deprotonated (*R*)- and (*S*)-HYTRA **83** reacts with enantiomerically pure 3-benzyloxybutanal **99** the (*R*)-configured acetate enolate attacks the aldehyde **99** (irrespective of its chirality) predominantly from the *Re* face so that, after hydrolysis, *anti* hydroxycarboxylic acid **100a** results. On the other hand, the (*S*)-configured enolate of **83** attacks the enantiomerically pure aldehyde preferentially from the *Si* side to give *syn* carboxylic acids **100b** with comparable selectivity, as shown in Scheme 1.22 [175].

In addition to this example of "acyclic" stereocontrol, the concept has also been applied to the cyclic chiral lithium enolate **101**; this also resulted in high diastereoselectivity when the enantiomeric enolates were combined with the chiral aldehydes **102**. As demonstrated by Eqs. (45) and (46), here again the stereochemical outcome is determined by the configuration of the enolate **101** [176].

Me, OLi, Me_3C, Bz, (*R*)-**101** + H, O

Me, O, Me_3C, N, H, H, OBz (45)

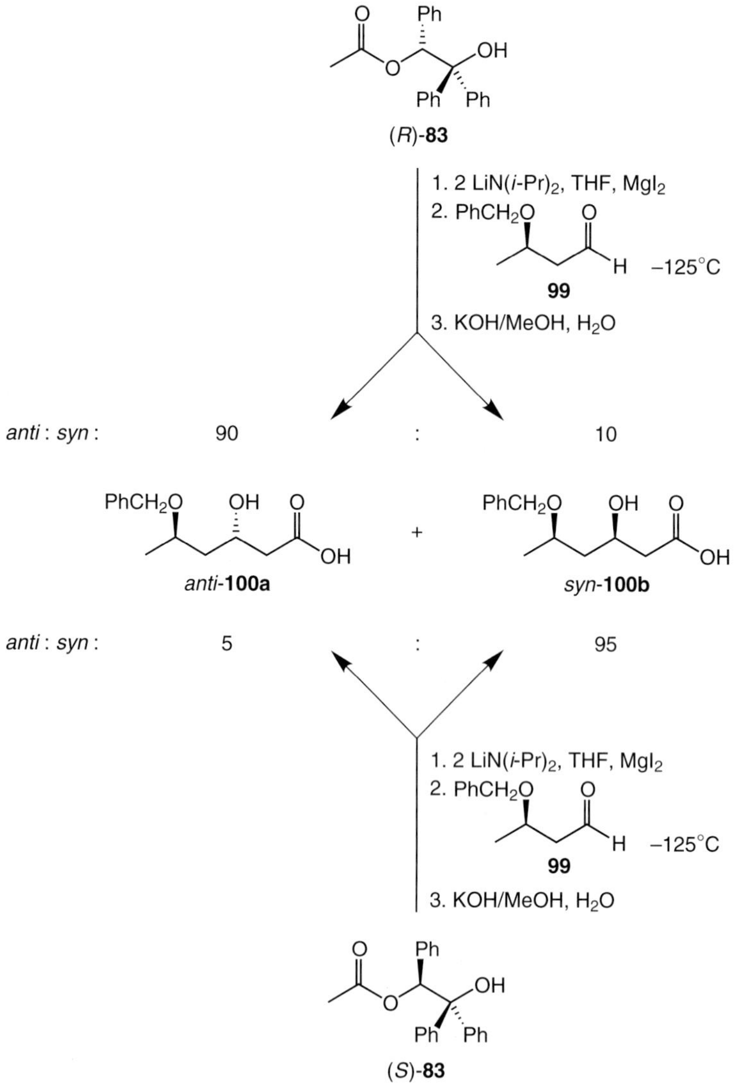

(*R*)-**83**

1. 2 $LiN(i\text{-}Pr)_2$, THF, MgI_2
2. $PhCH_2O$ ... H, −125°C (**99**)
3. KOH/MeOH, H_2O

anti : *syn* : 90 : 10

anti-**100a** + *syn*-**100b**

anti : *syn* : 5 : 95

1. 2 $LiN(i\text{-}Pr)_2$, THF, MgI_2
2. $PhCH_2O$ ... H, −125°C (**99**)
3. KOH/MeOH, H_2O

(*S*)-**83**

Scheme 1.22
"Enolate controlled" addition of doubly deprotonated (*R*) and (*S*) acetate **83** to (*R*)-benzyloxybutanal.

(46)

(*S*)-**101**

1.4.4
Addition of Achiral Enolates to Achiral Carbonyl Compounds in the Presence of Chiral Additives and Catalysts

Combination of achiral enolates with achiral aldehydes mediated by chiral ligands at the enolate counter-ion opens another route to non-racemic aldol adducts. Again, this concept has been extremely fruitful for boron, tin, titanium, zirconium and other metal enolates. It has, however not been applied very frequently to alkaline and earth alkaline metals. The main, inherent, drawback in the use of these metals is that the reaction of the corresponding enolate, which is not complexed by the chiral ligand, competes with that of the complexed enolate. Because the former reaction pathway inevitably leads to formation of the racemic product, the chiral ligand must be applied in at least stoichiometric amounts. Thus, any catalytic variant is excluded per se. Among the few approaches based on lithium enolates, early work revealed that the aldol addition of a variety of lithium enolates in the presence of (*S*,*S*)-1,4-(bisdimethylamino)-2,3-dimethoxy butane or (*S*,*S*)-1,2,3,4-tetramethoxybutane provides only moderate induced stereoselectivity, typical ee values being 20% [177]. Chelation of the ketone enolate **104** by the chiral lithium amide **103** is more efficient – the *β*-hydroxyl ketone *syn*-**105** is obtained in 68% ee and no *anti* adduct is formed (Eq. (47)) [178].

(47)

103

104

105
92%, *e.e.*: 68%

A variety of other chiral lithium amides, for example **106** and **108**, have been applied more recently to bring about enantioselective aldol additions. As shown in Eqs. (48) [179] and (49) [180], both simple diastereoselectivity and induced stereroselectivity can be induced by these reagents. In the latter reaction, the enolate itself becomes chiral, because of desymmetrization of ketone **107** on deprotonation.

OCMe$_3$ + LiN(i-Pr)$_2$ + **106**

1. PhCHO
2. Ac$_2$O

anti (76%)
e.e.: 94%

syn (7%)

(48)

107 + **108** + LiCl

CHO
95%

single diastereomer
e.e.: 90%

(49)

1.5 Conclusion

In a period longer than a century the aldol addition has proven itself to be an extremely useful "work horse" in organic synthesis. During the long history of this reaction the preformed-enolate technique was a breakthrough enabling particular stereoselectivity to be achieved. Although boron and titanium become increasingly favored as counter-ions, lithium and magnesium enolates found their place in the repertoire of organic chemistry. The high reactivity of these reagents combined with easy handling has made them a useful tool in the hands of synthetic chemists.

References

1 A. T. Nielsen, W. J. Houlihan, *Organic Reactions* **1968**, Vol. 16.
2 H. O. House, *Modern Synthetic Reactions*, 2nd ed., W. A. Benjamin, Menlo Park, 1972.
3 C. H. Heathcock in *Comprehensive Organic Synthesis*, Vol. 2 (Ed.: B. M. Trost), Pergamon, Oxford 1991, Chapter 1.5.
4 R. Kane, *J. Prakt. Chem.* **1838**, *15*, 129.
5 J. G. Schmidt, *Ber. Dtsch. Chem. Ges.* **1880**, *13*, 2342; **1881**, *14*, 1459.
6 L. Claisen, A. Claparède, *Ber. Dtsch. Chem. Ges.* **1881**, *14*, 349.
7 "... dasselbe ist zugleich Aldehyd und Alkohol,... es ist der erste Aldehyd eines Glycols": W. Wurtz, *Ber. Dtsch. Chem. Ges.* **1872**, *5*, 326.
8 For the thermochemistry of the aldol reaction, see: J. P. Guthrie, *Can. J. Chem.* **1974**, *52*, 2037; **1978**, *56*, 962; E. M. Arnett, F. J. Fisher, M. A. Nichols, A. A. Ribeiro, *J. Am. Chem. Soc.* **1989**, *111*, 748; J. P. Richard, R. W. Nagorski, *J. Am. Chem. Soc.* **1999**, *121*, 4763.
9 In the case of non-coordinating counter ions, the formation of the aldolate anion has been shown to be endothermic; cf. R. Noyori, I. Nishida, J. Sakata, M. Nishizawa, *J. Am. Chem. Soc.* **1980**, *102*, 1223.
10 V. Grignard, A. Vesterman, *Bull. Soc. Chim. Fr.* **1925**, *37*, 425; C. Weizmann, S. F. Garrard, *J. Chem. Soc.* **1920**, *117*, 324.
11 M. Häusermann, *Helv. Chim. Acta* **1951**, *34*, 1482.
12 J. B. Conant, N. Tuttle in R. Amus, *Organische Synthesen*, Vieweg u. Sohn, Braunschweig 1937, p. 192; cf. *Organikum*, Wiley–VCH, Weinheim, 21nd ed.; **2001**, p. 522.
13 For more recent applications of solid basic metal salts as catalysts in aldol reactions, see: J. F. Sanz, J. Oviedo, A. Márquez, J. A. Odriozola, M. Montes, *Angew. Chem. Int. Ed. Engl.* **1999**, *38*, 506; G. Zhang, H. Hattori, K. Tanabe, *Bull. Chem. Soc. Jap.* **1989**, *62*, 2070; K. Tanabe, G. Zhang, H. Hattori, *Appl. Catalysis* **1989**, *48*, 63; G. Zhang, H. Hattori, K. Tanabe, *Appl. Catalysis* **1988**, *36*, 189.
14 G. Wittig, *Top. Curr. Chem.* **1976**, *67*, 1.
15 G. Wittig, H. Reiff, *Angew. Chem. Int. Ed. Engl.* **1968**, *7*, 7.
16 T. Mukaiyama, *Organic Reactions* **1982**, *28*, 203.
17 A. Russell, R. L. Kenyon, *Org. Synth. Coll. Vol. III*, **1955**, 747.
18 C. R. Conard, M. A. Dolliver, *Org. Synth. Coll. Vol. II*, **1943**, 167.

19 J. G. Leuck, L. Cejka, *Org. Synth. Coll. Vol. I*, **1944**, 283.
20 E. P. Kohler, H. M. Chadwell, *Org. Synth. Coll. Vol. I*, **1944**, 78.
21 G. Kabas, *Tetrahedron* **1966**, *22*, 1213.
22 M. E. Jung, *Tetrahedron* **1976**, *32*, 3.
23 R. E. Gawley, *Synthesis* **1976**, 777.
24 R. B. Woodward, F. Sondheimer, D. Taub, K. Heusler, W. M. McLamore, *J. Am. Chem. Soc.* **1952**, *74*, 4223.
25 G. Büchi, B. Gubler, R. S. Schneider, J. Wild, *J. Am. Chem. Soc.* **1967**, *89*, 2776.
26 G. I. Poos, W. F. Johns, L. H. Sarett, *J. Am. Chem. Soc.* **1955**, *77*, 1026.
27 J. A. Marshall, A. E. Greene, *J. Org. Chem.* **1972**, *37*, 982.
28 T.-L. Ho, *Synth. Commun.* **1974**, *4*, 265 and references given therein.
29 W. S. Johnson, J. J. Korst, R. A. Clement, J. Dutta, *J. Am. Chem. Soc.* **1960**, *82*, 614.
30 For a short survey, see: J. Mulzer, H.-J. Altenbach, M. Braun, K. Krohn, H.-U. Reissig, *Organic Synthesis Highlights*, VCH, Weinheim **1991**, 232.
31 J. N. Collie, *J. Chem. Soc.* **1907**, *91*, 1806.
32 T. M. Harris, C. M. Harris, *Tetrahedron* **1977**, *33*, 2159.
33 S. G. Gilbreath, C. M. Harris, T. M. Harris, *J. Am. Chem. Soc.* **1988**, *110*, 6172.
34 G. Bringmann, *Tetrahedron Lett.* **1982**, *23*, 2009; G. Bringmann, J. R. Jansen, *Tetrahedron Lett.* **1984**, *25*, 2537.
35 Z. G. Hajos, D. R. Parrish, *J. Org. Chem.* **1974**, *39*, 1615; *Org. Synth.* **1985**, *63*, 26.
36 U. Eder, G. Sauer, R. Wiechert, *Angew. Chem. Int. Ed. Engl.* **1971**, *10*, 496.
37 G. Sauer, U. Eder, G. Haffer, G. Neef, R. Wiechert, *Angew. Chem. Int. Ed. Engl.* **1975**, *14*, 417; P. Buchschacher, A. Fürst, *Org. Synth.* **1985**, *63*, 37.
38 C. Agami, C. Puchot, H. Sevestre, *Tetrahedron Lett.* **1986**, *27*, 1501; C. Agami, *Bull. Soc. Chim. Fr.* **1988**, 499.
39 B. List, R. A. Lerner, C. F. Barbas III, *J. Am. Chem. Soc.* **2000**, *122*, 2395; A. B. Northrup, D. W. C. MacMillan, *J. Am. Chem. Soc.* **2002**, *124*, 6798.
40 F. C. Frostick, Jr., C. R. Hauser, *J. Am. Chem. Soc.* **1949**, *71*, 1350; M. Hammell, R. Levine, *J. Org. Chem.* **1950**, *15*, 162.
41 M. W. Rathke, *J. Am. Chem. Soc.* **1970**, *92*, 3222; C. R. Krüger, E. G. Rochow, *J. Organomet. Chem.* **1964**, *1*, 476.
42 M. W. Rathke, A. Lindert, *J. Am. Chem. Soc.* **1971**, *93*, 2318.
43 R. A. Olofson, C. M. Dougherty, *J. Am. Chem. Soc.* **1973**, *95*, 581, 582.
44 H. O. House, M. Gall, H. D. Olmstead, *J. Org. Chem.* **1971**, *36*, 2361.
45 C. H. Heathcock in *Modern Synthetic Methods* 1992, (Ed.: R. Scheffold), VHCA, VCH, Basel, Weinheim **1992**, 1.
46 C. H. Heathcock in *Comprehensive Organic Synthesis*, Vol. 2 (Ed.: B. Trost), Pergamon, Oxford **1993**, Chapter 1.6.
47 For comprehensive reviews, see: D. Seebach, *Angew. Chem. Int. Ed. Engl.* **1988**, *27*, 1624; C. Lambert, P. v. R. Schleyer in *Houben Weyl, Methoden der Organischen Chemie*, Vol. E 19d

(Ed.: M. Hanack), Thieme, Stuttgart **1993**, p. 75; L. M. Jackman, J. Bortiatynski in *Advances in Carbanion Chemistry*, Vol. 1 (Ed.: V. Snieckus), JAI Press, Greenwich, Connecticut **1992**, p. 45; P. G. Williard in *Comprehensive Organic Synthesis*, Vol. 1 (Ed.: B. M. Trost), Pergamon, Oxford **1991**, Chap. 1.1.

48 C. H. Heathcock, *Science* **1981**, *214*, 395.

49 D. A. Evans, J. V. Nelson, T. R. Taber, *Top. Stereochem.* **1982**, *13*, 1.

50 C. H. Heathcock, in *Comprehensive Carbanion Chemistry*, Part B (Eds.: E. Buncel, T. Durst), Elsevier, Amsterdam **1984**, Chap. 4.

51 C. H. Heathcock, in *Asymmetric Synthesis* Vol. 3, Part B (Ed.: J. D. Morrison), Academic Press, New York **1984**, Chap. 2.

52 S. Masamune, W. Choy, J. S. Petersen, L. R. Sita, *Angew. Chem. Int. Ed. Engl.* **1985**, *24*, 1.

53 M. Braun, *Angew. Chem. Int. Ed. Engl.* **1987**, *26*, 24.

54 M. Braun in *Advances in Carbanion Chemistry*, Vol. 1 (Ed.: V. Snieckus), JAI Press, Greenwich, Connecticut **1992**, p. 177.

55 B. M. Kim, S. F. Willians, S. Masamune in *Comprehensive Organic Synthesis*, Vol. 2 (Ed.: B. M. Trost), Pergamon, Oxford **1993**, Chap. 1.7; M. W. Rathke, P. Weipert, *ibid.*, Chap. 1.8.; I. Paterson, *ibid.*, Chap. 1.9; C. Gennari, *ibid.*, Chap. 2.4.

56 M. Braun, H. Sacha, *J. Prakt. Chem.* **1993**, *335*, 653.

57 M. Braun in *Houben Weyl, Methoden der Organischen Chemie*, Vol. E21b (Eds.: G. Helmchen, R. W. Hoffmann, J. Mulzer, E. Schaumann), Thieme, Stuttgart **1996** p. 1603.

58 R. Mahrwald, *Chem. Rev.* **1999**, *99*, 1097.

59 M. Sawamura, Y. Ito in *Catalytic Asymmetric Synthesis* (Ed.: I. Ojima), VCH, Weinheim **1993**, chapter 7.2.

60 C. Palomo, M. Oiarbide, J. M. García, *Chem. Eur. J.* **2002**, *8*, 37; *Chem. Soc. Rev.* **2004**, *33*, 65.

61 The stereochemical notation follows that proposed in the Houben–Weyl edition on "*Stereoselective Synthesis*", cf. G. Helmchen in *Houben–Weyl, Methoden der Organischen Chemie*, Vol. E21a (Eds.: G. Helmchen, R. W. Hoffmann, J. Mulzer, E. Schaumann), Thieme, Stuttgart **1996**, p. 1.

62 See textbooks on stereochemistry: e.g.: E. L. Eliel, S. H. Wiley, L. N. Mander, *Stereochemistry of Organic Compounds*, Wiley, New York **1994**.

63 D. Seebach, V. Prelog, *Angew. Chem. Int. Ed. Engl.* **1982**, *21*, 654.

64 S. Masamune, S. A. Ali, D. L. Snitman, D. S. Garvey, *Angew. Chem. Int. Ed. Engl.* **1980**, *19*, 557.

65 J. E. Dubois, M. Dubois, *Tetrahedron Lett.* **1967**, 4215; *J. Chem. Soc., Chem. Commun.* **1968**, 1567; *Bull. Soc. Chim. Fr.* **1969**, 3120, 3553; J. E. Dubois, P. Fellmann, C. R. *Acad. Sci. Sér. C* **1972**, *274*, 1307; Tetrahedron Lett. **1975**, 1225.

66 C. H. Heathcock, C. T. Buse, W. A. Kleschick, M. C. Pirrung, J. E. Sohn, J. Lampe, *J. Org. Chem.* **1980**, *45*, 1066.

67 C. H. Heathcock, M. C. Pirrung, *J. Org. Chem.* **1980**, *45*, 1727; C. H. Heathcock, M. C. Pirrung, S. H. Montgomery, J. Lampe, *Tetrahedron* **1981**, *37*, 4087.

68 P. Fellmann, J.-E. Dubois, *Tetrahedron* **1978**, *34*, 1349.

69 T. H. Chan, T. Aida, P. W. K. Lau, V. Gorys, D. N. Harpp, *Tetrahedron Lett.* **1979**, 4029.
70 R. Noyori, I. Nishida, J. Sakata, *J. Am. Chem. Soc.* **1981**, *103*, 2106.
71 See, for example, the stereochemical outcome of aldol addition performed with the lithium enolate of cyclohexanone, which definitely has (*E*) configuration, H. O. House, D. S. Crumrine, A. Y. Teranishi, H. D. Olmstead, *J. Am. Chem. Soc.* **1973**, *95*, 3310; M. Majewski, D. M. Gleave, *Tetrahedron Lett.* **1989**, *42*, 5681.
72 D. Ivanoff, A. Spassoff, *Bull. Soc. Chim. Fr.* **1931**, *49*, 371; D. Ivanoff, N. Nicoloff, *Bull. Soc. Chim. Fr.* **1932**, *51*, 1325.
73 H. E. Zimmerman, M. D. Traxler, *J. Am. Chem. Soc.* **1957**, *79*, 1920.
74 S. Masamune, T. Sato, B. M. Kim, T. Wollmann, *J. Am. Chem. Soc.* **1986**, *108*, 8279.
75 E. J. Corey, S. S. Kim, *J. Am. Chem. Soc.* **1990**, *112*, 4976.
76 R. Armstutz, W. B. Schweizer, D. Seebach, J. D. Dunitz, *Helv. Chim. Acta* **1981**, *64*, 2617; E. Juaristi, A. K. Beck, J. Hansen, T. Matt, T. Mukhopadhyay, M. Simson, D. Seebach, *Synthesis* **1993**, 1271.
77 D. A. Evans, L. R. McGee, *Tetrahedron Lett.* **1980**, *25*, 3975.
78 R. W. Hoffmann, K. Ditrich, S. Froech, D. Cremer, *Tetrahedron* **1985**, *41*, 5517.
79 R. W. Hoffmann, K. Ditrich, *Tetrahedron Lett.* **1984**, *25*, 1781; C. Gennari, L. Colombo, C. Scolastico, R. Todeschini, *Tetrahedron* **1984**, *40*, 4051.
80 R. Noyori, S. Murata, M. Suzuki, *Tetrahedron* **1981**, *37*, 3899; T. Mukaiyama, S. Kobayashi, M. Tamura, Y. Sagawa, *Chem. Lett.* **1987**, 491; E. Nakamura, M. Shimizu, I. Kuwajima, J. Sakata, K. Yokoyama, R. Noyori, *J. Org. Chem.* **1983**, *48*, 932.
81 S. S. Labadie, J. K. Stille, *Tetrahedron* **1984**, *40*, 2329.
82 D. A. Evans, L. R. McGee, *J. Am. Chem. Soc.* **1981**, *103*, 2876; T. Katuski, M. Yamaguchi, *Tetrahedron Lett.* **1985**, *26*, 5807.
83 R. W. Hoffmann, *Pure Appl. Chem.* **1988**, *60*, 123 and references given therein.
84 C. Gennari, R. Todeschini, M. G. Baretta, G. Favini, C. Scolastico, *J. Org. Chem.* **1986**, *51*, 612.
85 I. Kuwajima, E. Nakamura, *Acc. Chem. Res.* **1985**, *18*, 181.
86 R. Mahrwald, B. Ziemer, *Tetrahedron Lett.* **2002**, *43*, 4459.
87 S. E. Denmark, X. Su, Y. Nishigaichi, *J. Am. Chem. Soc.* **1998**, *120*, 12990.
88 J. Mulzer, G. Brüntrup, J. Finke, M. Zippel, *J. Am. Chem. Soc.* **1979**, *101*, 7723; J. Mulzer, M. Zippel, G. Brüntrup, J. Senger, J. Finke, *Liebigs Ann. Chem.* **1980**, 1108.
89 N. T. Anh, B. T. Thanh, *Nouv. J. Chim.* **1986**, *10*, 681.
90 Y. Yamamoto, K. Maruyama, *Tetrahedron Lett.* **1980**, *21*, 4607.
91 D. A. Evans, M. C. Kozlowski, J. A. Murry, C. S. Burgey, K. R. Campos, B. T. Connell, R. J. Staples, *J. Am. Chem. Soc.* **1999**, *121*, 669.
92 Y. Li, M. N. Paddon-Row, K. Houk, *J. Org. Chem.* **1990**, *55*, 481.

93 R. Leung-Toung, T. T. Tidwell, *J. Am. Chem. Soc.* **1990**, *112*, 1042.
94 Based on calculations, a similar coordination of the lithium cation has been postulated previously, cf. E. Kaufmann, P. v. R. Schleyer, K. N. Houk, Y. D. Wu, *J. Am. Chem. Soc.* **1985**, *107*, 5560; S. M. Bachrach, A. Streitwieser, Jr., *J. Am. Chem. Soc.* **1986**, *108*, 3946; G. Stork, R. L. Polt, Y. Li, K. N. Houk, *J. Am. Chem. Soc.* **1988**, *110*, 8360; S. M. Bachrach, J. P. Ritchie, *J. Am. Chem. Soc.* **1989**, *111*, 3134; E. Kaufmann, S. Sieber, P. v. R. Schleyer, *J. Am. Chem. Soc.* **1989**, *111*, 4005; A. E. Dorigo, K. Morokuma, *J. Am. Chem. Soc.* **1989**, *111*, 4635; see also ref. 92.
95 H. B. Bürgi, E. Shefter, J. D. Dunitz, *Tetrahedron* **1975**, *31*, 3089; H. B. Bürgi, J. D. Dunitz, E. Shefter, *J. Am. Chem. Soc.* **1973**, *95*, 5065.
96 S. E. Denmark, B. R. Henke, *J. Am. Chem. Soc.* **1991**, *113*, 2177.
97 C. H. Heathcock, C. T. White, J. J. Morrison, D. VanDerveer, *J. Org. Chem.* **1981**, *46*, 1296.
98 C. H. Heathcock, C. T. White, *J. Am. Chem. Soc.* **1979**, *101*, 7076.
99 C. H. Heathcock, S. Arseniyadis, *Tetrahedron Lett.* **1985**, *26*, 6009.
100 N. A. Van Draanen, S. Arseniyadis, M. T. Crimmins, C. H. Heathcock, *J. Org. Chem.* **1991**, *56*, 2499.
101 A. Choudhury, E. Thornton, *Tetrahedron* **1992**, *48*, 5701.
102 A. Choudhury, E. Thornton, *Tetrahedron Lett.* **1993**, *34*, 2221.
103 J. B. Goh, B. R. Lagu, J. Wurster, D. C. Liotta, *Tetrahedron Lett.* **1994**, *35*, 6029.
104 M. Braun, H. Sacha, *Angew. Chem. Int. Ed. Engl.* **1991**, *30*, 1318.
105 H. Sacha, D. Waldmüller, M. Braun, *Chem. Ber.* **1994**, *127*, 1959.
106 D. Seebach, A. R. Sting, M. Hoffmann, *Angew. Chem. Int. Ed. Engl.* **1996**, *35*, 2708.
107 D. Seebach, R. Naef, G. Calderari, *Tetrahedron* **1984**, *40*, 1313.
108 D. Seebach, T. Weber, *Helv. Chim. Acta* **1984**, *67*, 1650.
109 W. Amberg, D. Seebach, *Chem. Ber.* **1990**, *123*, 2413.
110 D. Seebach, J.-M. Lapierre, W. Jaworek, P. Seiler, *Helv. Chim. Acta* **1993**, *76*, 459.
111 I. Hoppe, D. Hoppe, R. Herbst-Irmer, E. Egert, *Tetrahedron Lett.* **1990**, *31*, 6859.
112 S. G. Davies, I. M. Dordor, P. Warner, *J. Chem. Soc., Chem. Commun.* **1984**, 956.
113 S. G. Davies, *Pure Appl. Chem.* **1988**, *60*, 13.
114 L. S. Liebeskind, M. E. Welker, *Tetrahedron Lett.* **1984**, *25*, 4341.
115 J. S. McCallum, L. S. Liebeskind in *Houben–Weyl, Methoden der Organischen Chemie*, Vol. E21b (Eds.: G. Helmchen, R. W. Hoffmann, J. Mulzer, E. Schaumann), Thieme, Stuttgart **1996**, p. 1667.
116 P. W. Ambler, S. G. Davies, *Tetrahedron Lett.* **1985**, *26*, 2129.

117 S. G. Davies, M. Wills, *J. Organomet. Chem.* **1987**, *328*, C 29.
118 S. G. Davies, D. Middlemiss, A. Naylor, M. Wills, *Tetrahedron Lett.* **1989**, *27*, 2971.
119 D. Seebach, V. Ehrig, M. Teschner, *Liebigs Ann. Chem.* **1976**, 1357.
120 C. Mioskowski, G. Solladié, *J. Chem. Soc., Chem. Commun.* **1977**, 162.
121 For reviews, see: G. Solladié in *Asymmetric Synthesis* Vol. 2, Part A (Ed.: J. D. Morrison), Academic Press, New York **1983**, p. 184.
122 C. Mioskowski, G. Solladié, *Tetrahedron* **1980**, *36*, 227.
123 G. Solladié in *Houben Weyl, Methoden der Organischen Chemie*, Vol. E21b (Eds.: G. Helmchen, R. W. Hoffmann, J. Mulzer, E. Schaumann), Thieme, Stuttgart **1996**, p. 1808.
124 E. J. Corey, L. O. Weigel, A. R. Chamberlin, H. Cho, D. H. Hua, *J. Am. Chem. Soc.* **1980**, *102*, 6613.
125 R. Annunziata, M. Cinquini, F. Cozzi, F. Montanari, A. Restelli, *J. Chem. Soc., Chem. Commun.* **1983**, 1138; *Tetrahedron* **1984**, *40*, 3815; for stereoselective aldol additions of 2-(arylsulfinyl)phenyl ketones, see: S. Nakamura, M. Oda, H. Yasuda, T. Toru, *Tetrahedron* **2001**, *57*, 8469.
126 S. S. Keito, K. Hatanaka, T. Kano, H. Yamamoto, *Angew. Chem. Int. Ed.* **1998**, *37*, 3378.
127 I. Ojima, H. B. Kwon, *J. Am. Chem. Soc.* **1988**, *110*, 5617.
128 M. Braun, R. Devant, *Tetrahedron Lett.* **1984**, *25*, 5031.
129 R. Devant, U. Mahler, M. Braun, *Chem. Ber.* **1988**, *121*, 397.
130 M. Braun, S. Gräf, S. Herzog, *Org. Synth.* **1993**, *72*, 32. For an alternative acylation of **82**, see: J. Macor, A. J. Sampognaro, P. R. Verhoest, R. A. Mack, *Org. Synth.* **2000**, *77*, 45; for a modified procedure of the preparation of **82**, see: A. Millar, L. W. Mulder, K. E. Mennen, C. W. Palmer, *Organic Preparations and Procedures Int.* **1991**, *23*, 173; *Chem. Abstr.* **1991**, *114*, 163667.
131 M. Braun, S. Gräf, *Org. Synth.* **1993**, *72*, 38.
132 S. Gräf, M. Braun, *Liebigs Ann. Chem.* **1993**, 1091.
133 M. Hashmi, S. Gräf, M. Braun, M. W. Anders, *Chem. Res. Toxicol.* **1996**, *9*, 361.
134 M. Braun, D. Waldmüller, *Synthesis* **1989**, 856.
135 M. Braun, C. Bauer, *Liebigs Ann. Chem.* **1991**, 1157.
136 M. Braun, J. Moritz, *Synlett* **1991**, 750.
137 W. R. Ewing, B. D. Harris, K. L. Bhat, M. M. Joullié, *Tetrahedron* **1986**, *42*, 2421.
138 P. Barbier, F. Schneider, U. Widmer, *Helv. Chim. Acta* **1987**, *70*, 1412; *Eur. Patent* 185359, 1986; P. Barbier, F. Schneider, U. Widmer, Hoffmann–La Roche and Co. AG; *Chem. Abstr.* **1987**, *106*, 102077; *Eur. Patent* 189577, 1986; *Chem. Abstr.* **1987**, *106*, 196249; *US Patent* 4931463, 1990; P. Barbier, F. Schneider, U. Widmer, Hoffmann–La Roche, Inc.; *Chem. Abstr.* **1991**, *114*, 82542; *Eur. Patent* 444482, 1991; R. Derunges, H. P. Maerki, H. Stalder, A. Szente, Hoffmann–La Roche AG; *Chem. Abstr.*, **1992**, *117*, 7786.
139 P. G. M. Wuts, S. R. Putt, *Synthesis* **1989**, 951.

140 R. Devant, H. E. Radunz, *Tetrahedron Lett.* **1988**, 2307; *German Patent* 3743225, 1989; R. M. Devant, H. E. Radunz, Merck GmbH; *Chem. Abstr.* **1990**, *112*, 36458.
141 J. E. Lynch, R. P. Volante, J. V. Wattley, I. Shinkai, *Tetrahedron Lett.* **1987**, *28*, 1385; *Eur. Patent* 251714, 1988; R. P. Volante, E. Corley, I. Shinkai, Merck and Co., Inc.; *Chem. Abstr.* **1988**, *108*, 150455; *US Patent* 4611081, 1986; J. E. Lynch, I. Shinkai, R. P. Volante, Merck and Co., Inc.; *Chem. Abstr.* **1987**, *106*, 18119.
142 T. Allmendinger, E. Felder, E. Hungerbühler, *Tetrahedron Lett.* **1990**, *31*, 7301; *Eur. Patent* 353732, 1990; T. Allmendinger, E. Hungerbühler, R. Lattmann, S. Ofner, W. Schilling, G. von Sprecher, E. Felder, Ciba–Geigy AG; *Chem. Abstr.* **1990**, *113*, 153046.
143 S. Wattanasin, H. D. Do, N. Bhongle, F. G. Kathawala, *J. Org. Chem.* **1993**, *58*, 1610.
144 D. Schinzer, A. Bauer, O. M. Bohm, A. Limberg, M. Cordes, *Chem. Eur. J.* **1999**, *5*, 2483; *Int. Patent* 9959985, 1999; K.-H. Altmann, A. Bauer, D. Schinzer; Novartis AG; *Chem. Abstr.* **2000**, *132*, 3284.
145 K. Nagasawa, H. Ishihara, Y. Zako, I. Shimizu, *J. Org. Chem.* **1993**, *58*, 2523.
146 *Jap. Patent* 08301811, 1996; S. Ri, M. Yamanashi, N. Shimizu; Kuraray Co; *Chem. Abstr.* **1997**, *126*, 131697.
147 C. T. Brain, A. Chen, A. Nelson, N. Tanikkul, E. J. Thomas, *Tetrahedron Lett.* **2001**, *42*, 1247.
148 M. Suzuki, Y. Yanagawa, H. Iwasaki, H. Kanda, K. Yanagihara, H. Matsumoto, Y. Ohara, Y. Yazaki, R. Sakoda, *Bioorg. Med. Chem. Lett.* **1999**, *9*, 2977.
149 J. M. Weiss, H. M. R. Hoffmann, *Tetrahedron: Asymmetry* **1997**, *8*, 3913.
150 S. Mills, R. Desmond, R. A. Reamer, R. P. Volante, I. Shinkai, *Tetrahedron Lett.* **1988**, *29*, 281.
151 *US Patent* 5216015, 1993; D. G. McGarry, F. A. Volz, J. R. Regan, M. N. Chang; Rohne–Poulenc Rorer Pharmaceuticals Inc.; *Chem. Abstr.* **1993**, *119*, 226236.
152 H. Jendralla, E. Baader, W. Bartmann, G. Beck, A. Bergmann, E. Granzer, B. v. Kerekjarto, K. Kesseler, R. Krause, W. Schubert, G. Wess, *J. Med. Chem.* **1990**, *33*, 61; H. Jendralla, E. Granzer, B. v. Kerekjarto, R. Krause, U. Schacht, E. Baader, W. Bartmann, G. Beck. A. Bergmann, K. Kesseler, G. Wess, L.-J. Chen, S. Granata, J. Herchen, H. Kleine, H. Schüssler, K. Wagner, *J. Med. Chem.* **1991**, *34*, 2962; *German Patent* 3722807, 1989; G. Wess, W. Bartmann, G. Beck, E. Granzer, Hoechst AG; *Chem. Abstr.* **1989**, *110*, 231121; *Eur. Patent* 324347, 1989; E. Baader, H. Jendralla, B. Kerekjarto, G. Beck, Hoechst AG; *Chem. Abstr.* **1990**, *112*, 21003; *German Patent* 3826814, 1990; G. Beck, W. Bartmann, G. Wess, E. Granzer, Hoechst AG; *Chem. Abstr.* **1991**, *114*, 23976.
153 O. Tempkin, S. Abel, C.-P. Chen, R. Underwood, K. Prasad, K.-M. Chen, O. Repic, T. J. Blacklock, *Tetrahedron* **1997**, *53*, 10659.
154 D. V. Patel, R. J. Schmidt, E. M. Gordon, *J. Org. Chem.*

1992, *57*, 7143; *German Patent* 3805801, 1988; J. J. Wright, S. Y. Sit, Bristol–Myers Co.; *Chem. Abstr.* **1989**, *110*, 114836.
155 B. D. Roth, C. J. Blankley, A. W. Chucholowski, E. Ferguson, M. L. Hoefle, D. F. Ortwine, R. S. Newton, C. S. Sekerke, D. R. Sliskovic, C. D. Stratton, M. M. Wilson, *J. Med. Chem.* **1991**, *34*, 357; *Eur. Patent* 409281, 1991; B. D. Roth, Warner–Lambert Co.; *Chem. Abstr.* **1991**, *115*, 29107; see also: *German Patent* 3932887, 1990; N. S. Watson, C. Chan, B. C. Ross, Glaxo Group Ltd.; *Chem. Abstr.* **1991**, *114*, 164567. For a synthesis of the [14C] side chain-labeled compound, see: T. H. Lee, P. W. K. Woo, *J. Labelled Compounds and Radiopharmaceuticals* **1999**, *42*, 129. – For a review, see: B. D. Roth, *Progress in Med. Chem.* **2002**, *40*, 1.
156 D. R. Sliskovic, C. J. Blankley, B. R. Krause, R. S. Newton, J. A. Picard, W. H. Roark, B. D. Roth, C. Sekerke, M. K. Shaw, R. L. Stanfield, *J. Med. Chem.* **1992**, *35*, 2095.
157 *Eur. Patent* 424929, 1991; H. Natsugari, H. Ikeda, Takeda Chemical Industries, Ltd.; *Chem. Abstr.* **1991**, *115*, 114373.
158 *Int. Patent*, 1991; M. Matsuo, T. Manabe, H. Okumura, H. Matsuda, N. Fujii, Fujisawa Pharmaceutical Co. Ltd; *Chem. Abstr.* **1992**, *116*, 151782.
159 K. Prasad, K. M. Chen, O. Repic, G. E. Hardtmann, *Tetrahedron: Asymmetry* **1990**, *1*, 703.
160 D. J. Cram, F. A. Abd Elhafez, *J. Am. Chem. Soc.* **1952**, *74*, 5828.
161 D. J. Cram, D. R. Wilson, *J. Am. Chem. Soc.* **1963**, *85*, 1245.
162 For a discussion, see: E. L. Eliel in *Asymmetric Synthesis*, J. D. Morrison, Ed., Vol. 2, Part A, p. 125, Academic, New York **1983**; J. Mulzer, H.-J. Altenbach, M. Braun, K. Krohn, H.-U. Reissig, *Organic Synthesis Highlights*, p. 3, VCH, Weinheim **1991**.
163 M. Chérest, H. Felkin, N. Prudent, *Tetrahedron Lett.* **1968**, 2199.
164 Review: M. T. Reetz, *Angew. Chem. Int. Ed. Engl.* **1984**, *23*, 556.
165 M. Braun in *Houben–Weyl, Methoden der Organischen Chemie*, Vol. E21b (Eds.: G. Helmchen, R. W. Hoffmann, J. Mulzer, E. Schaumann), Thieme, Stuttgart **1996**, p. 1713.
166 R. B. Woodward, E. Logusch, K. P. Nambiar, K. Sakan, D. E. Ward, B.-W. Au-Yeung, P. Balaram, L. J. Browne, P. J. Card, C. H. Chen, R. B. Chênebert, A. Fliri, K. Frobel, H.-J. Gais, D. G. Garratt, K. Hayakawa, W. Heggie, D. P. Hesson, D. Hoppe, I. Hoppe, J. A. Hyatt, D. Ikeda, P. A. Jacobi, K. S. Kim, Y. Kobuke, K. Kojima, K. Krowicki, V. J. Lee, T. Leutert, S. Malchenko, J. Martens, R. S. Matthews, B. S. Ong, J. B. Press, T. V. Rajan Babu, G. Rousseau, H. M. Sauter, M. Suzuki, K. Tatsuta, L. M. Tolbert, E. A. Truesdale, I. Uchida, Y. Ueda, T. Uyehara, A. T. Vasella, W. C. Vladuchick, P. A. Wade, R. M. Williams, H. N.-C. Wong, *J. Am. Chem. Soc.* **1981**, *103*, 3210.
167 A. I. Meyers, J. P. Hudspeth, *Tetrahedron Lett.* **1981**, *22*, 3925.
168 C. H. Heathcock, S. D. Young, J. P. Hagaen, M. C. Pirrung, C. T. White, D. VanDerveer, *J. Org. Chem.* **1980**, *45*, 3846.

169 M. T. Reetz, M. W. Drewes, A. Schmitz, *Angew. Chem. Int. Ed. Engl.* **1987**, *26*, 1141.

170 M. T. Reetz, *Pure Appl. Chem.* **1988**, *60*, 1607.

171 D. A. Evans, J. R. Gage, *Tetrahedron Lett.* **1990**, *31*, 6129.

172 E. J. Corey, G. A. Reichard, *J. Am. Chem. Soc.* **1992**, *114*, 10677.

173 J.-M. Escudier, M. Baltas, L. Gorrichon, *Tetrahedron Lett.* **1991**, *32*, 5345.

174 D. E. Ward, M. Sales, C. C. Man, J. Shen, P. Sasmal, C. Guo, *J. Org. Chem.* **2002**, *67*, 1618.

175 U. Mahler, R. M. Devant, M. Braun, *Chem. Ber.* **1988**, *121*, 2035.

176 D. Seebach, E. Juaristi, D. D. Miller, C. Schickli, T. Weber, *Helv. Chim. Acta* **1987**, *70*, 237.

177 D. Seebach, H. O. Kalinowski, B. Bastani, G. Crass, H. Daum, N. Dörr, N. DuPreez, V. Ehrig, W. Langer, C. Nüsser, H.-A. Oei, M. Schmitt, *Helv. Chim. Acta* **1977**, *60*, 301. For more recent applications of this method, see: Y. Nomura, M. Iguchi, H. Doi, K. Tomioka, *Chem. Pharm. Bull.* **2002**, *50*, 1131.

178 A. Ando, T. Shioiri, *Tetrahedron* **1989**, *45*, 4969.

179 M. Uragami, K. Tomioka, K. Koga, *Tetrahedron: Asymmetry* **1995**, *6*, 701.

180 M. Majewski, P. Nowak, *J. Org. Chem.* **2000**, *65*, 5152.

2
The Development of Titanium Enolate-based Aldol Reactions

Arun K. Ghosh and M. Shevlin

2.1
Introduction

Asymmetric aldol reactions are very important carbon–carbon bond-forming reactions in organic synthesis. Applications of aldol reactions have been widespread, particularly in the synthesis of complex and bioactive natural products. An aldol reaction can create two new stereogenic centers depending upon the choice of enolate substituent and the aldehyde or ketone. Furthermore, as shown in Scheme 2.1, the reaction can generate four possible diastereomers when X contains a chiral center – two *syn* (**3** and **4**) and two *anti* diastereomers (**5** and **6**). Significant advances in asymmetric aldol reactions has led to control of aldol products with high enantio- and diastereoselectivity [1]. The field of stereoselective synthesis of *syn* aldol products is particularly sophisticated. Important methods are also emerging for stereoselective *anti* aldol reactions.

The use of chiral auxiliaries [2] has been an important concept in terms of generating one diastereomer in a stereopredictable manner. The ready availability of both enantiomers of a chiral auxiliary and easy removal and recovery are very important. *syn* and *anti* diastereoselectivity depends on several factors including enolate geometry, related metal ions, and reaction conditions. In general, *Z* enolates provide *syn* aldol products with high selectivity and *E* enolates afford *anti* aldol products. This result can be rationalized on the basis of a cyclic Zimmerman–Traxler transition state [3]. As shown in transition state **7** for *Z* enolates (Scheme 2.2) the enolate substituent is generally pseudoaxial and the aldehyde substituent occupies a pseudoequatorial position, providing a *syn* product, **8**. *E* enolates adopt transition state **9**, in which the enolate and aldehyde substituents are pseudoequatorial and leads to the *anti* aldol product **10**. Beside chair-like transition-state models, boat-like transition-state models have been proposed to explain stereoselectivity. It should be noted that strategies have been developed over the years to form the *anti* product from the *Z* enolate, as will be discussed in this chapter.

Modern Aldol Reactions. Vol. 1: Enolates, Organocatalysis, Biocatalysis and Natural Product Synthesis.
Edited by Rainer Mahrwald

ISBN: 3-527-30714-1

syn-aldol

anti-aldol

Scheme 2.1
Generation of four possible diastereomers when *x* contains a chiral center.

Diastereoselective aldol reactions have been extensively utilized in the synthesis of complex natural products, including macrolides [4] and ionophores [5]. In this context, iterative approaches are often exploited to append propionate units one at a time. This approach leads to double stereodifferentiation [6] in which the reactant pairs can be either "matched" or "mismatched". The chirality of the two reactants reinforce each other if they are matched. As a result, the diastereoselectivity is often higher than would

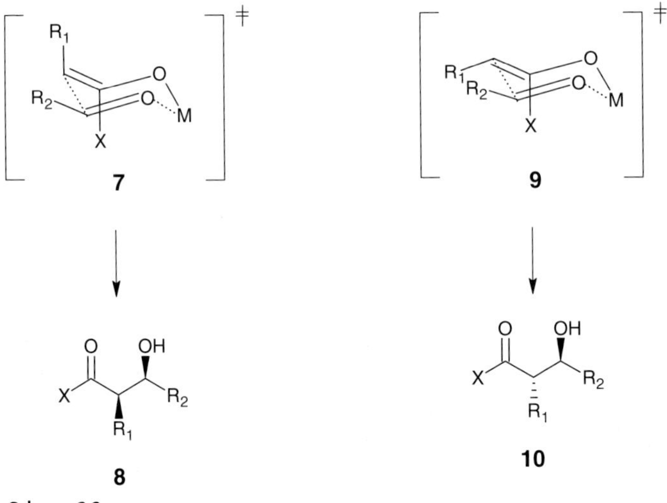

Scheme 2.2
Zimmerman–Traxler transition states.

be expected from either reagent alone. In they are mismatched, the chirality of the reagents has a detrimental effect, leading to a lower selectivity than would have been expected.

The metal ion associated with the enolate has pronounced effect on stereoselectivity. Numerous titanium enolate-based asymmetric aldol methods have provided convenient access to aldol products in enantiomerically pure form. The titanium enolate aldol reaction has tremendous synthetic potential, because titanium reagents are readily available and inexpensive. This chapter will focus on the development of a variety of titanium enolate aldol reactions.

The first titanium enolate aldol reactions were reported by De Kimpe and coworkers [7], who reported the self-condensation and elimination of α-chloroketones with $TiCl_4$ and pyridine.

2.2
Additions of Enolates to Ketones

Although aldol additions to aldehydes are robust methods and the corresponding theory is well developed, aldol additions to ketones are still largely unexplored, possibly because of the additional complexity of differentiating sterically between the two different alkyl groups of ketones. The difficulty of aldol additions to ketones is apparent from the lack of asymmetric methods available for addition of titanium enolates to ketones. The notion of *syn* and *anti* products also tends to break down when dealing with non-symmetrical ketones with similar substituents.

Tanabe and coworkers reported the first instances of additions of titanium enolates to ketones [8]. They reported moderate to excellent yields for several reactions. When the two substituents on the ketone were sufficiently sterically differentiated, good to excellent *syn* diastereoselectivity was observed. Representative examples of these crossed-aldol reactions are shown in Table 2.1. The initial method involved a catalytic (5 mol%) amount of trimethylsilyl triflate in the reaction mixture; it was suggested this generates a trichlorotitanium triflate species in situ which is more effective at enolization. This theory was, however, called into question by reports of additions of enolates to ketones with trimethylsilyl chloride as catalyst, which also provides moderate to good yields [9]. Additions of enolates of phenyl esters or phenyl thioesters to ketones were also shown to proceed in good yield and with *syn* diastereoselectivity in the absence of silyl additive [10].

Oshima and coworkers reported reactions of aldehyde enolates with ketones [11]. They used enolates generated from silyl enol ether **14** with methyllithium and transmetalated with titanium tetrabutoxide. Representative examples are shown in Table 2.2. Selectivity was minimal, except for 1,1,1-trichloroacetone (entry 5), for which the *anti* isomer was formed almost exclusively.

Tab. 2.1
Additions of enolates to ketones.

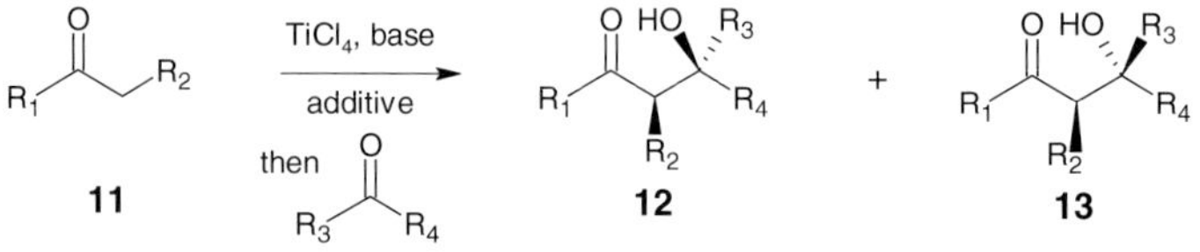

Entry	R_1	R_2	R_3	R_4	Base	Additive	Yield (%)	12:13	Ref.
1	Ph	Me	Ph	Me	Bu_3N	TMSOTf	95	100:0	8
2	Ph	Me	Ph	CH_2Cl	Bu_3N	TMSOTf	91	100:0	8
3	Et	Me	Ph	Me	Bu_3N	TMSOTf	84	84:16	8
4	Et	Me	Ph	Et	Bu_3N	TMSOTf	92	72:28	8
5	Pr	Et	Ph	Me	Bu_3N	TMSOTf	72	100:0	8
6	Pr	Et	n-C_7H_{15}	Me	Bu_3N	TMSOTf	60	60:40	8
7	$TBSOCH_2$	H	Ph	CH_2Cl	Bu_3N	TMSCl	94	–	9
8	PhO	Me	Et	Ph	Et_3N	–	83	64:36	10
9	PhO	Me	CH_2Cl	Ph	Et_3N	–	77	68:32	10
10	PhS	Me	Et	Et	Et_3N	–	77	–	10
11	PhS	Me	Et	Ph	Et_3N	–	98	77:23	10

2.3 Addition of Enolates Without α-Substituents to Aldehydes

The development of asymmetric aldol reactions involving enolates with no α-substituent has been hindered by low asymmetric induction compared with the corresponding propionate aldol reactions. It is postulated that this difficulty is because of the lack of stereochemical constraints from the enolate portion of the Zimmerman–Traxler [3] transition-state model.

Tab. 2.2
Reaction of aldehyde enolates with ketones.

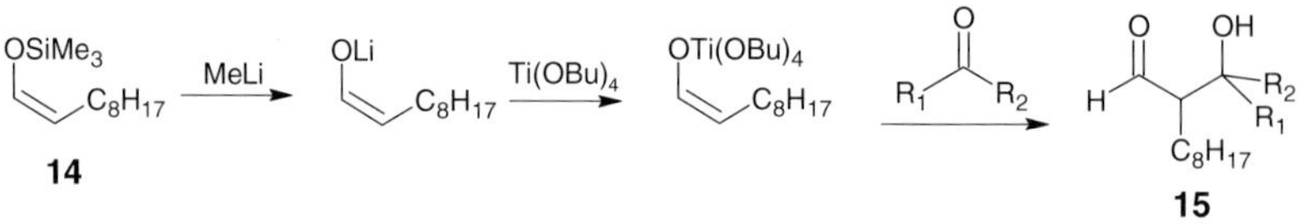

Entry	R_1	R_2	Yield (%)	Ratio[a]
1	HC≡C	Me	85	59:41
2	PhC≡C	Me	86	52:48
3	BuC≡C	Me	81	55:45
4	$Me_3SiC{\equiv}C$	n-C_7H_{15}	74	51:49
5	CCl_3	Me	72	>99:1[b]

[a] Product stereochemistry not assigned
[b] *anti:syn*

Tab. 2.3
Asymmetric acetate aldol reactions.

16 → a) BuLi, THF, 0° b) Lewis acid c) PhCHO → 17 + 18

R	*Lewis Acid*	*7:8*
Me	Ti(O*i*-Pr)$_3$Cl	83:17
CH_2OH	$TiCl_4$	72:28

2.3.1
Stereoselective Acetate Aldol Reactions Using Chiral Auxiliaries

The early work in the field of titanium enolate acetate aldol reactions was conducted by Braun in a general investigation of acetate aldol reactions [12]. The enolates were generated from chiral acetamide **16** by transmetalation of the lithium enolate with triisopropoxytitanium chloride or titanium tetrachloride, as shown in Table 2.3. They reported moderate selectivity for the reaction with benzaldehyde.

Yan and coworkers developed titanium enolate acetate aldol reactions as an extension of their boron acetate enolate methodology [13, 14]. Good yields and diastereoselectivity were reported when using camphor-derived *N*-acyloxazolidinethione **19** (Scheme 2.3, Table 2.4, entries 1–7, and Figure 2.1). The high selectivities were attributed to additional chelation afforded by the thiocarbonyl of the chiral auxiliary in transition state assembly **20**, shown in Scheme 2.3. The corresponding camphor-derived oxazolidinone acetate imide provided no stereocontrol, supporting the chelation control hypothesis.

Shortly thereafter, acetate aldol reactions using camphor-derived imidazolidinone **27** were reported by Palomo and coworkers [15]. They reported moderate yields and enantioselectivity for a variety of unsaturated and aliphatic aldehydes (Table 2.4, entries 8–12). Interestingly, enantioselectivity for unsaturated aldehydes was opposite that for aliphatic aldehydes. Also, enantioselectivity reported for titanium was completely opposite that of the corresponding lithium enolate reactions.

Recently, Phillips and Guz reported a titanium enolate acetate aldol reaction based on a valine-derived *N*-acyloxazolidinethione **30** [16]. The titanium enolate is directly generated from $TiCl_4$, (−)-sparteine, and *N*-methylpyrrolidone by use of Crimmins' procedure, as discussed in Section 2.4.1.2.1. Their highly hindered chiral auxiliary can be synthesized in three steps from commercially available starting material. They reported good yields and good to excellent diastereoselectivity for a range of aldehydes,

Tab. 2.4
Survey of acetate aldol reactions employing chiral auxiliaries.

23 $\xrightarrow[\text{R}_2\text{CHO}]{\text{Lewis acid}}$ **24** + **25**

Entry	R_1	R_2	*Lewis Acid*	*Yield (%)*	*24:25*	*Ref.*
1	**26**	*n*-Pr	$TiCl_4$	85	95:5	13
2	**26**	*i*-Pr	$TiCl_4$	86	94:6	13
3	**26**	(*E*)-$CH_3CH{=}CH$	$TiCl_4$	86	93:7	13
4	**26**	Ph	$TiCl_4$	91	91:9	13
5	**26**	(*E*)-PrCH=CH	$TiCl_4$	87	90:10	14
6	**26**	(*E*)-$Me_3SiCH{=}CH$	$TiCl_4$	80	90:10	14
7	**26**	(*E*)-PhSCH=CH	$TiCl_4$	77	90:10	14
8	**27**	Ph	$TiCl_4$	65	88:12	15
9	**27**	(*E*)-PhCH=CH	$TiCl_4$	70	86:14	15
10	**27**	*i*-Pr	$TiCl_4$	85	17:83	15
11	**27**	*t*-Bu	$TiCl_4$	74	47:53	15
12	**27**	$PhCH_2CH_2$	$TiCl_4$	37	42:58	15
13	**28**	Ph	Ti(O*i*-Pr)$_3$Cl	89	35:65	19
14	**29**	Ph	Ti(O*i*-Pr)$_3$Cl	62	70:30	19

as shown in Table 2.5. Once again, the extra chelation afforded by the thiocarbonyl moiety might be responsible for their excellent results. It has, furthermore, been noted that diastereoselectivity depends critically on the exact stoichiometry of the reagents. The stereochemical outcome has been rationalized on the basis of the coordinated chair model **33** or dipole-minimized boat transition model **34**, shown in Figure 2.2.

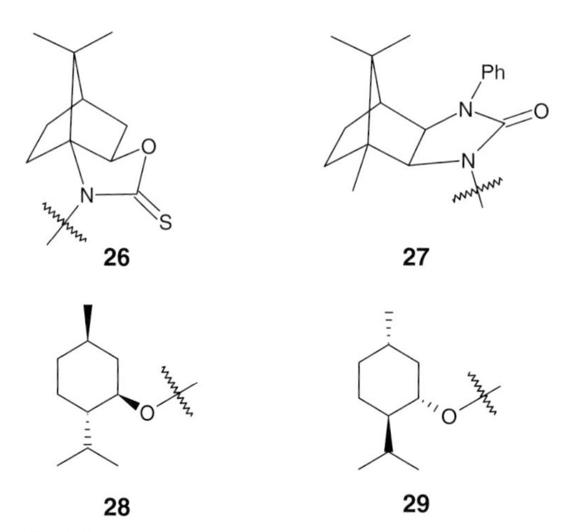

Fig. 2.1
Chiral auxiliaries used in the asymmetric acetate aldol reactions described in Table 2.4.

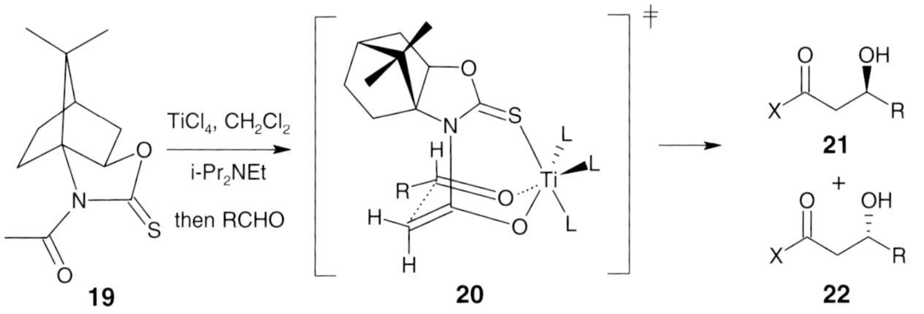

Scheme 2.3
Camphor-derived asymmetric acetate aldol reactions.

Urpí and Vilarassa investigated an asymmetric acetate aldol reaction in conjunction with the synthesis of Macrolactin A [17]. They reported good yield and excellent stereoselectivity for a handful of unsaturated aldehydes using a valine-derived thiazolidinethione chiral auxiliary.

2.3.2
Stereoselective Acetate Aldol Reactions Involving Chiral Titanium Ligands

Duthaler and coworkers demonstrated that stereoselectivity in the titanium enolate acetate aldol reaction could also be induced by chiral ligands on titanium [18]. Their cyclopentadienylbis(1,2:5,6-di-*O*-isopropylidene-α-D-

Tab. 2.5
Sterically hindered oxazolidinethione-derived asymmetric acetate aldol reactions.

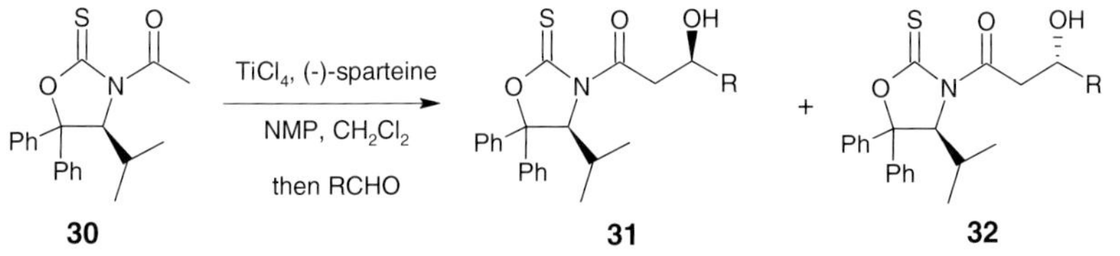

Entry	*R*	*Yield (%)*	*31:32*
1	$PhCH_2CH_2$	83	95:5
2	*n*-Pr	78	95:5
3	Et	90	93:7
4	*n*-Bu	77	95:5
5	$(CH_3)_2CHCH_2$	82	96:4
6	*i*-Pr	83	92:8
7	*n*-C_6H_{13}	78	95:5
8	(*E*)-$CH_3CH{=}CH$	85	99:1
9	$PMBOCH_2$	55	97:3
10	$TBDPSOCH_2CH_2$	56	99:1
11	Ph	86	85:15

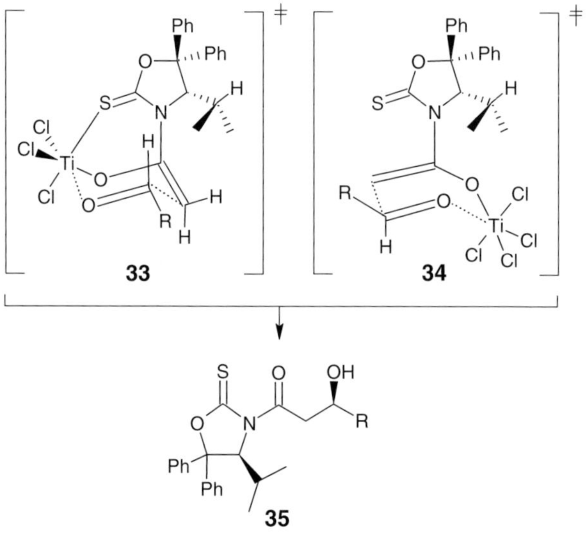

Fig. 2.2
Transition-state models of the asymmetric acetate aldol reaction.

glucofuranose-3-*O*-yl)chlorotitanate Lewis acid provided excellent enantioselectivity and moderate to good yields with *tert*-butyl acetate reacting with a wide range of aliphatic and unsaturated aldehydes (Table 2.6, entries 1–7, and Figure 2.3). The chiral ligands are, furthermore, commercially available and can be recovered after the reaction.

This work was extended by Rutledge to an investigation of the effects of the double stereodifferentiation of chiral ligands on titanium and a chiral auxiliary derived from (+)- or (−)-menthol (**45** and **46**). (Table 2.6, entries 8–11, and Figure 2.4) [19]. Similar work by Fringuielli examined the effects of chiral ligands [20] and auxiliaries [21] on the basis of (+)-2-carane **43** and (+)-3-carane **44** (Table 2.6, entries 12–17, and Figure 2.3). Although double stereodifferentiation can result in impressive stereoselectivity, these methods suffer from requiring stoichiometric amounts of both chiral auxiliary and chiral ligands and yields are usually modest.

2.3.3 Alternative Approaches to Acetate Aldol Adducts

Because of the inherent difficulty of inducing chirality in the acetate enolate reaction, alternative approaches have been developed. A general approach is to synthesize α-substituted aldols and then reductively remove the α-substituent. Yan reported a one-step bromination–aldolization which provided α-bromo aldols in excellent yield and diastereoselectivity (Scheme 2.4) [22]. They then demonstrated that the α-bromo substituent could be reduc-

Tab. 2.6
Stereoselective acetate aldol reactions involving chiral titanium Lewis acids.

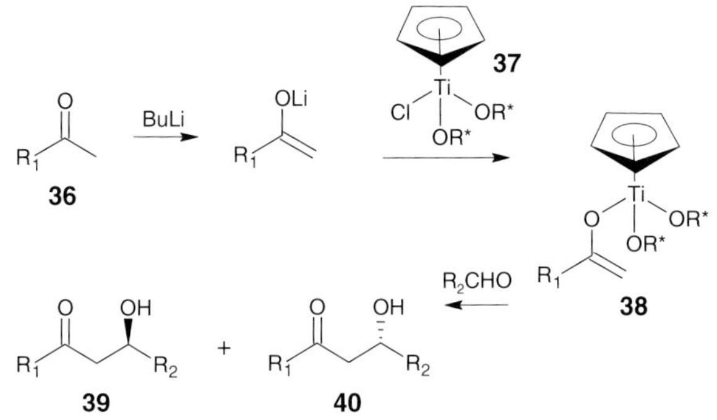

Entry	R_1	R_2	*OR**	*Yield (%)*	*39:40*	*Ref.*
1	*t*-BuO	*n*-Pr	**41**	51	97:3	18
2	*t*-BuO	*n*-C_7H_{15}	**41**	87	98:2	18
3	*t*-BuO	*i*-Pr	**41**	66	98:2	18
4	*t*-BuO	*t*-Bu	**41**	80	96:4	18
5	*t*-BuO	CH_2=CMe	**41**	81	98:2	18
6	*t*-BuO	Ph	**41**	69	98:2	18
7	*t*-BuO	2-furyl	**41**	62	95:5	18
8	**45**	Ph	**41**	51	96:4	19
9	**46**	Ph	**41**	54	98:2	19
10	**45**	Ph	**42**	40	30:70	19
11	**46**	Ph	**42**	40	21:79	19
12	HO	Ph	**43**	54	82:18	20
13	HO	Ph	**44**	52	83:17	20
14	*t*-BuO	Ph	**43**	68	81:19	20
15	*t*-BuO	Ph	**44**	70	75:25	20
16	**45**	Ph	**44**	76	90:10	21
17	**47**	Ph	**44**	70	96:4	21

tively removed under mild conditions using aluminum amalgam to provide acetate aldols in good yield and enantioselectivity [23].

Urpí and co-workers reported aldol-like reactions of titanium enolates with acetals by use of a valine-derived *N*-acetylthiazolidinethione **52** [24]. They reported moderate to good stereoselectivity and moderate yield for a variety of aliphatic and unsaturated acetals, as shown in Table 2.7. In these reactions, the enolate presumably reacts with the oxycarbenium ion generated by Lewis acid activation of the acetals.

Recently, Ghosh and Kim developed an alternative procedure for acetate aldol reactions utilizing ester-derived titanium enolate aldol reactions [25]. As shown in Scheme 2.5, reaction of chloroacetate **55** with $TiCl_4$ and diisopropylethylamine provided the titanium enolate, which upon reaction with monodentate aldehydes gave a highly *anti*-diastereoselective aldol product

Fig. 2.3
Chiral ligands used in the stereoselective acetate aldol reactions described in Table 2.6.

56 in excellent yield. Addition of 2 equiv. acetonitrile is critical to the observed diastereoselectivity.

Interestingly, when the aldol reaction is conducted with bidentate aldehydes, a variety of *anti* aldolates are obtained in excellent diastereoselectivity and yield. The reason for this reversal of diastereoselectivity will be discussed in Section 2.4.2.2.1. Reductive removal of chlorine provided convenient access to acetate aldol products. Representative aldol reactions with mono- and bidentate aldehydes are shown in Table 2.8.

2.4
Addition of Enolates with α-Substituents to Aldehydes

As described previously, when enolates have α-substituents, new α- and β-chiral centers are formed. Thus, it is possible to form aldol products in which the α- and β-substituents are in a *syn* or *anti* relationship to each

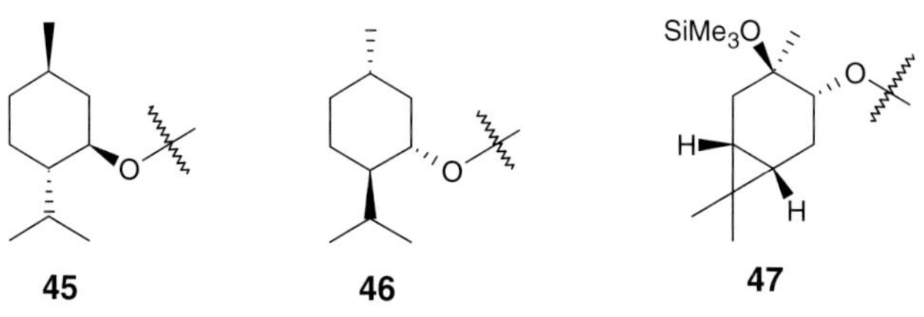

Fig. 2.4
Chiral auxiliaries used in the stereoselective acetate aldol reactions described in Table 2.6.

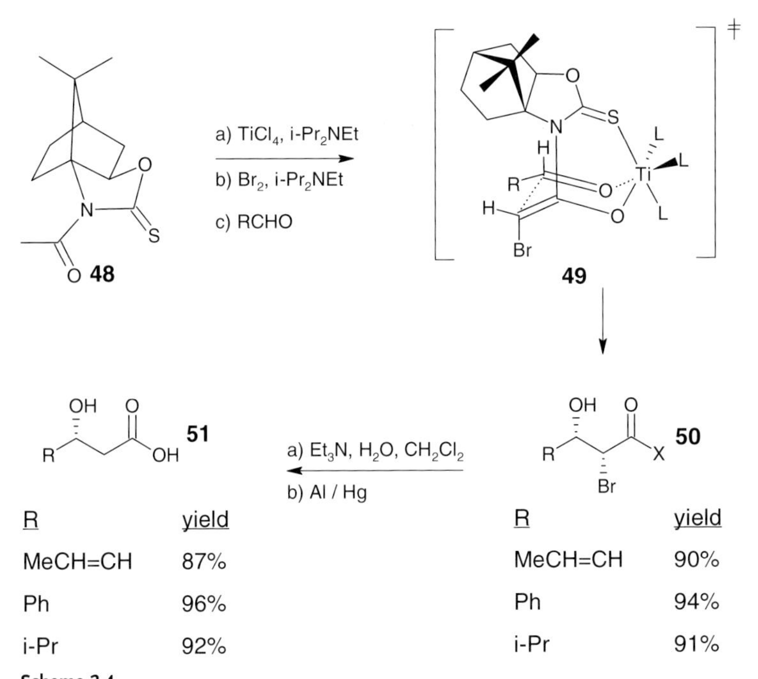

Scheme 2.4
Bromination–aldol reaction and reductive debromination–deprotection.

Tab. 2.7
Acetate aldol adducts from acetals.

52 → 53 + 54

1) $TiCl_4$, i-Pr_2NEt
2) $RCH(OMe)_2$, LA

Entry	*R*	*Lewis Acid*	*Yield (%)*	*53:54*
1	Ph	$BF_3 \cdot Et_2O$	77	88:12
2	4-$MeOC_6H_4$	$BF_3 \cdot Et_2O$	87	93:7
3	3-$MeOC_6H_4$	$BF_3 \cdot Et_2O$	75	88:12
4	4-ClC_6H_4	$BF_3 \cdot Et_2O$	81	88:12
5	(*E*)-PhCH=CH	$BF_3 \cdot Et_2O$	77	82:18
6	Me	$SnCl_4$	57	76:24
7	*n*-Pr	$SnCl_4$	62	73:27
8	*i*-Bu	$SnCl_4$	70	79:21
9	*i*-Pr	$SnCl_4$	60	71:29

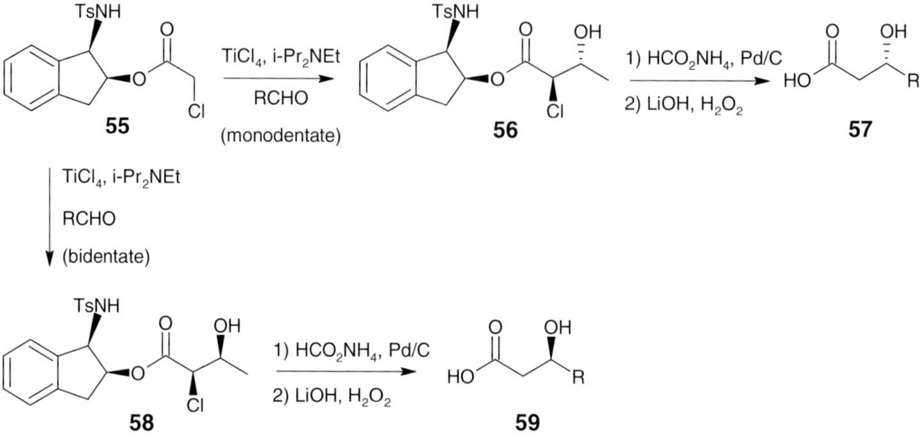

Scheme 2.5
Asymmetric chloroacetate aldol reactions.

other. Methods have been developed to provide both diastereoselectivity and enantioselectivity in the aldol products. These developments are not mere extensions of previous methods. In fact, many titanium enolate-based asymmetric transformations provide stereoselectivity and efficiency which cannot be obtained by use of other methods.

2.4.1
Syn Diastereoselectivity

In general, most enolates are formed in the *Z* configuration. When the metal involved in the reaction is acidic enough to coordinate to both the enolate and the aldehyde of the aldol reaction, *syn* adducts result via a six-membered cyclic Zimmerman–Traxler [3] transition state. This results in inherent preference for titanium enolate aldol reactions to form *syn* aldols,

Tab. 2.8
Chloroacetate aldol reactions.

Entry	*Aldehyde*	*Additive*	*Yield (%)*	*58:56*
1	*i*-PrCHO	–	82	25:75
2	*i*-PrCHO	MeCN (2.2 equiv.)	66	2:98
3	*i*-PrCHO	NMP (2.2 equiv.)	71	6:94
4	*i*-BuCHO	MeCN (2.2 equiv.)	88	<1:99
5	*i*-BuCHO	NMP (2.2 equiv.)	90	<1:99
6	PhCHO	NMP (2.2 equiv.)	47	4:96
7	$BnOCH_2CHO$	–	86	>99:1
8	$BnO(CH_2)_2CHO$	–	79	96:4

and several excellent methods have been developed to that end. One particular important feature of titanium enolate-based *syn* aldol reactions is the opportunity to develop chelation-controlled reactions. Several very interesting methods have been developed that exploit this unique ability of titanium. In asymmetric *syn* aldol reactions utilizing chiral oxazolidinone or oxazolidinethiones one can generate either *syn* aldol diastereomer from the same chiral auxiliary system.

2.4.1.1 Synthesis of *syn* Aldols in Racemic Form

2.4.1.1.1 Reactions of Ketones

Aldol reactions of metal enolates containing α-substituents have been investigated extensively. Reetz reported *syn*-selective aldol reactions of titanium enolates of a variety of cyclic ketones and aldehydes providing diastereoselective *syn* aldols in the racemic form [26]. As shown in Table 2.9, formation of titanium enolates from ketones was achieved by transmetalation of the corresponding lithium enolates. They reported good to excellent selectivity for reactions of cyclic ketones with aldehydes. Whereas *Z* enolates generally furnish *syn* aldol products under kinetic conditions, formation of *syn* aldols from cyclic enolates seems to be difficult. Interestingly, titanium enolates of cyclic ketones, which can only form *Z* enolates, provided high *syn* diastereoselectivity. Good selectivity in the reactions of acyclic ketones

Tab. 2.9
Aldol reactions of cyclic ketones with aldehydes.

O, $(CH_2)n$, **60** —LDA→ OLi, $(CH_2)n$ —$Ti(Oi\text{-}Pr)_3Cl$ or $Ti(NEt_2)_3Br$→ $OTiL_3$, $(CH_2)n$ —RCHO→ O, OH, R, $(CH_2)n$ **61** + O, OH, R, $(CH_2)n$ **62**

Entry	*R*	*n*	*Lewis Acid*	*61:62*
1	Ph	3	$Ti(NEt_2)_3Br$	85:15
2	Ph	4	$Ti(NEt_2)_3Br$	97:3
3	Ph	4	$Ti(NMe_2)_3Br$	92:8
4	Ph	4	$Ti(Oi\text{-}Pr)_3Cl$	86:15
5	*i*-Pr	4	$Ti(Oi\text{-}Pr)_3Cl$	96:4
6	Ph	5	$Ti(NEt_2)_3Br$	90:10
7	Ph	6	$Ti(Oi\text{-}Pr)_3Cl$	91:9

a) LDA
b) Ti(Oi-Pr)$_3$Cl
c) R$_2$CHO

63 → 64 + 65

R_1	R_2	64 : 65
Et	Ph	89 : 11
Et	t-Bu	81 : 19
t-Bu	Ph	87 : 13
Ph	Ph	87 : 13

Scheme 2.6
Aldol reactions of acyclic ketones with aldehydes.

with aldehydes has been observed, as shown in Scheme 2.6. Interestingly, *syn* aldol product preference occurred almost independently of enolate configuration; stereochemical rationale has not been forthcoming.

The aldol reactions of titanium enolates generated in situ were reported by Harrison [27] to give excellent yield and selectivity for *syn* aldol products, as shown in Table 2.10. However, methyl ketones tended to eliminate under the reaction conditions and provided α,β-unsaturated ketones. Reactions with propiophenone and benzaldehyde provided excellent yields of aldolates, with *syn* aldols being the major product (95:5 ratio). The stereochemical outcome was rationalized by Zimmerman–Traxler transition state model **67**.

Evans used a combination of titanium tetrachloride and diisopropylethylamine, which proved to be efficient and general for many different kinds of substrates [28]. The *syn* aldol diastereoselectivity was comparable with that of boron-mediated processes. Isolated yields with titanium enolates are considerably higher than from boron enolates. Furthermore, *syn*-selectivity

Tab. 2.10
syn-Selective aldol reactions of directly generated titanium enolates.

66 —($TiCl_4$, Et_3N; ArCHO)→ [67]‡ → 68 + 69

Entry	*Ar*	*Yield (%)*	*68:69*
1	Ph	91	95:5
2	p-MeC_6H_4	94	96:4
3	p-$MeOC_6H_4$	95	89:11
4	o-$MeOC_6H_4$	96	87:13
5	p-$NO_2C_6H_4$	98	87:13

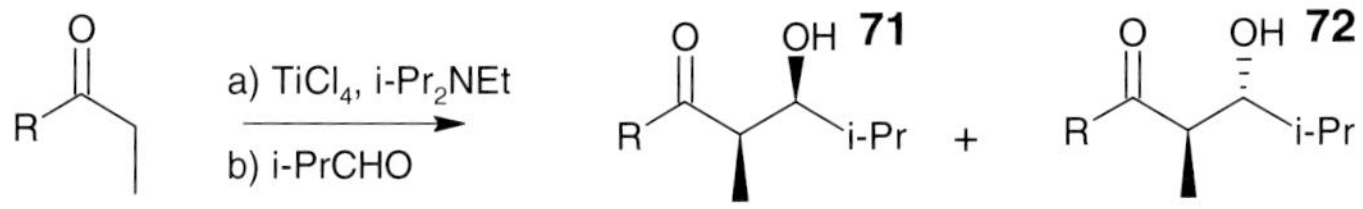

R	Yield	71 : 72
Et	95	98 : 2
i-Pr	95	93 : 7

Scheme 2.7
Aldol reactions of directly generated titanium enolates.

depends on the size of the amine bases. The enolates exist as aggregated complexes, as was evidenced by NMR studies. Representative examples are shown in Scheme 2.7.

2.4.1.1.2 **Reactions of Esters and Thiol Esters**

Syn-selective aldol reactions involving directly generated thiol ester enolates were reported to give moderate to good selectivity and moderate yield, as shown in Table 2.11 [29]. Modest *anti*-selectivity was obtained in reactions of 2-pyridylthiopropionate (entry 7), suggesting the 2-pyridyl moiety could be coordinating with titanium.

The reactions of α-thio-substituted ester enolates have been reported to give moderate to excellent selectivity and moderate to good yield, as shown in Table 2.12 (entries 1–6) [29]. Interestingly, for α-substituted propionates (entries 4–6), selectivity was good to excellent for the product in which the

Tab. 2.11
syn-Selective aldol reactions of thiol esters.

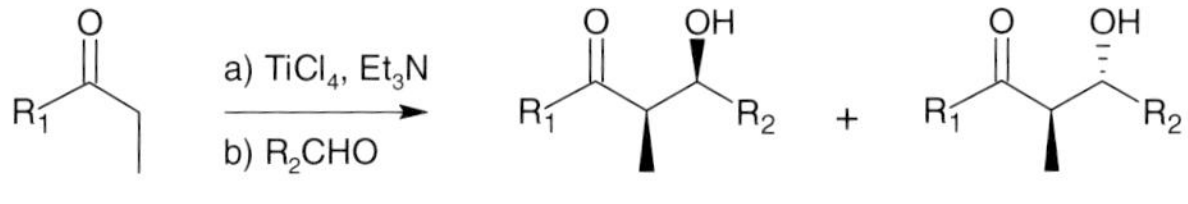

Entry	R_1	R_2	*Yield (%)*	*74:75*
1	PhS	Ph	70	85:15
2	PhS	*n*-Pr	68	78:22
3	PhS	*i*-Pr	65	89:11
4	*t*-BuS	Ph	77	86:14
5	*t*-BuS	*n*-Pr	72	69:31
6	*t*-BuS	*i*-Pr	75	71:29
7	2-pyridyl-S	*i*-Pr	69	29:71
8	*o*-$MeOC_6H_4S$	Ph	71	87:13
9	C_6F_5S	Ph	40	95:5

Tab. 2.12
syn-Selective aldol reactions with α-heteroatomic substituents.

76 → 77 + 78

Entry	*R₁*	*R₂*	*R₃*	*R₄*	*Additive*	*Yield (%)*	*77:78*	*Ref.*
1	EtO	*n*-BuS	H	Ph	–	80	94:6	29
2	EtO	*n*-BuS	H	*n*-Pr	–	74	76:24	29
3	EtO	*n*-BuS	H	*i*-Pr	–	73	84:16	29
4	EtO	*n*-BuS	Me	Ph	–	90	95:5	29
5	EtO	*n*-BuS	Me	*n*-Pr	–	71	88:12	29
6	EtO	*n*-BuS	Me	*i*-Pr	–	76	88:12	29
7	MeO	PhSe	H	*i*-Pr	Ph_3P	86	90:10	30
8	MeO	PhSe	H	Ph	Ph_3PO	92	97:3	30
9	MeO	PhSe	H	$PhCH_2CH_2$	Ph_3PO	92	88:12	30
10	MeO	PhSe	H	(*E*)-PhCH=CH	Ph_3PO	81	>98:2	30
11	MeO	PhSe	H	*n*-C_5H_{11}	Ph_3PO	83	95:5	30
12	EtO	PhSe	Me	Ph	Ph_3P	93	>98:2	31
13	EtO	PhSe	Me	*i*-Pr	Ph_3P	92	>98:2	31
14	Me	PhSe	H	Ph	–	91	>98:2	31
15	Me	PhSe	H	*i*-Pr	–	75	>98:2	31
16	Me	PhSe	H	PhCH=CH	–	74	96:4	31
17	Me	PhSe	H	$PhCH_2CH_2$	–	94	93:7	31
18	Me	PhSe	H	*n*-$C_6H_{13}C{\equiv}C$	–	97	94:6	31
19	Et	PhSe	H	Ph	–	71	>98:2	31
20	*i*-Pr	PhSe	H	$PhCH_2CH_2$	–	92	>98:2	31

α-sulfur substituent and β-hydroxyl group are *syn* to each other. This might indicate additional chelation to titanium.

Reactions of α-selenoacetate enolates have been reported to give excellent selectivity and yield in the presence of triphenylphosphine or triphenylphosphine oxide, as shown in Table 2.12 (entries 7–11) [30]. This work was extended to α-selenopropionate esters (Table 2.12, entries 12 and 13) and α-selenoketones (Table 2.12, entries 14–20) [31]. As for α-sulfur substituents, the α-selenopropionate ester enolates gave excellent yield and selectivity for aldols in which the α-heteroatom and β-hydroxyl group were *syn* to each other. These papers highlighted the importance of the presence of 1 equiv. bulky phosphine or amine to coordinate to titanium to improve the yield and the selectivity.

2.4.1.1.3 **Aldol Reactions of Aldehyde Hydrazones**

In contrast with the aldol reactions of ketones or carboxylic acid derivatives, those of aldehyde enolates typically proceed with little to no selectivity. This

Tab. 2.13
syn-Selective aldol-like reactions of hydrazones with aldehydes.

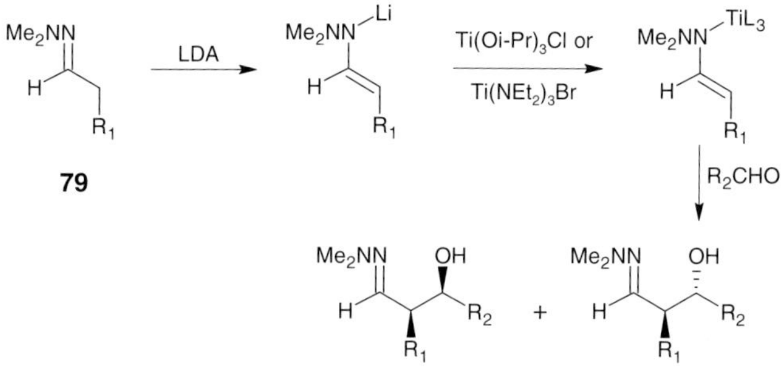

Entry	R_1	R_2	*Lewis Acid*	*Yield (%)*	*80:81*
1	Me	Ph	$Ti(O\textit{i}\text{-}Pr)_3Cl$	80	91:9
2	Me	Ph	$Ti(NEt_2)_3Br$	61	85:15
3	Ph	Ph	$Ti(O\textit{i}\text{-}Pr)_3Cl$	95	98:2
4	Ph	$p\text{-}NO_2C_6H_4$	$Ti(O\textit{i}\text{-}Pr)_3Cl$	40	98:2
5	Ph	Me	$Ti(O\textit{i}\text{-}Pr)_3Cl$	95	98:2
6	Ph	Me	$Ti(NEt_2)_3Br$	50	90:10
7	Ph	*i*-Pr	$Ti(O\textit{i}\text{-}Pr)_3Cl$	78	98:2
8	*i*-Pr	Ph	$Ti(O\textit{i}\text{-}Pr)_3Cl$	78	94:6
9	Me	Me	$Ti(O\textit{i}\text{-}Pr)_3Cl$	61	95:5
10	Me	*t*-Bu	$Ti(O\textit{i}\text{-}Pr)_3Cl$	70	93:7

could be because of lack of steric bulk to influence the cyclic six-membered Zimmerman–Traxler transition state. For this reason there are few useful reactions that directly utilize aldehyde enolates, even though the term aldol originally referred to a *β*-hydroxyaldehyde. This issue was resolved by using titanium enolates generated from *N*,*N*-dimethylhydrazones by transmetalation with $Ti(O\textit{i}\text{-}Pr)_3Cl$ or $Ti(NEt_2)_3Br$ [32]. The reaction furnished racemic *syn* aldols with excellent selectivity and moderate to excellent yield, as shown in Table 2.13. Reactions with cyclic ketone hydrazones also provided excellent *syn* diastereoselectivity and yields, as shown in Scheme 2.8.

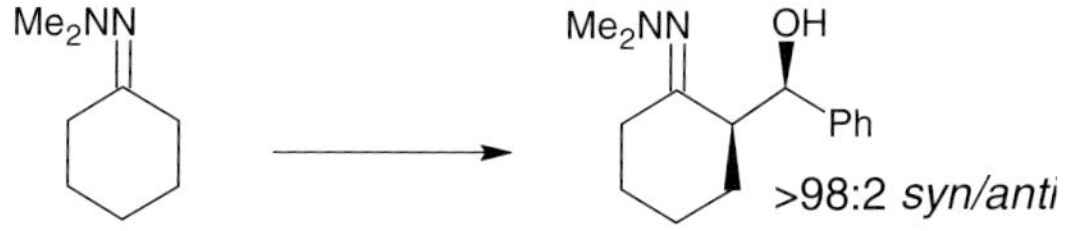

Scheme 2.8
syn-Selective aldol-like reaction of cyclic hydrazone.

2.4.1.2 Synthesis of Optically Active *syn* Aldols Using Chiral Auxiliaries

Asymmetric aldol reactions utilizing chiral auxiliaries or templates have emerged as one of the most reliable methods in organic synthesis. Both *syn*- and *anti*-selective aldol reactions have been developed over the years. The field of asymmetric *syn* aldol reactions has been largely advanced by Evans since his development of dibutylboron enolate aldol chemistry based on amino acid-derived chiral oxazolidinones. This method requires expensive dibutylboron triflate, however, and the amino acid-derived chiral auxiliary is only readily available in one enantiomer and thus only provides one enantiomer of the *syn* aldol. Several methods developed on the basis of titanium enolates provide convenient access to both "Evans" and "non-Evans" *syn* aldol products.

2.4.1.2.1 Amino Acid-derived Oxazolidinone and Related Auxiliaries

Oxazolidinone-, oxazolidinethione-, oxazolidineselone-, and thiazolidinethione-based enolates react with aldehydes via the well-established six-membered Zimmerman–Traxler [3] chair-like transition state. Exhaustive studies and analysis by Crimmins have established the theoretical basis of these reactions [33]. These transition states can proceed without chelation between carbonyl or thiocarbonyl (**84**) or with an additional chelation to titanium (**85**), as shown in Scheme 2.9. To proceed via the chelated transition structure **85**, one of the ligands on titanium (typically chloride) must be displaced by the carbonyl or thiocarbonyl group. Although these groups are not sufficiently nucleophilic to completely displace this ligand on their own,

"Evans" *syn* **86** "non-Evans" *syn* **87**

Scheme 2.9
Asymmetric *syn* aldol transition-state models.

eq. $Ti(Oi\text{-}Pr)_3Cl$	89 : 90
3	3 : 92
1	77 : 16

base	91 : 92	yield
$i\text{-}Pr_2NEt$	94 : 6	84
TMEDA	98 : 2	83

Scheme 2.10
syn-Selective aldol reactions involving oxazolidinone chiral auxiliaries.

the ligand can be easily abstracted with a second equivalent of titanium. A consequence of this is that these substrates will occasionally give mixtures of products resulting from incomplete conversion of **84** to **85**. Also, addition of chelating ligands, extra equivalents of amine bases, or even some solvent molecules tend to disfavor transition state **85**, because of their preferential chelation to the titanium enolate.

The initial adaptation of amino acid-derived oxazolidinone chiral auxiliaries to titanium enolate aldol reactions was conducted by Thornton [34–36]. The procedure involved transmetalation of the lithium enolate with $Ti(O\textit{i}\text{-}Pr)_3Cl$ in THF or ether. By varying the amount of titanium, the ratio of products **89**:**90** could be varied, as shown in Scheme 2.10. They attributed this observation to quenching of lithium interference, but it could also be a matter of switching between chelated and non-chelated transition structures with excess titanium. They observed that a stoichiometric amount of THF gave rise to titanium chelation [35]. By choosing the more powerfully chelating THF they could favor formation of the *syn* product **89** whereas the weakly chelating ether favored *syn* product **90**. Because both ethereal solvents can chelate to titanium, however, their products were always complex mixtures of both *syn* and *anti* aldols. The ethereal solvents were necessary because of solubility problems.

These complications were alleviated when Evans demonstrated that titanium enolates could be directly generated in dichloromethane with amine bases [28]. Stereoselection and reaction yields are comparable with those of boron enolate-based *syn* aldol reactions. Amine bases seemed to have a noticeable effect on stereoselectivity. As shown in Scheme 2.10, the use of TMEDA improved selectivity compared with *i*-Pr_2NEt; this suggests the reaction proceeded through non-chelated transition state **84**. Furthermore, stoichiometry of aldehydes (2 equiv.) is also critical for complete conversion.

Crimmins et al. developed amino acid-derived oxazolidinethione [37] and thiazolidinethione [38] chiral auxiliaries and demonstrated their utility in titanium enolate aldol reactions. Depending upon the amount and nature of amine bases and the stoichiometry of $TiCl_4$, "Evans" or "non-Evans" *syn* aldol products can be provided with excellent diastereoselectivity and isolated yields. Reactions involving oxazolidinethione auxiliaries led to "Evans" *syn* aldol product **94** when 1.0 equiv. $TiCl_4$, 1.1 equiv. aldehyde, and 2.5 equiv. (−)-sparteine were used, as shown in Table 2.14, entries 1–3. Rationalizing that the extra amine base or excess aldehyde could be acting as a ligand on titanium, they demonstrated that 1.05 equiv. $TiCl_4$, 1.1 equiv. aldehyde, 1.0 equiv. (−)-sparteine, and 1.0 equiv. *N*-methylpyrrolidinone also gave product **94**, the latter conditions being more economical and simpler to work up. (−)-Sparteine was discovered to have a dramatic rate-enhancement effect on these aldol reactions, but it was demonstrated that its chiral architecture did not lead to significant asymmetric induction. Using oxazolidinethione auxiliaries, the chelation-controlled "non-Evans" product **95**

Tab. 2.14
Oxazolidinethione based *syn* aldol reactions.

A: a) 1.0 eq. $TiCl_4$ b) 2.5 eq. (-)-sparteine c) RCHO
B: a) 2 eq. $TiCl_4$ b) 1.1 eq. i-Pr_2NEt c) RCHO

93 → **94** + **95**

Entry	*Method*	*R*	*Yield (%)*	*94:95:anti*
1	A	*i*-Pr	70	99:1:0
2	A	Ph	89	97:2:1
3	A	MeCH=CH	65	97:2:1
4	B	*i*-Pr	87	0:95:5
5	B	Ph	88	1:98:1
6	B	MeCH=CH	81	0:95:5

Scheme 2.11
Conversion of oxazolidinethione auxiliary to other functionality.

could be formed by using 2 equiv. $TiCl_4$ and only 1 equiv. diisopropylethylamine, as shown in Table 2.14, entries 4–6.

In a typical procedure, 2 mmol $TiCl_4$ was added dropwise to a solution of 1 mmol oxazolidinethione **93** in 6 mL CH_2Cl_2 at 0 °C and stirred for 5 min. *i*-Pr_2NEt (1.1 mmol) was added dropwise and the dark red solution stirred for 20 min at 0 °C and then cooled to −78 °C. Aldehyde (1.1 mmol) was added dropwise and stirred for 1 h. One intrinsic feature of oxazolidinethiones is that the chiral auxiliary could be readily converted to Weinreb amide **97** by reaction with imidazole and methoxylamine salt, as shown in Scheme 2.11. Similarly, sodium borohydride reduction gave alcohol **98** and DIBALH reduction provided the corresponding aldehyde in excellent yield.

To demonstrate the full utility of the (−)-sparteine-mediated enolization, they also showed that oxazolidinone chiral auxiliaries could be reacted with either 1.1 equiv. $TiCl_4$ and 2.5 equiv. (−)-sparteine or 1.05 equiv. $TiCl_4$, 1.0 equiv. (−)-sparteine, and 1.0 equiv. *N*-methylpyrrolidinone to yield product **94**.

Thiazolidinethione-derived chiral auxiliaries have similar reactivity and selectivity, as shown in Table 2.15. Because of the increased nucleophilicity of the thiazolidinethione ring, chelation-controlled reaction through transition state **85** enabled preferential formation of the "non-Evans" *syn* product **101**. One equivalent of *i*-Pr_2NEt, TMEDA, or (−)-sparteine and 1 equiv. $TiCl_4$ provided **101** diastereoselectively. Interestingly, when the reaction was performed with 2 equiv. of TMEDA or (−)-sparteine, "Evans" *syn* aldol adduct **100** was obtained diastereoselectively. These aldol products can be converted to a variety of other functionality under mild conditions. Other oxazolidinethiones and thiazolidinethiones have resulted in comparable diastereoselectivity and yields.

Similar reactions utilizing oxazolidineselone chiral auxiliaries were developed by Silks and coworkers [39]. They demonstrated good to excellent

Tab. 2.15
Thiazolidinethione-based *syn* aldol reactions.

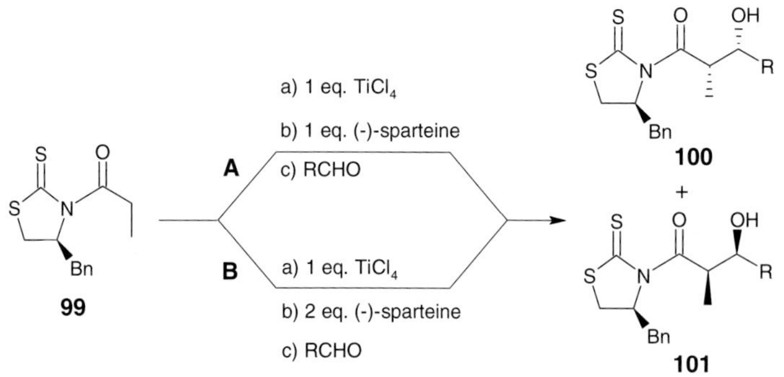

Entry	*Method*	*R*	*Yield (%)*	*100:101*
1	A	*i*-Pr	60	2:98
2	A	Ph	52	<1:99
3	B	*i*-Pr	75	97:3
4	B	Ph	62	>99:1

yields and excellent stereoselectivity in reactions of *N*-propionyl- and *N*-benzyloxyacetyloxazolidineselone with a variety of aldehydes, as shown in Table 2.16. The acylated oxazolidineselone can be prepared in a one-pot procedure from oxazoline **102** via lithiation, addition of elemental selenium, and quenching with the appropriate acyl chloride. The selenocarbonyl also has utility as a chiral probe via ^{77}Se NMR. The selone chiral auxiliary can also be converted to other functionality similar to oxazolidinethione **96** and thiazolidinethione **99**.

2.4.1.2.2 **Camphor-derived Chiral Auxiliaries**

After adaptation of amino acid-derived oxazolidinone chiral auxiliaries to titanium enolate aldol reactions, Thornton and coworkers went on to develop a camphor-derived chiral auxiliary [40]. Moderate to good selectivity was observed for reactions with a variety of aldehydes using camphorquinone-derived *N*-propionyl oxazolidinone **106**, as shown in Table 2.17. The carbonyl of the oxazolidinone is not a good enough nucleophile to completely displace chloride to form chelated transition structure **107**, so the non-chelated transition state assembly **108** is always a competing pathway. Stereodifferentiation experiments with (*R*)- and (*S*)-2-benzyloxypropanal were also investigated. *syn* Diastereoselectivity was, however, moderate (53% de) compared with lithium enolate-based reactions which provided high selectivity (85% de) for the (*R*) isomer. Reactions with (*S*)-benzyloxypropanal resulted in mismatched aldehyde and enolate selectivity and a 42:52 ratio of *syn* adducts.

Tab. 2.16
Oxazolidineselone-based *syn* aldol reactions.

Entry	R_1	R_2	R_3	*Yield (%)*	*104:105*
1	*i*-Pr	BnO	*i*-Pr	72	75:25
2	Bn	Me	*i*-Pr	86	>99:1
3	Bn	Me	MeCH=CH	86	>99:1
4	Bn	Me	Ph	91	>99:1

Tab. 2.17
Camphor-derived *syn* aldol reactions.

Entry	*R*	*109:110:anti*
1	Et	76:13:11
2	Ph	79:2:19
3	*i*-Pr	86:2:12
4	MeCH=CH	49:4:49

Tab. 2.18
Camphor-derived *syn*-selective aldol reactions.

Entry	*R*	*Yield (%)*	*114:115*
1	Me	84	>99:1
2	*i*-Pr	70	>99:1
3	Ph	79	>99:1
4	MeCH=CH	70	99:1

To improve the π-facial and *syn* selectivity of the camphor-derived chiral auxiliary, Ahn et al. developed a chiral oxazinone derived from ketopinic acid (**111**) [41]. The explanation for the low diastereoselectivity of camphor-derived auxiliary **106** was that the steric influence of the proximal bridge-head methyl group was insufficient for π-facial selectivity. It seemed that the *syn*-7-methyl group was too far away from the acyl moiety. In oxazinone-based chiral auxiliary **111** one enolate face is in proximity to the camphor skeleton and, as a result, steric bias was significantly enhanced over the **106**-derived system. Moderate to good yields and near complete *syn* diastereoselectivity for a range of aldehydes have been reported, as shown in Table 2.18. The steric influence of their auxiliary is apparently strong enough to overcome inherent weakness in titanium chelation; near complete selectivity for chelation-controlled product **114** arising from transition state assembly **112** was reported.

Yan et al. developed camphor-derived oxazolidinethione chiral auxiliary **116** from ketopinic acid [42, 43]. The opposing location of the *N*-acyl group in this auxiliary when compared with **106** and **111** leads to formation of the other *syn* product **120** via chelation control. Good yields and excellent selectivity were reported for a range of aldehydes, as shown in Table 2.19. Be-

Tab. 2.19
Camphor-derived *syn*-selective aldol reactions.

Entry	*R*	*Yield (%)*	*119:120*
1	*i*-Pr	85	2:98
2	Ph	84	3:97
3	MeCH=CH	86	<1:99

cause the thiocarbonyl is a good ligand for titanium, the reaction proceeds through chelated transition state assembly **118**. The reactions with the corresponding *N*-bromoacyl derivatives also provided excellent *syn* diastereoselectivity and isolated yields [22]. The observed *syn* stereochemistry is consistent with chelation-controlled model **118**.

2.4.1.2.3 Aminoindanol and Amino Acid-derived Chiral Auxiliaries

Ghosh et al. developed highly diastereoselective ester-derived titanium enolate-based *syn* aldol reactions [44]. Chiral sulfonamide **121** was readily prepared by tosylation of optically active aminoindanol followed by reaction with propionyl chloride in pyridine. As shown in Scheme 2.12, the corresponding titanium enolate of **121** was generated with 1.2 equiv. $TiCl_4$ and 3.8 equiv. *i*-Pr_2NEt at 0 °C to 23 °C for 1 h. These conditions provided the *Z* enolate which, upon reacting with a variety of bidentate titanium-complexed aldehydes (2 equiv. aldehyde/2.2 equiv. $TiCl_4$), afforded good to excellent yields of *syn* aldol products **125** with high diastereoselectivity, as shown in Table 2.20. On the other hand, when the reaction was carried out with monodentate aldehydes, *anti* aldol product **126** was obtained.

The stereochemical outcome of these reactions has been rationalized by use of a chelation-controlled model. In this model, the titanium enolate

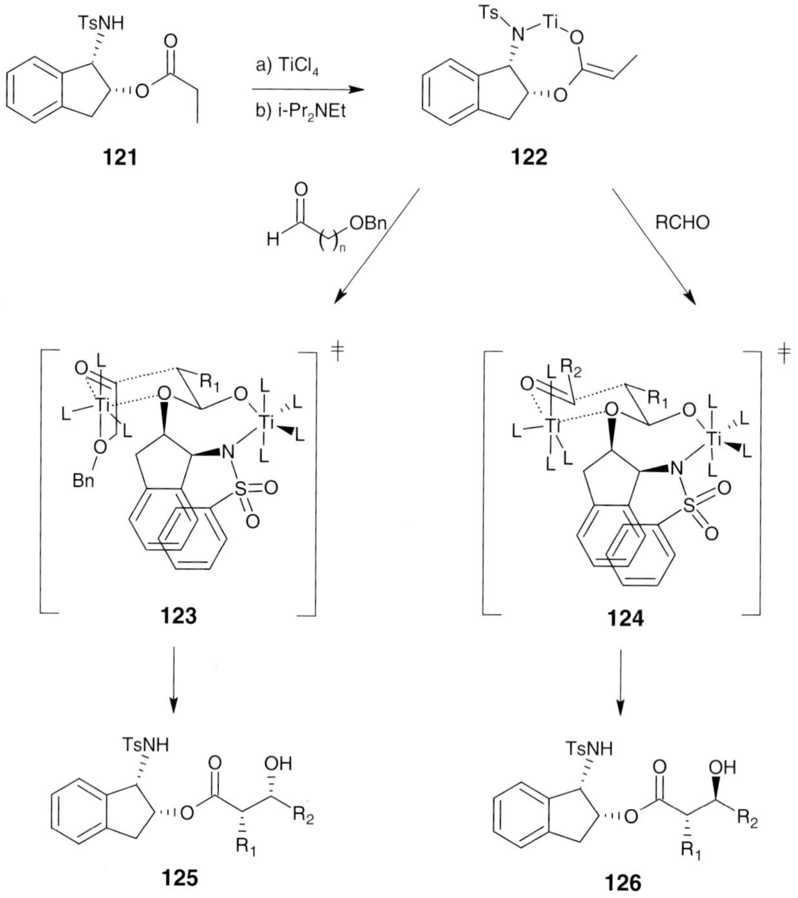

Scheme 2.12
Aminoindanol-derived asymmetric aldol reactions.

is a seven-membered metallocycle and is assumed to have a chair-like conformation. It has been postulated that *anti* aldol diastereoselectivity was obtained from monodentate aldehydes via a Zimmerman–Traxler-like model **124** whereas reactions with bidentate aldehydes presumably proceed through **123**. The oxyaldehyde side-chain is oriented pseudo-axially for effective metal chelation. As evidenced, this methodology provides excellent diastereoselectivity with a range of esters. The enhanced selectivity for benzyloxyacetaldehyde and benzyloxypropionaldeyde is because of five- or six-membered chelation. Reaction of benzyloxybutyraldehyde results in slightly reduced the *syn* diastereoselectivity. This might be because of less favorable seven-membered chelation.

On the basis of the possible transition state assembly **123**, Ghosh and coworkers further speculated that the α-chiral center or the indane ring

Tab. 2.20
Aminoindanol-based *syn* aldol reactions.

Entry	R_1	R_2	*Yield (%)*	*125:126*
1	Me	$BnOCH_2$	84	98:2
2	Me	$BnO(CH_2)_2$	51	98:2
3	Bn	$BnOCH_2$	84	99:1
4	Bn	$BnO(CH_2)_2$	51	99:1
5	Bn	$BnO(CH_2)_3$	55	94:6
6	*i*-Bu	$BnOCH_2$	83	99:1
7	*i*-Bu	$BnO(CH_2)_2$	56	99:1

may not be necessary for *syn* selectivity for bidentate oxyaldehydes [45]. Indeed, the corresponding phenylalanine-derived chiral auxiliary **127** (R = Bn) resulted in good *syn* selectivity with different bidentate oxyaldehydes, as shown in Table 2.21. They also investigated the effect of the *β*-chiral substituent and discovered that use of valine-derived chiral auxiliary **127** (R = *i*-Pr) resulted in good yields and excellent *syn* selectivity for a wide range of mono- and bidentate aldehydes [46].

Reaction of enolate **124** with a variety of monodentate aldehydes provided aldol products with good to excellent *anti* diastereoselectivity; this will be

Tab. 2.21
Amino acid-derived *syn* aldol reactions.

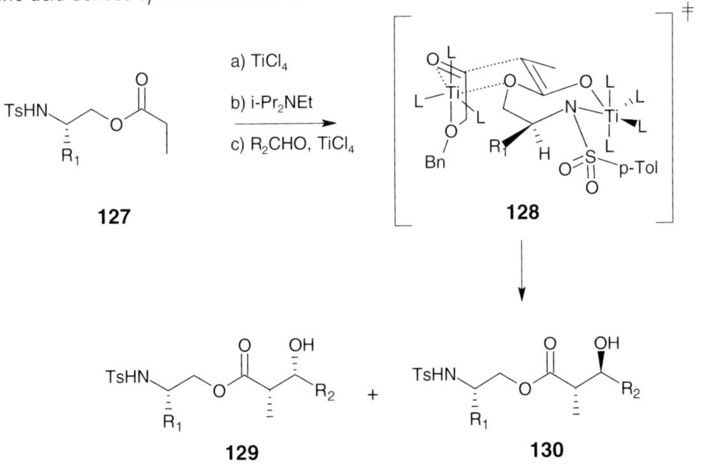

Entry	R_1	R_2	*Yield (%)*	*129:130*
1	Bn	$BnOCH_2$	80	98:2
2	Bn	$BnO(CH_2)_2$	81	97:3
3	*i*-Pr	*i*-Bu	93	95:5
4	*i*-Pr	PhCH=CH	89	96:4

Tab. 2.22
Effect of $TiCl_4$ stoichiometry on diastereoselectivity.

TsNH, O, 131; a) $TiCl_4$ b) i-Pr_2NEt c) PhCH=CHCHO, $TiCl_4$ → 132 + 133 (TsNH, O, OH, Ph)

Entry	*$TiCl_4$ Equiv.*	*Yield (%)*	*132:133*
1	0.0	6	0:100
2	1.0	68	2:98
3	2.0	85	16:84
4	3.0	94	75:25
5	4.0	94	93:7
6	5.0	95	94:6

discussed in detail in Section 2.4.2.2. Subsequent investigation revealed that the stoichiometry of $TiCl_4$ required for aldehyde activation is critical to the observed selectivity [47]. As shown in Table 2.22, with increasing quantities of $TiCl_4$, the *syn:anti* product ratio and reaction yields improved. There was, however, a dramatic reversal of diastereoselectivity when 3 equiv. $TiCl_4$ was used for complexation with cinnamaldehyde (2 equiv.). Reaction of 2 equiv. cinnamaldehyde precomplexed with 5 equiv. $TiCl_4$ provided diastereoselective *syn* aldol product **132** in excellent yield.

The scope and generality of this methodology were then examined with a variety of aldehydes, as shown in Table 2.23. The stereochemical outcome can be rationalized by using open-chain transition state model **136**, which is favored by use of increasing amounts of $TiCl_4$ and furnishes *syn* aldol **138** as the major product. One of the most important features of ester-derived titanium enolate aldol reactions is that, depending upon the choice of aldehyde and the stoichiometry of $TiCl_4$, one can generate *syn* or *anti* aldol adducts diastereoselectively from the same chiral auxiliary. Ready availability of either enantiomer of *cis*-aminoindanol also provides convenient access to both diastereomers of *syn* or *anti* aldols in optically active form.

2.4.1.2.4 **Other Chiral Auxiliaries**

Other chiral auxiliaries used in *syn* aldol reactions are illustrated in Figure 2.5. Xiang et al. developed *N*-tosylnorephedrine-based chiral auxiliary **139**

Tab. 2.23
syn-Selective aminoindanol-derived aldol reactions.

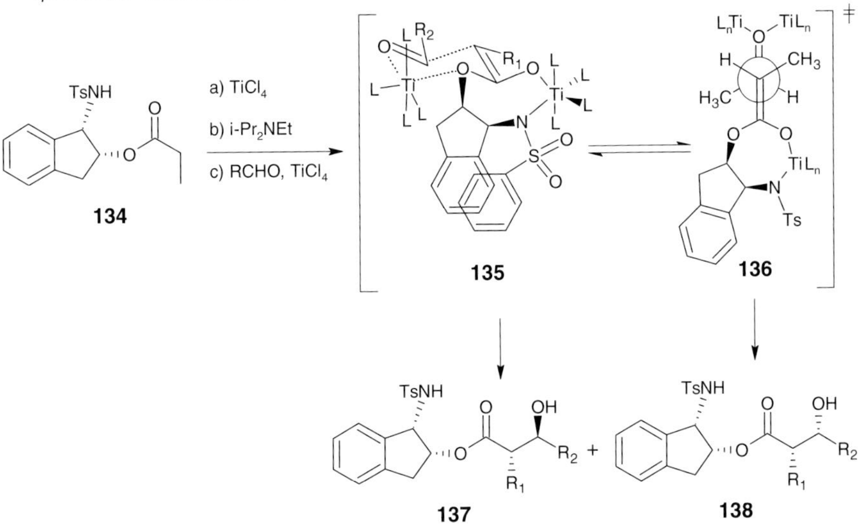

Entry	*R*	*Yield (%)*	*137:138*
1	Et	80	13:87
2	*i*-Pr	82	13:87
3	*t*-Bu	65	1:99
4	Ph	91	7:93
5	$Me_2C{=}CH$	65	11:89

[48]. It was shown that **139**-based esters could enolize in the presence of $TiCl_4$ and amine bases. Although the auxiliary favored *syn* products, the reactions yielded complex mixtures of both *syn* and *anti* products.

Ahn and coworkers developed stilbenediamine-derived 1,2,5-thiadiazolidine-1,1-dioxide-based chiral auxiliary **140** and demonstrated its utility in *syn* aldol reactions [49]. Excellent selectivity and yield were observed for a variety of aldehydes, as shown in Table 2.24, entries 1–3. The cyclic sulfamide auxiliary is novel in that it is bifunctional and C_2-symmetric, so only

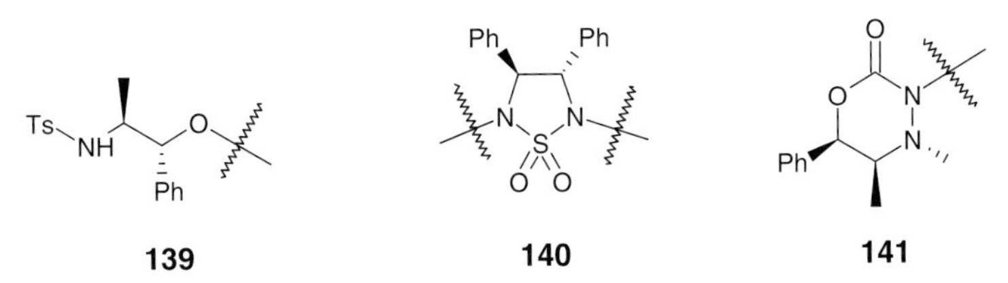

Fig. 2.5
Other chiral auxiliaries used in *syn* aldol reactions.

Tab. 2.24
syn Aldol reactions utilizing other chiral auxiliaries.

142 —RCHO→ 143 + 144

Entry	*X*	*R*	*Yield (%)*	*143:144*	*Ref.*
1	**140**	Ph	91	96:4	49
2	**140**	*i*-Pr	93	97:3	49
3	**140**	(*E*)-MeCH=CH	89	95:5	49
4	**141**	*p*-$MeOC_6H_4$	93	96:4	50
5	**141**	*t*-Bu	70	75:25	50

half an equivalent of chiral material is needed. The first aldol fragment can be removed by simple treatment with sodium methoxide in THF. To remove the second aldol fragment it was necessary to protect the free sulfamide with a Boc group. The second aldol product could then be removed by a method analogous to the first. The favored product **143** is derived from chelation control, because of the excellent chelating ability of the sulfone in the chiral auxiliary.

Hitchcock and coworkers developed ephedrine-derived 3,4,5,6-tetrahydro-2*H*-1,3,4-oxadiazin-2-one-based chiral auxiliary **141** and demonstrated its utility in aldol reactions [50]. It was discovered that the aldehyde had to be present during enolization for reaction to occur, because of difficulties in enolization. Use of aromatic aldehydes resulted in good yield and good to excellent selectivity, as did aliphatic aldehydes without α-hydrogen atoms. This method is not useful for aldehydes bearing α-hydrogen atoms, because of self-condensation.

2.4.1.3 Synthesis of Optically Active *syn* Aldols Using Chiral Titanium Ligands

Duthaler and coworkers used carbohydrate–titanium complexes for synthesis of optically active *syn*-β-hydroxy-α-amino acids [51]. These *syn*-α-aminoaldols were obtained in moderate yield and excellent *syn* diastereoselectivity, as shown in Table 2.25. Transmetalation of the lithium enolate of glycine ester derivative **145** with chiral titanium complex **146** provided a titanium enolate which upon reaction with a wide variety of aldehydes provided *syn*-β-hydroxy-α-amino esters **148**. Subsequent hydrolysis and *N*-protection gave α-aminoaldols **149**.

Duthaler and coworkers also reported asymmetric *syn* aldol methodology based on their titanium complex **146** [52]. Heathcock demonstrated the capacity of 2,6-dimethylphenyl propionate-derived lithium enolates to undergo addition to a range of aldehydes affording racemic *anti* aldol adduct **151**

Tab. 2.25
syn Aldol reactions of glycine derivatives.

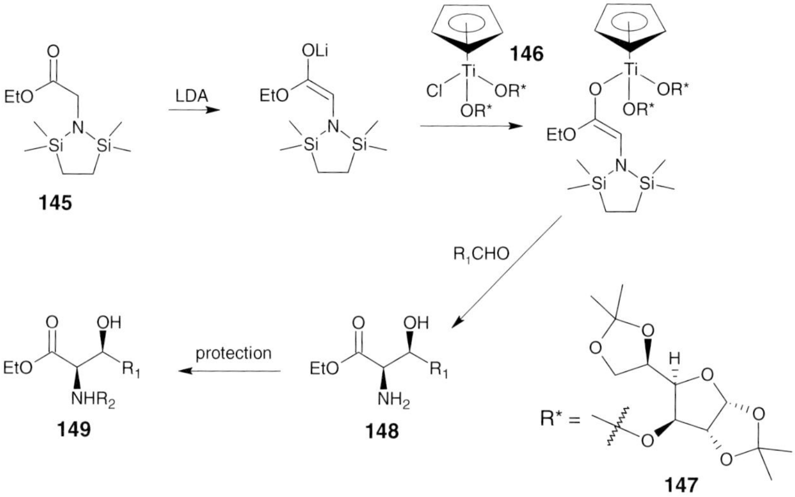

Entry	R_1	R_2	*Yield (%)*	*de (%)*
1	Me	Boc	53	>98
2	$CH_2{=}CMe$	CHO	61	99
3	Ph	Boc	60	>96
4	*t*-BuO_2C	CHO	66	>96

with excellent diastereoselectivity. The *E* enolate presumably reacted with aldehydes via a cyclic transition state to form *anti* aldol adducts **151**. Transmetalation of the lithium enolate of **150** with **146** and reaction with a variety of aldehydes provided *syn* aldols in moderate to good yield and excellent stereoselectivity, as shown in Table 2.26. It was observed that the kinetically generated *E* enolates were responsible for the observed stereoselectivity. Equilibration to the more stable *Z*-enolates gave *anti* products, as discussed in Section 2.4.2.3. To rationalize the stereochemical outcome of these *syn* aldol products, boat-like transition state **152** was proposed. There are also other examples of *syn* aldol product formation from titanium *E*-enolates [53].

Mahrwald reported aldol reactions of ketone enolates with aldehydes in which the reaction was conducted with equimolar amounts of titanium(IV) alkoxides and α-hydroxy acids [54]. This provided aldol products with high *syn* diastereoselectivity, as shown in Table 2.27. Among a variety of alkoxides and α-hydroxy acids examined, the use of Ti(O*t*-Bu)$_2$-BINOL and (*R*)-mandelic acid resulted in high *syn* diastereoselectivity and aldol products were obtained in enantiomerically enriched form.

Other α-hydroxy acids such as tartaric acid or lactic acid resulted in very low enantioselectivity (18–24%). The influence of the chirality of BINOL

Tab. 2.26
syn Aldol reactions using chiral titanium enolates.

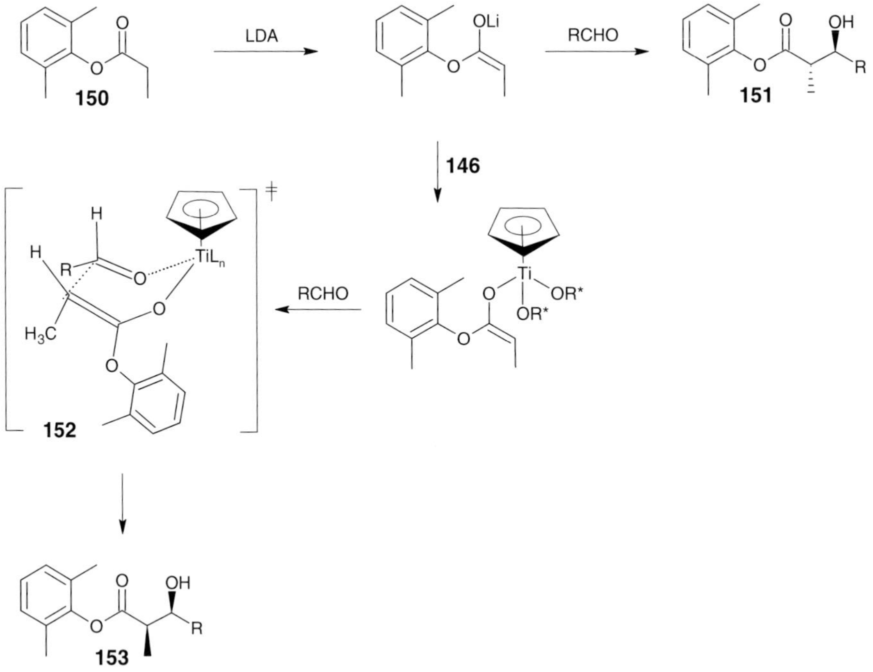

Entry	*R*	*Yield (%)*	*153:151*
1	*i*-Pr	76	94:6
2	CH_2=CMe	61	96:4
3	Ph	82	96:4
4	Pr	87	92:8

Tab. 2.27
syn-Selective aldol reactions.

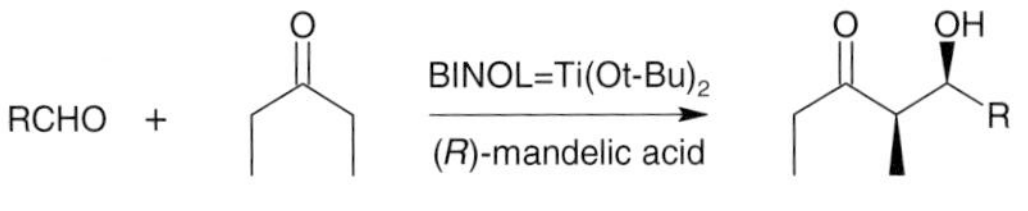

Entry	*R*	*Yield (%)*	*syn:anti*	*ee (syn) (%)*
1	Ph	71	95:5	93
2	*t*-Bu	55	85:15	87
3	PhC≡C	72	73:27	94
4	*i*-Pr	48	89:11	83

Tab. 2.28
syn-Selective aldol reactions of chiral α-silyloxyketones.

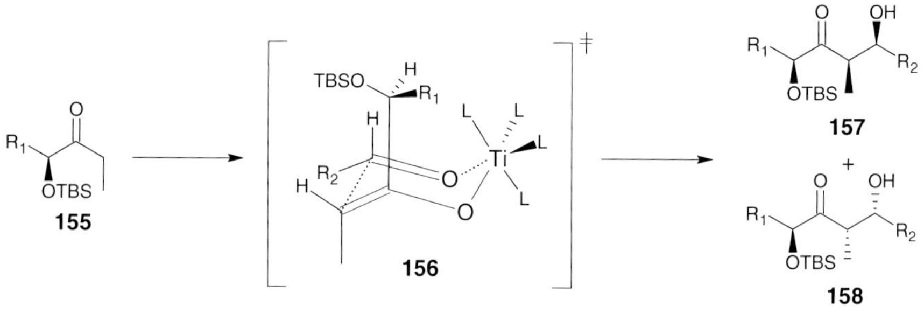

Entry	R_1	R_2	*Method*[a]	*Yield (%)*	*157:158*	*Ref.*
1	Cyclohexyl	Ph	A	–	99:1	55
2	Cyclohexyl	*i*-Pr	A	–	>99:1	55
3	Cyclohexyl	Ph	B	–	99:1	56
4	Cyclohexyl	*i*-Pr	B	–	99:1	56
5	Me	*i*-Pr	C	90	97:3	57
6	Bn	*i*-Pr	C	85	97:3	57

[a] Method A: LDA, $Ti(Oi\text{-}Pr)_3Cl$, R_2CHO; Method B: LDA, $Ti(Oi\text{-}Pr)_4$, R_2CHO; Method C: $TiCl_4$, $i\text{-}Pr_2NEt$, R_2CHO

was insignificant. Although the mechanism of this reaction is unclear, ligand exchange between *t*-BuOH and α-hydroxy acids is evident from NMR analysis and might be necessary for the *syn* diastereoselectivity and enantioselectivity observed.

2.4.1.4 Synthesis of Optically Active *syn* Aldols with Chiral Enolates

Thornton and Siegel have reported that reactions of the titanium enolates of chiral α-silyloxyketones resulted in excellent *syn* diastereoselectivity, as shown in Table 2.28, entries 1–4 [55, 56]. Use of tetrakisisopropoxytitanium enolates also afforded excellent *syn* diastereoselectivity. In work similar to that with oxazolidinone chiral auxiliaries, these enolates were generated as lithium enolates then transmetalated with the appropriate titanium Lewis acid. Large excesses of titanium were necessary for good stereoselectivity. They also noted that carboxylic acid derivatives could be obtained by deprotection and oxidative cleavage. Chiral auxiliary methods are, however, more efficient at providing adducts of this nature.

Urpí and coworkers demonstrated that a variety of directly generated titanium enolates of α-silyloxyketones reacted with aldehydes to give *syn* aldols with excellent yield and selectivity, as shown in Table 2.28, entries 5–6 [57]. The selectivity of these reactions can be explained by the transition state assembly **156**. Between the *O*-benzyl derivative and the OTBS protected ketones, the latter provided excellent yield and *syn–syn* diastereoselectivity.

Tab. 2.29
Reactions of β-ketoimide enolates.

a) $TiCl_4$
b) $i\text{-}Pr_2NEt$
c) RCHO

159 → [160]‡ → 161 + 162

Entry	*R*	*Yield (%)*	*161:162*
1	*i*-Pr	86	>99:1
2	$CH_2{=}CMe$	64	95:5
3	Et	86	>99:1
4	Ph	81	96:4

The steric bulk of the silyloxy group prohibits it from effectively chelating titanium, so the products arise from the non-chelated transition state.

Much work in the field of aldol reactions of ketones was performed by Evans to enable the synthesis of polypropionate natural products. They demonstrated that β-ketoimides like **159** were selectively and completely enolized at the C_4 position rather than the potentially labile methyl-bearing C_2 position, most probably because steric factors prohibited alignment of the carbonyl groups necessary to activate the C_2 proton. As shown in Table 2.29, it was demonstrated that these compounds would react with aldehydes to provide *syn–syn* product **161**, via titanium enolates, with good yield and excellent selectivity, and the corresponding *syn–anti* product **162** could be favored by use of a tin enolate reaction [58]. They invoked the chelated transition state assembly **160** to explain the product stereochemistry observed, in which the C_2-methyl group directs diastereofacial selectivity. Interestingly, reduction with $Zn(BH_4)_2$ provided the *syn* diol diastereoselectively.

Evans also investigated the stereochemical influence of two adjacent stereogenic centers in the titanium enolate-based aldol reaction [28]. As shown in Scheme 2.13, asymmetric induction of the enolate resulted from the influence of the α-stereocenter; the β-stereogenic center has very little effect.

Subsequently, Evans examined double-stereodifferentiating titanium enolate aldol reactions [59]. Both the aldehyde and the enolate contained α and β stereogenic centers. In aldol reactions between these substrates the enolate can adopt either a matched or mismatched relationship with the aldehyde. Several possible scenarios were investigated. As shown in Scheme

Scheme 2.13
syn-Selective aldol reactions of chiral ketones.

2.14, when the ketone enolate and aldehyde were chirally fully matched (**168**) or partially mismatched (**169** and **170**), *syn* diastereoselectivity was quite good. The diastereoselectivity was poor only in the completely mismatched case (**171**).

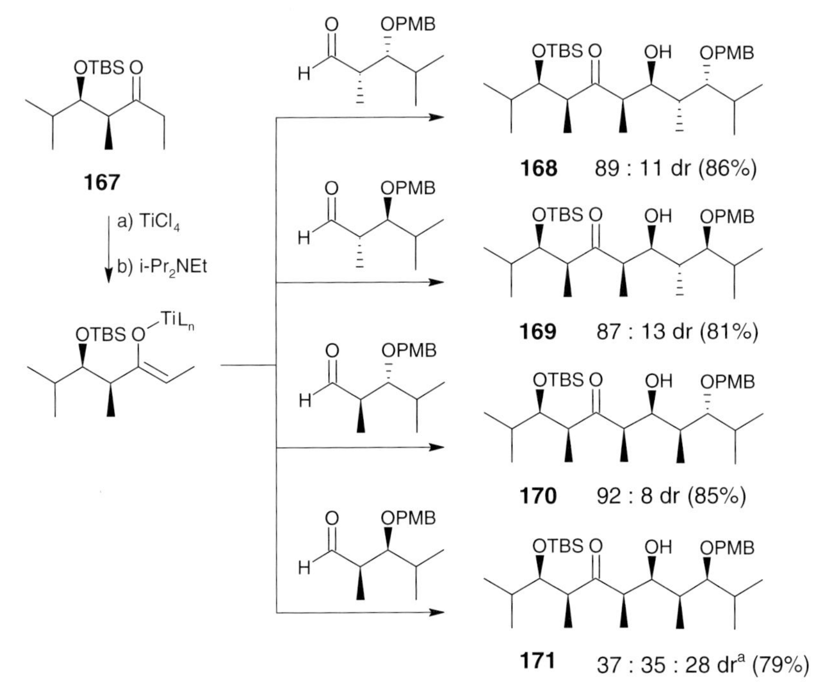

Scheme 2.14
Double stereodifferentiation in *syn* aldol reaction. ([a]28% of an *anti* diastereomer also isolated.)

2.4.2
Anti Diastereoselectivity

As can be seen from the developments described above, the control of both relative and absolute acyclic stereochemistry in a variety of *syn* aldol reactions can now be achieved highly stereoselectively. Both boron and titanium enolate-based *syn* aldol reactions have gained widespread popularity and are frequently used in synthesis. Whereas *anti*-α-alkyl-β-hydroxycarbonyl units are inherent to numerous bioactive natural products, there are relatively few effective synthetic processes that are convenient, operationally simple, and afford high diastereoselectivity for a wide range of aldehydes. Early examples of *anti*-selective aldol reactions, reported by Meyers in 1984, were based upon oxazoline-derived boron enolates [60]. Several other methods based upon metal enolates other than titanium have subsequently been developed. In this chapter, however, we will focus on titanium enolate-based methods.

2.4.2.1 Synthesis of *anti* Aldols in Racemic Form

Procter et al. reported a novel *anti*-selective aldol process using titanium enolates of *N*-propionylpyrrolidine [61]. The aldol products were obtained in good yield and *anti* diastereoselectivity. The aldolates were produced in racemic form, however. As shown in Table 2.30, formation of the titanium enolate of **172** was achieved by transmetalation of the lithium enolate with Cp_2TiCl_2 in THF. The resulting titanium enolate reacted with a range of aldehydes. Interestingly, the lithium enolate of **172** has stereoselectivity and the corresponding zirconium enolate afforded *syn* diastereoselectivity.

Tab. 2.30
anti-Selective aldol reactions of *N*-propionylpyrrolidine.

172 —a) LDA; b) Cp_2TiCl_2→ $OTiCp_2Cl$ enolate —RCHO→ 173 + 174

Entry	*R*	*Yield (%)*	*173:174*
1	Et	65	79:21
2	*i*-Pr	64	87:13
3	Ph	68	98:2
4	MeCH=CH	77	90:10

Tab. 2.31
anti Aldol reactions of phenyl benzyloxythioacetate with respective aldehydes.

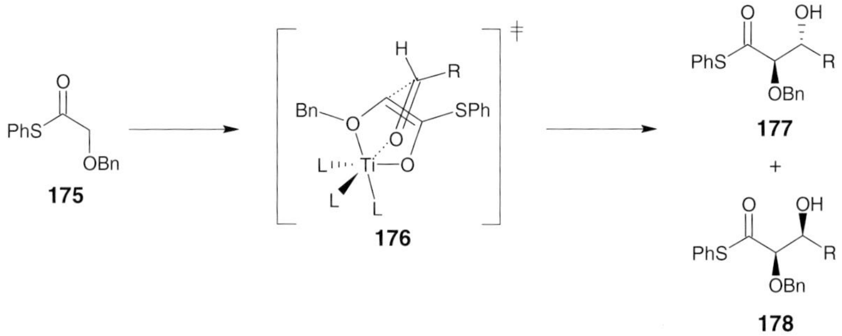

Entry	*R*	*Yield (%)*	*177:178*
1	Me	99	97:3
2	*n*-Pr	97	98:2
3	*t*-Bu	81	98:2
4	Ph	99	98:2

Cinquini and Cozzi reported the synthesis of racemic *anti*-α,β-dihydroxy thiolesters by aldol reaction of aldehydes with α-benzyloxythiolester-derived chlorotitanium enolates [62]. The thiolester was enolized by treatment with $TiCl_4$ and Et_3N in CH_2Cl_2 at −78 °C and the resulting enolate was reacted with a variety of aldehydes at −78 °C providing *anti* aldol products with good to excellent *anti* diastereoselectivity, as shown in Table 2.31. The stereochemical outcome can be explained by boat-like model **176** in which the thiolester formed an *E* enolate presumably by chelation through the benzyloxy group.

Kazmaier reported the synthesis of several racemic α-aminoaldols in which the α-amino group and the β-hydroxyl group are *anti* [63]. Reactions of *N*-benzyloxycarbonyl amino acid ester enolates with a variety of aldehydes afforded moderate to good yields and good to excellent diastereoselectivity, as shown in Table 2.32. Titanium enolates were generated by deprotonation of **179** with LDA in THF followed by the addition of 2 equiv. $Ti(O\textit{i}\text{-}Pr)_3Cl$ in THF at −78 °C.

2.4.2.2 Synthesis of Optically Active *anti* Aldols by Use of Chiral Auxiliaries

2.4.2.2.1 Aminoindanol and Related Chiral Auxiliaries

Ghosh and coworkers developed novel highly diastereoselective *anti* aldol methods based on ester-derived titanium enolate aldol reactions [64–66]. The utility of the *cis*-1-toluenesulfonamido-2-indanol-derived chiral auxiliary was demonstrated. This chiral auxiliary is readily prepared from commercially available enantiomerically pure *cis*-1-amino-2-indanol. Both enantio-

Tab. 2.32
Synthesis of sterically demanding α-aminoaldols.

Entry	*R_1*	*R_2*	*Yield (%)*	*180:181*
1	*i*-Pr	*t*-Bu	70	97:3
2	*i*-Pr	*i*-Pr	87	92:8
3	Me	*i*-Pr	87	92:8
4	Et	*i*-Pr	85	95:5

mers of the chiral auxiliary are, furthermore, readily available, enabling synthesis of either enantiomer of the *anti* aldol. As shown in Scheme 2.15, ester **183** was readily enolized with titanium tetrachloride and diisopropylethylamine. First it was treated with $TiCl_4$ in CH_2Cl_2 at 0 °C. After stirring at 23 °C for 15 min, *i*-Pr_2NEt was added. Evans reported that alkyl esters cannot be enolized with $TiCl_4$ and amine bases [28]. Smooth enolization of sulfonamidoesters is presumably because of internal chelation with the sulfonamido group, as was recently documented by Xiang et al. [48].

Scheme 2.15
Enolization of *N*-tosylaminoindanol esters.

Scheme 2.16
Aminoindanol-derived *anti* aldol reaction.

Although treatment of enolate **184** with isovaleraldehyde produced no reaction, addition of the enolate to a solution of isovaleraldehyde precomplexed with $TiCl_4$ provided the *anti* aldol product **187** in 97% yield as a single isomer (by HPLC and ^1H NMR) as shown in Scheme 2.16.

Reaction with several other aldehydes also produced the *anti* isomer as the major product. It should be noted that of four possible diastereomers, only one *syn* and one *anti* product were observed in this reaction. The *anti*:*syn* selectivity ranged from 85:15 to >99:1 and yields ranged from 41–97%, as shown in Table 2.33. Whereas benzaldehyde resulted in very little to no selectivity under these conditions, in the presence of an additive there was a dramatic improvement in *anti* diastereoselectivity. Thus, aldol reaction of the enolate of **189** with benzaldehyde (2 equiv.) precomplexed with $TiCl_4$ (2.2 equiv.) in the presence of CH_3CN additive (2.2 equiv.) afforded *anti* al-

Tab. 2.33
anti-Selective asymmetric aldol reactions.

Entry	*R*	*Yield (%)*	*190:191*
1	*i*-Bu	97	>99:1
2	Me	50	85:15
3	Pr	74	95:5
4	*i*-Pr	91	85:15
5	MeCH=CH	41	95:5
6	PhCH=CH	63	99:1
7	Ph	85	45:55

192 → **193** (20.5 : 5.8 : 1 dr): a) $TiCl_4$; b) i-Pr_2NEt; c) i-BuCHO, $TiCl_4$

194 → **195** (1 : 1 dr): a) $TiCl_4$; b) i-Pr_2NEt; c) i-BuCHO, $TiCl_4$

196 → **197** (70 : 30 dr): a) $TiCl_4$; b) i-Pr_2NEt; c) i-BuCHO, $TiCl_4$

Scheme 2.17
Structure–reactivity relationship studies in *anti* aldol reactions.

dol product stereoselectively (96:4) in good yield. These conditions, however, did not improve the yield or selectivity for reactions with aliphatic aldehydes. The *anti* selectivity can be rationalized by using a novel Zimmerman–Traxler type transition state **186**. In this transition state, both the methyl group of the enolate and the alkyl group of the aldehyde adopt pseudo-equatorial positions, leading to the *anti* product.

Several structural features have been shown to be critical for the high selectivities observed. First, reaction of dimethylphenylglycinol-derived auxiliary **192** with isovaleraldehyde provided a 20.5:5.8:1 ratio of isomers, as shown in Scheme 2.17. The major isomers are thought to be the two *anti* diastereomers. Removal of the indane aromatic ring resulted in complete loss of diastereoselectivity. Thus, reaction of the corresponding 1,2-aminocyclopentanol derivative **194** with isovaleraldehyde resulted in a 1:1 mixture of *syn* and *anti* diastereomers. The use of the mesylate derivative **196** also resulted in a large loss of selectivity. The reaction of *N*-mesylaminoindanol derivative **196** with isovaleraldehyde provided a 70:30 mixture of *anti* and *syn* diastereomers. These results suggest a possible π-stacking interaction between the two aromatic rings. The result of this interaction would be to help stabilize the conformation shown in enolate **184**, although this interaction has not been rigorously established.

Double stereodifferentiation experiments with matched chiral aldehyde **199** provided *anti* aldol **200** as a single diastereomer, as shown in Scheme 2.18 [67].

Scheme 2.18
Double stereodifferentiation in aminoindanol-derived *anti* aldol reaction.

On the basis of these structure–reactivity studies, Ghosh and Kim subsequently speculated that the planarity of the acenaphthene ring in conjunction with its aromaticity might further enhance π-stacking interactions with the arylsulfonamide functionality [68]. In this context, an effective synthesis of both enantiomers of *cis*-2-amino-1-acenaphthenol from acenaphthylene was developed. Enolization of **202** using $TiCl_4$ and *i*-Pr_2NEt provided a single enolate, presumably the *Z* enolate. Reaction of this enolate with isovaleraldehyde precomplexed with $TiCl_4$ resulted in significantly reduced *anti* diastereoselectivity (78:22), however. Similar to the aminoindanol chiral auxiliary, reactions with the acenaphthene-derived auxiliary produced only one *anti* and one *syn* diastereomer. Interestingly, aldol reaction of the enolate of **202** with aldehydes (2 equiv.) precomplexed with $TiCl_4$ (2.2 equiv.) in the presence of CH_3CN (2.2 equiv.) resulted in excellent *anti* diastereoselectivity and yield for a wide range of aldehydes, as shown in Table 2.34.

2.4.2.2.2 Oxazolidinethione and Oxazolidineselone Chiral Auxiliaries

Silks and coworkers reported *anti*-selective aldol reactions in conjunction with their investigations on chiral selenium compounds [39]. They discovered that bidentate oxyaldehydes precomplexed with $TiCl_4$ led to *anti* aldol products in good yield and with excellent stereoselectivity, as shown in Table 2.35, entries 1 and 2. Presumably, this chelation-controlled reversal of *syn*/*anti* selectivity is a phenomenon similar to that which Ghosh and coworkers exploited to produce *syn* aldols from aminoindanol chiral auxiliaries. Both benzyl and isopropyl oxazolidineselone chiral auxiliaries have comparable *anti* diastereoselectivity.

Crimmins and McDougall demonstrated that protected glycolyloxazolidinethiones underwent aldol reactions with aldehydes to provide *anti* aldol products [69]. Moderate to good yields and good to excellent *anti* selectivity were observed for a wide range of reactions, as shown in Table 2.35, entries 3–8. To achieve *anti* selectivity, it is necessary to complex the aldehydes with $TiCl_4$ before addition of the enolate. Comparing their results with those of Silks, it seems that the α-oxy substituent on the enolate is a less powerful *anti* director than an oxygen substituent on the aldehyde.

Tab. 2.34
Acenaphthene-derived *anti* aldol reactions with respective aldehydes.

201 → 202; a) $TiCl_4$ b) $i\text{-}Pr_2NEt$ c) RCHO, $TiCl_4$, CH_3CN → 203 + 204

Entry	*R*	*Yield (%)*	*203:204*
1	*i*-Bu	95	97:3
2	Et	92	92:8
3	*i*-Pr	95	96:4
4	cyclohexyl	84	99:1
5	Ph	93	93:7
6	$PhCH_2CH_2$	97	95:5

2.4.2.3 Synthesis of Optically Active *anti* Aldols by Use of Chiral Titanium Ligands

Duthaler reported the synthesis of optically active *anti* aldols by use of the chiral $CpTi(DAGO)_2$ complex [70]. They reported moderate yields and

Tab. 2.35
anti-Selective aldol reactions involving oxazolidinethione and oxazolidineselone auxiliaries.

205 → (R_2CHO) → 206 + 207 + 208

Entry	*X*	*R_1*	*R_2*	*Yield (%)*	*206:207:208*	*Ref.*
1	Se	Me	$BnOCH_2$	100	>99:0:1	39
2	Se	Me	$BuOCH_2$	81	>99:0:1	39
3	S	$OCH_2CH{=}CH_2$	Me	84	94:6:0	69
4	S	$OCH_2CH{=}CH_2$	*i*-Pr	61	87:13:0	69
5	S	$OCH_2CH{=}CH_2$	Ph	56	65:24:11	69
6	S	OBn	$CH_3(CH_2)_4$	64	88:12:0	69
7	S	OBn	$CH{=}CH_2$	48	74:26:0	69
8	S	OMe	Me	62	86:4:10	69

Tab. 2.36
Stereoselective synthesis of *anti* aldols using chiral titanium Lewis acids.

a) LDA
b) $CpTi(DAGO)_2Cl$
c) RCHO

209 → **210** + **211**

Entry	R_2	*Yield (%)*	*210:211*
1	Pr	74	89:11
2	*t*-Bu	59	83:17
3	CH_2=CMe	50	54:46
4	Ph	73	23:77

good diastereoselectivity for a range of aldehydes when reacted with 2,6-dimethylphenyl propionate derived titanium enolate, as shown in Table 2.36, entries 1 and 2. The method provided no selectivity (entry 3) or *syn* selectivity (entry 4), however, when the aldehyde was unsaturated and substituted. The corresponding *N*-propionyl-1,3-oxazolidin-2-one provided good selectivity with isobutyraldehyde (88:12 *anti/syn*) but poor yield, because of the formation of byproducts.

2.5 Natural Product Synthesis via Titanium Enolate Aldol Reactions

Numerous applications of titanium enolate aldol reactions have been reported for the synthesis of natural products and bioactive compounds of pharmaceutical interest. These syntheses were performed by utilizing the titanium enolate methods described above as key steps. There are many transformations that simply cannot be achieved by use of conventional aldol chemistry. Furthermore, the use of inexpensive titanium reagents, operational simplicity, and functional group tolerance make this chemistry very attractive in synthesis. The following applications in synthesis further exemplify the increasing importance of titanium aldol chemistry.

2.5.1 Lactone Natural Products

Asymmetric *syn* and *anti* aldol reactions have been used for synthesis of the following natural products containing lactone moieties. A titanium enolate-based aldol reaction has also been utilized in the aldol dehydration sequence during the synthesis of pyrone natural products mxyopyronin A and B. The following syntheses of lactone-containing natural products highlight the variety of aldol transformations.

2.5.1.1 Tetrahydrolipstatin

Ghosh and Fidanze reported a synthesis of tetrahydrolipstatin (**217**) in which the *anti* stereochemistry of the β-lactone moiety is set by using asymmetric *anti* aldol methodology, discussed in Section 2.4.2.2.1 [71]. Tetrahydrolipstatin is a saturated analog of the natural product lipstatin, isolated from *Streptomyces toxytricini*. It is a potent and irreversible inhibitor of pancreatic lipase and has been marketed in several countries as an anti-obesity agent under the trade name Xenical. Aldol reaction between the enolate of **212**, generated with $TiCl_4$ and *i*-Pr_2NEt, and cinnamaldehyde precomplexed with Bu_2BOTf gave **213** in 60% yield as a 6.1:1 mixture of *anti* and *syn* diastereomers, as shown in Scheme 2.19. Aldol adduct **213** was converted to the benzyl ester **214**, which was reduced selectively to the *anti* 1,3-diol by use of Evans' procedure [72] and protected as the triisopropylsilyl ether **215**. Hydrogenation of the *O*-benzyl group over Pearlman's catalyst and lactonization with phenylsulfonyl chloride in pyridine gave the β-lactone **216** which, on deprotection and reaction with *N*-formyl-l-leucine under Mitsunobu conditions, afforded (−)-tetrahydrolipstatin.

2.5.1.2 Myxopyronins A and B

Panek and coworkers developed syntheses of myxopyronin A (**220a**) and myxopyronin B (**220b**) in which an aldol reaction was used to append the

Scheme 2.19
Synthesis of (−)-tetrahydrolipstatin.

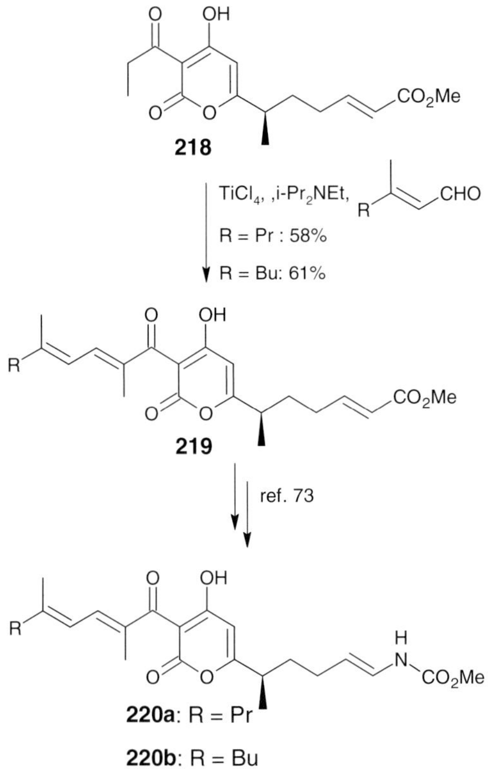

Scheme 2.20
Myxopyronin A and B synthesis.

hydrophobic side chain [73]. The myxopyronins are natural products isolated from the gliding bacterium *Myxococcus fulvus* MX f50. These molecules are bacterial growth inhibitors, because of their capacity to selectively inhibit bacterial RNA polymerase. Aldol condensation of ketone **218**-derived titanium enolate and subsequent elimination provided diene **219**, as shown in Scheme 2.20. Attachment of the other side chain provided myxopyronins A and B.

2.5.1.3 **Callystatin A**

Crimmins and King reported a total synthesis of callystatin A (**228**) using their aldol methodology (as discussed in Section 2.4.1.2.1) to construct three of the four chiral centers in the molecule [74]. Callystatin A is a natural product, isolated from the marine sponge *Callyspongia truncata*, which has potent in-vitro toxicity against KB cell lines ($IC_{50} = 0.01$ ng mL^{-1}). Phenylalanine-derived *N*-propionyloxazolidethione **221** was subjected to enolization with $TiCl_4$ and (−)-sparteine using Crimmins' procedure and reacted with (*S*)-2-methylbutanal to provide *syn* aldol adduct **222** in 83% yield with 98% selectivity, as shown in Scheme 2.21. Protection of the alco-

Scheme 2.21
Synthesis of (−)-callystatin A.

hol as the TBS ether, reductive removal of the chiral auxiliary, and Swern oxidation furnished aldehyde **223**. Aldol reaction under similar conditions with oxazolidethione **221** gave the all-*syn* product **224** in 81% yield with 98% selectivity; this was then converted to **225**. Wittig olefination with fragment **226** (synthesized from (*S*)-glycidol) gave olefin **227** which, on removal of the OTMS and isopropyl acetal protecting groups, perruthenate oxidation, and deprotection of the OTBS group gave (−)-callystatin A.

2.5.1.4 **AI-77-B**

Ghosh and coworkers developed a convergent synthesis of (−)-AI-77-B (**237**) in which all five stereogenic centers were set by asymmetric synthesis [75].

As in their previous work [76], four of those chiral centers were set using the aminoindanol-based *syn* and *anti* aldol methods discussed in Sections 2.4.1.2.3 and 2.4.2.2.1, respectively. AI-77-B is a natural product isolated from *Bacillus pumilus* with potent antiulcerogenic and anti-inflammatory properties. *Anti* aldol reaction of **229**-derived titanium enolate with 4,4,4-trichlorobutryaldehyde gave adduct **230** in 90% yield with 90% de; this was subsequently transformed to isocoumarin fragment **231**, as shown in Scheme 2.22.

Aldol reaction of **232**-derived titanium enolate with benzyloxyacetaldehyde

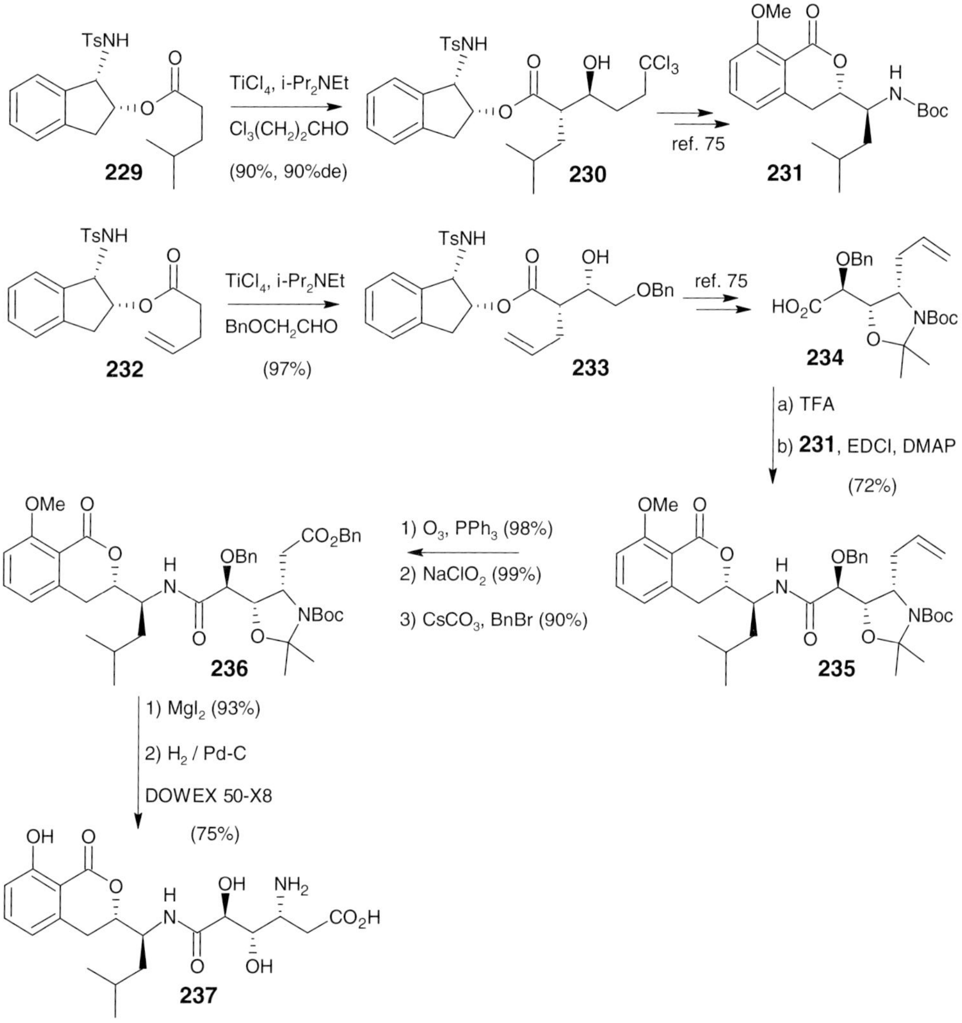

Scheme 2.22
AI-77B synthesis.

provided *syn* adduct **233** in 97% yield as a single product; this was transformed into carboxylic acid **234** via Curtius rearrangement, stereoselective homologation after Dondoni's procedure [77] and functional group manipulation. Deprotection of **231** and coupling with EDCI and DMAP gave amide **235**. The allyl group was converted to benzyl ester **236** which, on demethylation with MgI_2 and concomitant hydrogenolysis of the *O*-benzyl groups and isopropylidine deprotection, gave AI-77B. Key steps included stereoselective ester-derived asymmetric *syn* and *anti* aldol reactions, a regioselective Diels–Alder reaction, and Dondoni homologation.

2.5.2 Macrolide Natural Products

Titanium enolate aldol reactions have been used in the syntheses of several of important antitumor macrolides and of the immunosuppressive macrolide rapamycin. Duthaler's titanium enolate acetate aldol reaction was used in the synthesis of epothilone 490. A variety of titanium enolate-based *syn* aldol reactions highlight the scope and utility of this technique.

2.5.2.1 Epothilone 490

Danishefsky and coworkers reported a concise, convergent synthesis of epothilone 490 (**243**) with a key late-stage acetate aldol reaction [78]. Epothilone 490 is a natural product isolated from the cellulose-degrading bacterium *Sorangium cellulosum* which has potent Taxol-like microtubule-stabilizing induced cancer cytotoxicity. Known vinyl iodide **238** [79] was subjected to Stille coupling and acetylation to give acetate ester **239**. Reaction with previously synthesized aldehyde **240** [80] using to Duthaler's titanium enolate aldol method (discussed in Section 2.3.2) gave aldol **241** in 85% yield as a single diastereomer (Scheme 2.23). Deprotection of the OTroc group using zinc in acetic acid followed by ring closing metathesis using the Grubbs second generation catalyst **242** yielded epothilone 490 in 64% yield.

2.5.2.2 Cryptophycin B

Ghosh and Bischoff reported an efficient and convergent synthesis of cryptophycin B (**249**) using their aminoindanol-derived *syn* aldol methodology (discussed in Section 2.4.1.2.3) [81]. Cryptophycin B (**249**), a marine natural product isolated from *Nostoc sp.* GSV 224, has potency at the picogram level against KB cells. Unsaturated ester **244**-derived titanium enolate was subjected to an aldol reaction with 3-benzyloxypropionaldehyde to give *syn* product **245** in 98% yield as a single diastereomer, as shown in Scheme 2.24. Reductive removal of the chiral auxiliary then deoxygenation of the primary alcohol and protection of the remaining secondary alcohol furnished silyl ether **246**. Conversion of the benzyloxy functionality to the α,β-unsaturated ester moiety was accomplished by selective OBn deprotection, PCC oxidation, and Horner–Emmons olefination to yield ester **247** which

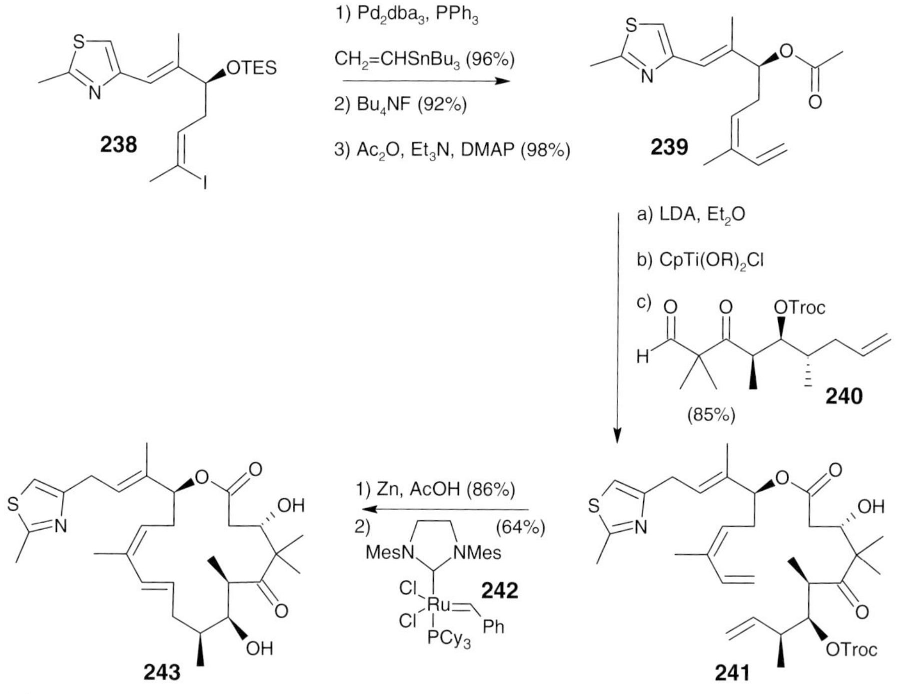

Scheme 2.23
Epothilone 490 synthesis.

was converted to compound **249**. Exposure to trifluoroacetic acid to remove the N-Boc and *O-tert*-butyl protecting groups, then Yamaguchi cycloamidation gave the macrocycle which, on selective epoxidation with dimethyldioxirane, gave cryptophycin B in 22% overall yield over fourteen steps.

2.5.2.3 **Amphidinolide T1**

Ghosh and Liu reported the first total synthesis of amphidinolide T1 (**258**), setting four of the seven stereogenic centers with their asymmetric aldol methodology (discussed in Section 2.4.1.2.3) [82]. Amphidinolide T1 is a marine natural product isolated from *Amphidinium sp.* with significant antitumor properties against a variety of cell lines. Aldol reaction of ester **250** with 3-benzyloxypropionaldehyde gave *syn* adduct **251** in 90% yield as a single diastereomer, as shown in Scheme 2.25. Ester **251** was then converted into tetrahydrofuran derivative **252** via Wittig olefination and olefin cross-metathesis.

Aldol reaction of ester **253** with benzyloxyacetaldehyde gave exclusively *syn* aldol adduct **254** in 93% yield; this was transformed into alkene **255**. Subsequent transformation yielded tetrahydrofuran derivative **256**. This cyclic bromoether serves as masked functionality for the labile exocyclic

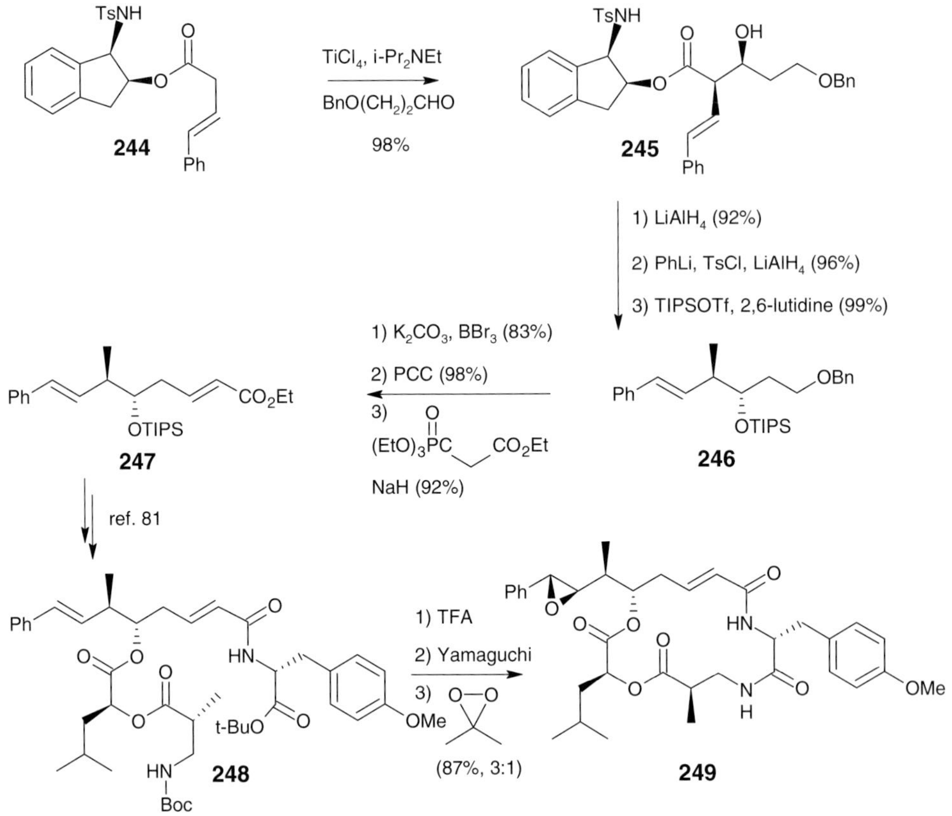

Scheme 2.24
Cryptophycin B synthesis.

methylene group in the final product. Stereoselective oxocarbenium ion-mediated alkylation using a modification of Ley's procedure [83] with DTBMP and $AlCl_3$ gave coupled product **257** in 73% yield as a single diastereomer. Deprotection of the alcohol and ester moieties followed by Yamaguchi macrolactonization, and treatment with zinc and ammonium chloride afforded amphidinolide T1. Key steps included stereoselective aminoindanol-derived asymmetric aldol reactions, efficient olefin cross-metathesis, stereoselective oxocarbenium ion-mediated anomeric alkylation, and the use of the cyclic bromomethyl ether as a novel exo-methylene group surrogate.

2.5.2.4 **Rapamycin**

Danishefsky and coworkers published a total synthesis of rapamycin (**260**) using a novel aldol macrocyclization reaction as the key step [84]. Rapamycin is a natural product with immunosuppresive properties. The conclusion

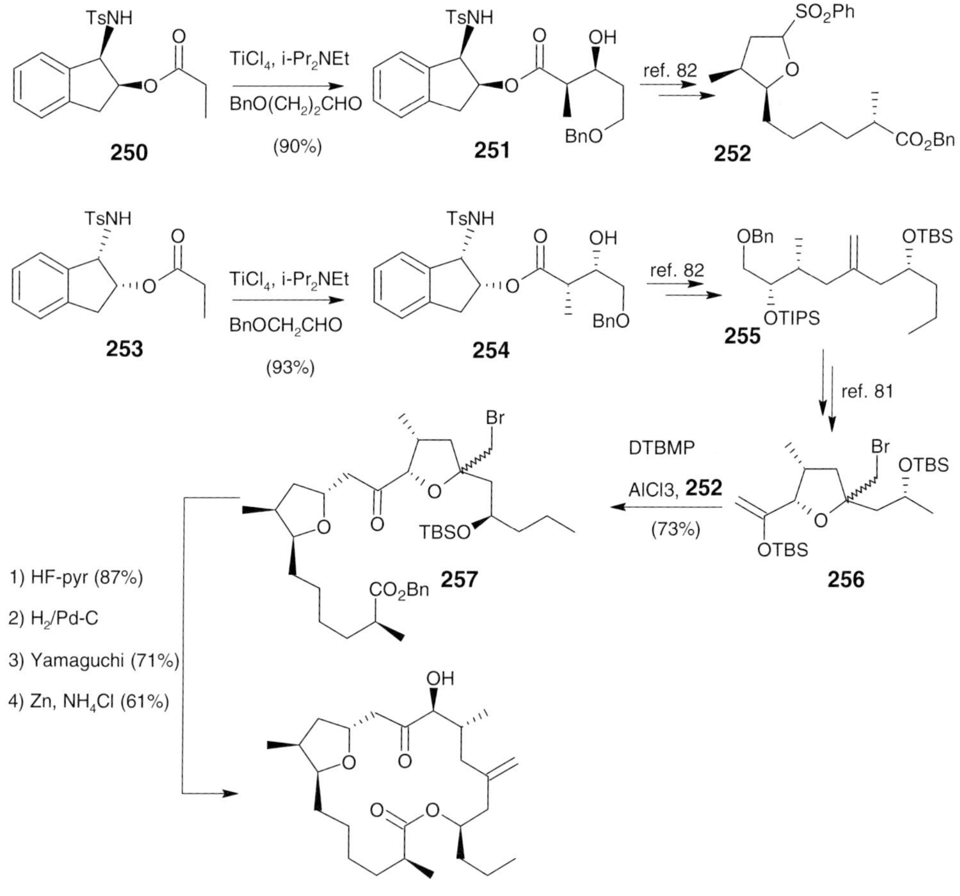

Scheme 2.25
Amphidinolide T1 synthesis.

of the synthesis treats late-stage intermediate aldehyde **259** with isopropoxytitanium trichloride in the presence of triethylamine to generate the cyclized product in 11% yield, with 22% of another isomer (possibly the *syn* aldol product), as shown in Scheme 2.26. Deprotection of the TIPS ether afforded rapamycin.

2.5.2.5 Spongistatins 1 and 2

Crimmins and coworkers published a convergent synthesis of spongistatin 1 (**263a**) and spongistatin 2 (**263b**), employing their phenylalanine-derived asymmetric aldol methodology (discussed in Section 2.4.1.2.1) to set one of the stereocenters [85]. Spongistatins 1 and 2 are natural products with sub-nanomolar growth inhibition of several NCI chemoresistant tumor types

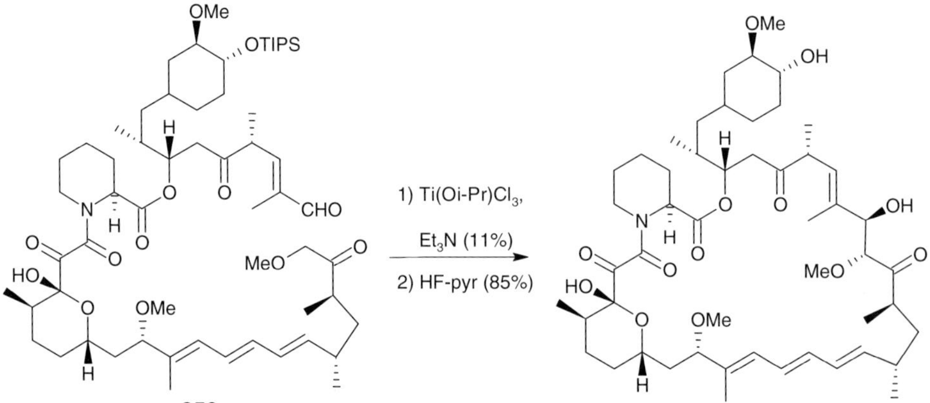

Scheme 2.26
Rapamycin synthesis.

including human melanoma, lung, brain, and colon cancers. Aldol reaction of aldehyde **261** with *N*-propionyloxazolidinethione then reductive removal of the chiral auxiliary furnished diol **262** in 74% yield as a 96:4 ratio of stereoisomers, as shown in Scheme 2.27. This fragment was then incorporated into synthetic spongistatins 1 and 2.

2.5.3
Miscellaneous Natural Products

Several other natural products have been synthesized by using titanium enolate-based aldol methods. Many of these syntheses utilize ketone enolate aldol reactions to establish *syn* stereochemistry. Duthaler's *anti* aldol reaction was used in the synthesis of tautomycin. Use of Evan's ketone–aldol reaction was nicely exemplified in syntheses of denticulatin B and membrenone C.

2.5.3.1 Tautomycin

Chamberlin and coworkers reported a convergent synthesis of tautomycin (**270**), employing Duthaler's *anti* aldol methodology, as discussed in Section 2.4.2.3, to set four of its stereogenic centers [86]. Tautomycin, a serine/threonine selective protein pyrophosphatase inhibitor is selective for PP1 over PP2A. Reaction of aldehyde **265** with the enolate of **264**, generated with LDA and transmetalated with Duthaler's reagent (R = diacetoneglucose) afforded compound **266** in 80% yield as an 8:1 mixture of *anti* and *syn* aldols, as shown in Scheme 2.28. Protection of the alcohol as the silyl ether and conversion of the ester gave iodide **267**, which was converted to spirocyclic aldehyde **268**. Aldehyde **268** which was reacted with the titanium

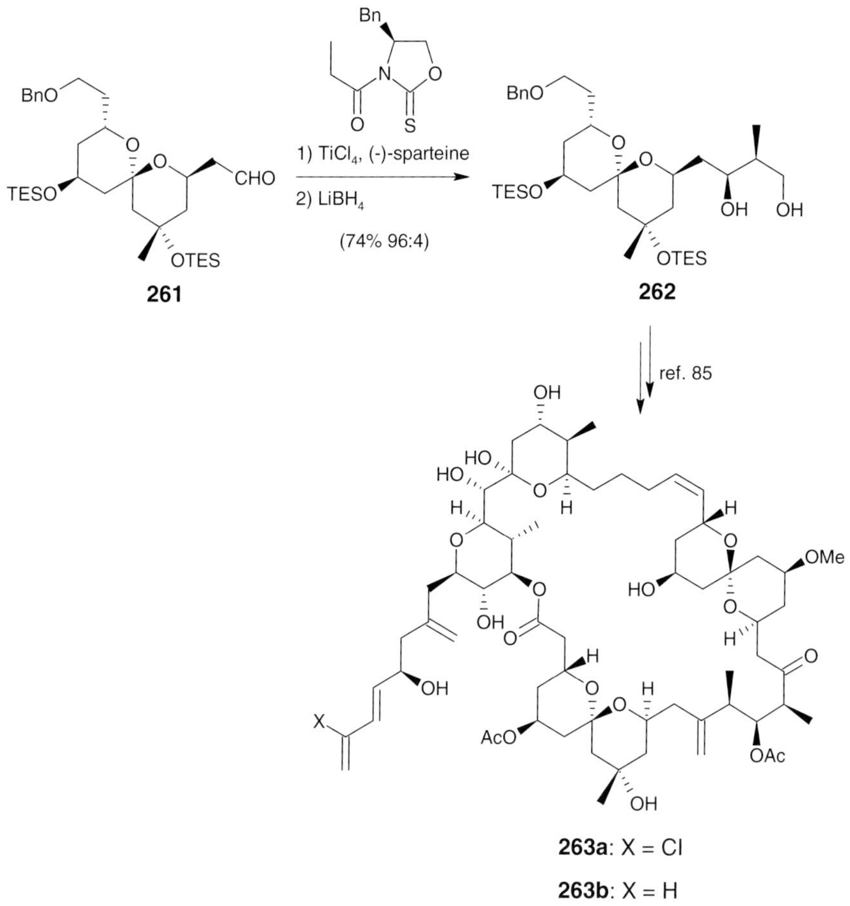

Scheme 2.27
Spongistatin 1 and 2 synthesis.

enolate of ester **264** using Duthaler's procedure to give *anti* aldol adduct **269** in 67% yield as a 7:1 mixture of the *anti* and *syn* isomers. Reaction of **269** provided synthetic tautomycin.

2.5.3.2 **Crocacin C**

Chakraborty and coworkers reported a synthesis of crocacin C (**276**) using asymmetric aldol methodology developed by Crimmins (Section 2.4.1.2.1) to set two of the four chiral centers [87]. Crocacin C is a natural product isolated from myxobacterium *Chondromyces crocatus* with potent growth inhibition of Gram-positive bacteria, fungi, and yeasts. Titanium enolate aldol reaction of phenylalanine-derived *N*-propionyloxazolidinethione **271** with cinnamaldehyde gave the *syn* aldol adduct **272** in 89% yield as a single dia-

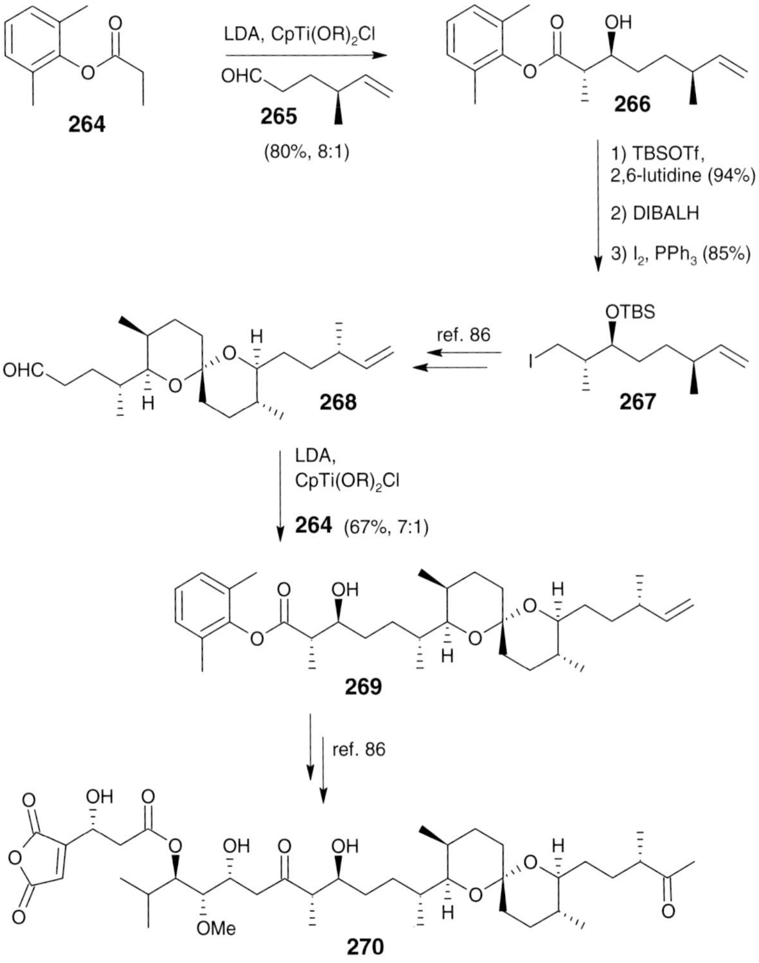

Scheme 2.28
Tautomycin synthesis.

stereomer; this was converted to allylic alcohol **273**, as shown in Scheme 2.29. Sharpless asymmetric epoxidation provided epoxide **274** which was opened regio- and stereoselectively with lithium dimethylcuprate to provide diol **275** in 86% yield. Diol **275** was then transformed to the final product crocacin C.

2.5.3.3 Stigmatellin A

Enders and coworkers developed a synthesis stigmatellin A (**281**) using aldol reactions to set two of the four chiral centers [88]. Stigmatellin A is a natural product isolated from bacterium *Stigmatella aurantica* and is one

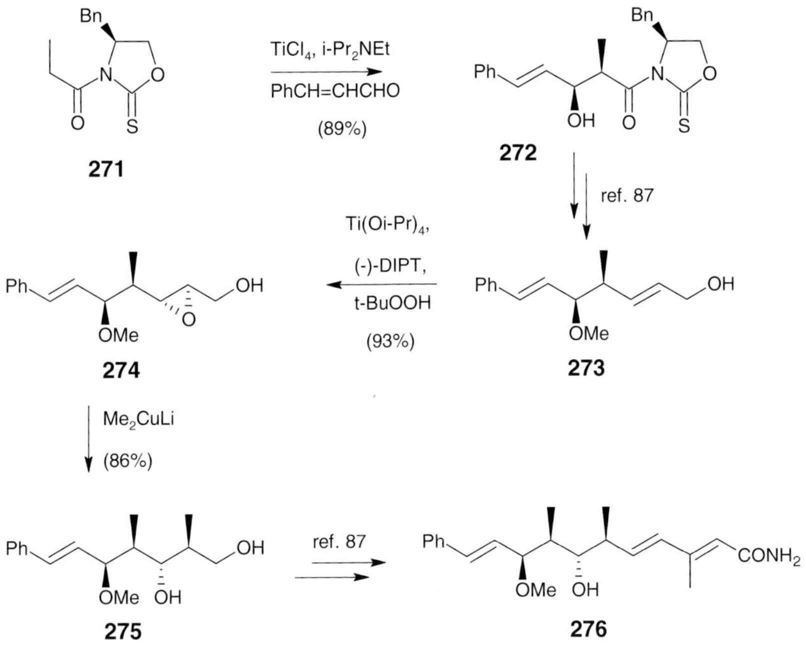

Scheme 2.29
Crocacin C synthesis.

of the most powerful electron-transport inhibitors in chloroplasts and mitochondria. Hydrazone **277** was alkylated with 3-(*p*-methoxyphenoxy)propyl iodide using their previously developed SAMP/RAMP methodology to provide hydrazone **278** in 80% yield with 98% de, as shown in Scheme 2.30 [89]. The chiral auxiliary was removed and the resulting ketone was subjected to an aldol reaction with benzyloxyacetaldehyde to give *syn* aldol adduct **279** in 64% yield as a 2:1 mixture of isomers. Ketone **279** was subjected to selective *anti* reduction using Evans' procedure [72] and methylated to yield compound **280**, which upon removal of the *p*-methoxyphenyl and benzyloxy groups gave stigmatellin A.

2.5.3.4 Denticulatin B

Paterson and Perkins developed a synthesis of denticulatin B (**285**) and its isomer denticulatin A using a late-stage aldol reaction [90]. Denticulatin B is a marine natural product isolated from the mollusk *Siphoneria denticulata*. Aldol coupling of ketone **282** and aldehyde **283** provided the *syn* aldol adduct **284** in 90% yield as a mixture of diastereomers, as shown in Scheme 2.31. Swern oxidation of the hydroxyl groups and deprotection and cyclization with HF–pyridine afforded denticulatin B in 20% overall yield over nine steps.

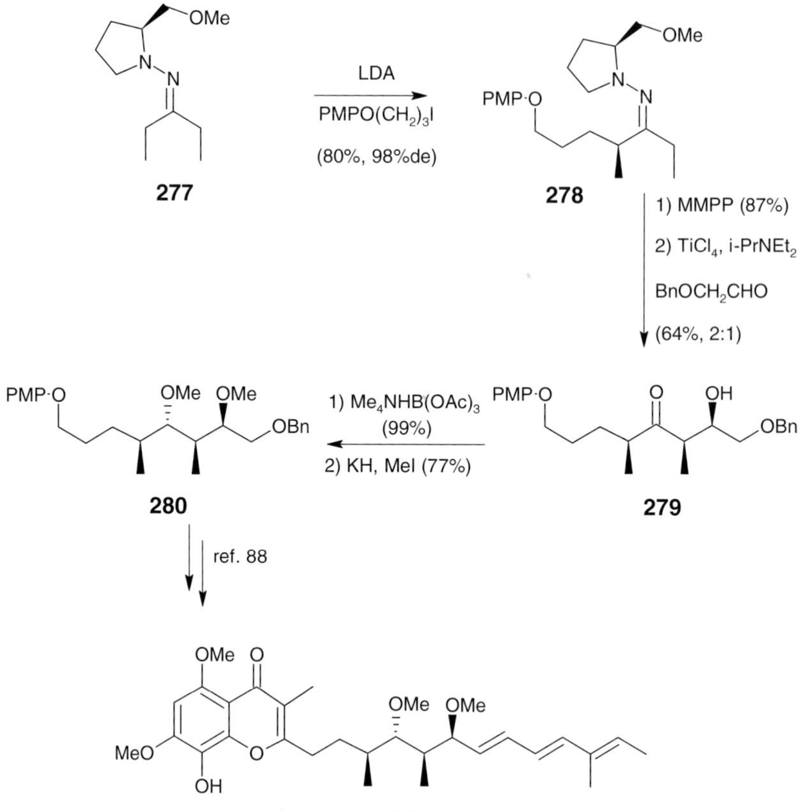

Scheme 2.30
Stigmatellin A synthesis.

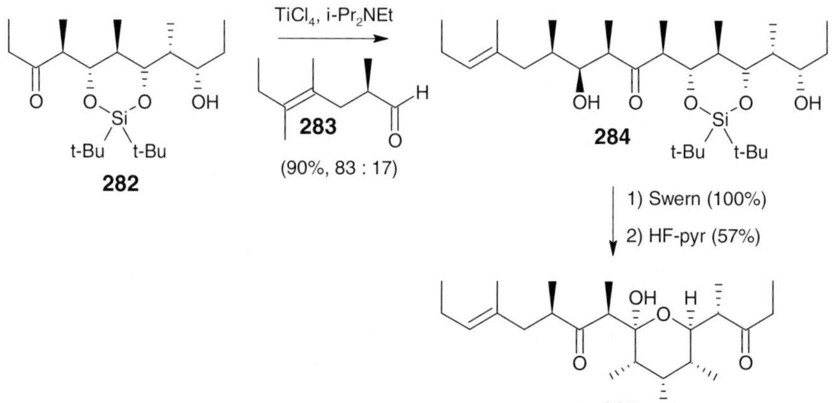

Scheme 2.31
Denticulatin B synthesis.

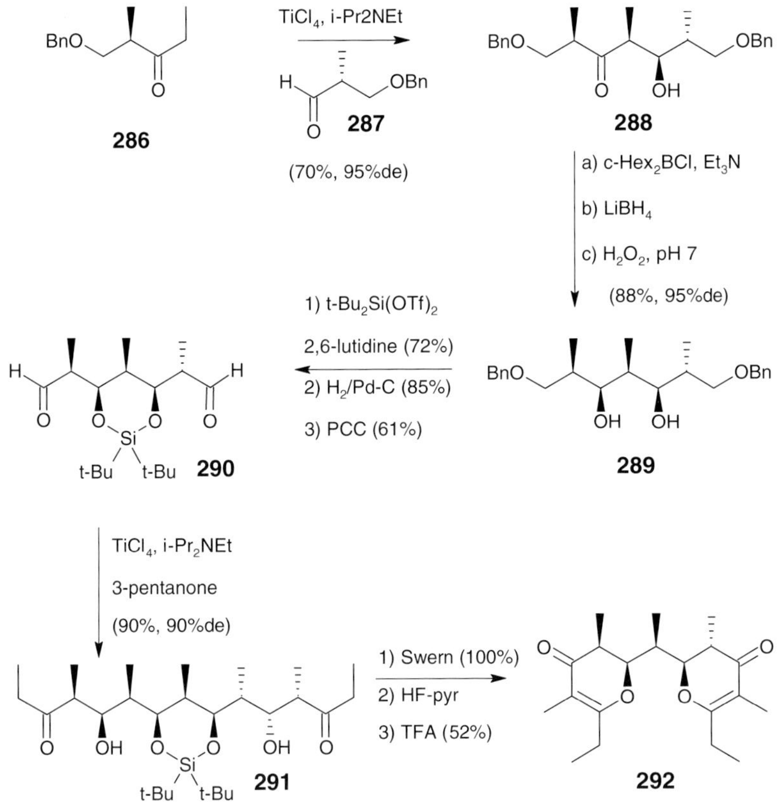

Scheme 2.32
Membrenone C synthesis.

2.5.3.5 Membrenone C

Perkins and Sampson developed syntheses of membrenone C (**292**) and its isomers to establish the absolute configuration of the natural product [91]. Membrenone C is a natural product isolated from the skin of a Mediterranean mollusk. Aldol reaction of the stereochemically matched chiral ketone **286** and protected (*R*)-3-hydroxy-2-methylpropionaldehyde **287** gave aldol adduct **288** in 70% yield and 95% de, as shown in Scheme 2.32. *Syn*-selective reduction of the ketone was accomplished by using a modification of Narasaka's procedure [92] to give diol **289** in 88% yield and 95% de. Protection of the diol as the cyclic silyl acetal followed by hydrogenolysis of the *O*-benzyl groups and PCC oxidation afforded dialdehyde **290**. Double aldol reaction with 3-pentanone gave compound **291** with nine contiguous chiral centers in 90% yield and 90% de. The synthesis was completed by Swern oxidation of the free hydroxyl groups, removal of the silyl protecting group, and cyclization in the presence of trifluoroacetic acid.

2.6
Typical Experimental Procedures for Generation of Titanium Enolates

2.6.1
Experimental Procedures

Titanium Enolate Formation by Transmetalation, Synthesis of *syn* Aldols (Thornton's Procedure, Section 2.4.1.2.1). The lithium enolate was generated at −78 °C with LDA in ether. Ti(O*i*-Pr)$_3$Cl (1–3 equiv.) was added dropwise with stirring. The clear solution became brown–orange and was left to warm to −40 °C over 1 h and then cooled to −78 °C. Aldehyde (1.1 equiv.) was added rapidly by syringe and the reaction was left to warm to −40 °C over 3 h. The reaction was quenched with saturated NH_4F and the layers were separated. The aqueous layer was extracted three times with ether. All organic layers were combined and dried over $MgSO_4$, followed by vacuum filtration and rotary evaporation to dryness. Purification by flash chromatography provided the major product.

Ester-derived Titanium Enolate by Transmetalation, Synthesis of *syn* Aldols (Duthaler's Procedure, Section 2.4.1.2.2). BuLi (6.2 mmol) is added at −20 °C to a solution of *i*-Pr$_2$NH (1 mL, 7.07 mmol) in 30 mL of ether under argon. After 15 min the reaction mixture is cooled to −78 °C and a solution of 2,6-dimethylphenyl propionate (1.0 g, 5.61 mmol) in 10 mL ether is added dropwise and stirred for 1.5 h. An ethereal solution of chlorocyclopentadienylbis(1,2:5,6-di-*O*-isopropylidene-α-D-glucofuranos-3-*O*-yl)titanium (0.088 M, 80 mL; 7.04 mmol, 1.25 equiv.) is added carefully via a cannula under argon pressure. After stirring for 24 h the aldehyde (7.29 mmol, 1.3 equiv.) is added and the reaction is monitored by TLC. The reaction mixture is quenched with 2 g NH_4Cl and 10 mL 1:1 THF–H_2O. After stirring for 2 h at 0 °C the precipitated titanium salts are separated by filtration and washed with ether. The filtrate is washed with 20 mL 1 M HCl, 10 mL satd $NaHCO_3$, and brine. The aqueous washings are re-extracted with 2 × 50 mL EtOAc. The combined organic extracts are dried over $MgSO_4$ and the solvent is removed. The crude products are either separated directly by chromatography or by first stirring for 1 h with 200 mL 0.1 M HCl, extracted with ether (3 × 100 mL), and washed with sat. $NaHCO_3$ and brine, removing glucose as the water soluble 1,2-acetonide.

Ester-derived Titanium Enolate for *anti* Aldol Reactions (Ghosh's Procedure, Section 2.4.2.2.1). *N*-tosylaminoindanol ester was reacted with $TiCl_4$ in CH_2Cl_2 at 0–23 °C for 15 min followed by addition of *i*-Pr$_2$NEt (4 equiv.) at 23 °C and stirring of the resulting brown solution for 2 h. The titanium enolate was then added to the representative aldehyde (2 equiv.) precomplexed with $TiCl_4$ (2.2 equiv.) at −78 °C and the mixture was stirred at −78 °C for 2 h before quenching with aqueous NH_4Cl. The aqueous layer was extracted with CH_2Cl_2. The combined organic extracts were washed with brine, dried over Na_2SO_4, and concentrated under reduced pressure. The crude products were purified by flash chromatography on silica gel.

Oxazolidinethione-derived Titanium Enolate for *syn* Aldol Reactions (Crimmins's Procedure, Section 2.4.1.2.1). To a dry round-bottomed flask under nitrogen was added 0.250 g (1.0 mmol) oxazolidinethione in 6 mL CH_2Cl_2. The solution was cooled to 0 °C and $TiCl_4$ (1.05 mmol, 0.115 mL) was added dropwise and the solution was stirred for 5 min. To the yellow slurry or suspension was added (−)-sparteine (2.5 mmol). The dark red enolate was stirred for 20 min at 0 °C. Freshly distilled aldehyde (1.1 mmol) was added dropwise and the reaction stirred for 1 h at 0 °C. The reaction was quenched with half-saturated NH_4Cl and the layers were separated. The organic layer was dried over Na_2SO_4, filtered, and concentrated. Purification of the crude material by column chromatography afforded the main diastereomer.

2.6.2 Alternative Approaches to Titanium Enolate Generation

Although the direct generation of titanium enolates is typically the most useful method of generating titanium enolates for aldol reactions, other methods have been described. Grubbs and Stille reported that titanium enolates could be generated by reaction of biscyclopentadienyltitanium alkylidene complexes and acyl halides [93]. Oshima and coworkers reported the formation of titanium enolates from α-iodoketones with allylsilane and titanium tetrachloride [94]. Mukaiyama and coworkers reported the generation of titanium enolates from α-bromoketones on treatment with $TiCl_2$ and copper powder [95].

2.7 Conclusion

Titanium enolate aldol reactions have been shown to be very effective for control of relative and absolute stereochemistry in acetate aldol and both *syn* and *anti* aldol reactions. The use of readily available and inexpensive titanium reagents make these methods convenient for large-scale synthesis. The synthetic potential of a variety of aldol reactions has been demonstrated by highlighting the synthesis of numerous bioactive complex natural products. The significance of enantio- and diastereoselection in synthesis, particularly in this pharmaceutical age, ensures that titanium enolate aldol reactions will remain an important part of organic synthesis for years to come. There is no doubt that unprecedented success has been achieved in the development of a variety of titanium enolate aldol reactions in the past decade. Much new potential and other exciting possibilities remain to be explored, however. We hope this chapter will stimulate further research and developments in titanium enolate-based aldol reactions and their use in organic synthesis.

References

1 Lim, B.-M.; Williams, S. F.; Masamune, S. *Comprehensive Organic Synthesis*; Trost, B. M.; Fleming, I. Eds; Pergamon Press: Oxford, **1991**, Vol. 2, Heathcock, C. H., ed., Chapter 1.7, 239.

2 The term 'chiral auxiliary' was first introduced by Eliel. Eliel, E. L.; *Tetrahedron* **1974**, *30*, 1503 and references cited therein.

3 Zimmerman, H. E.; Traxler, M. D. *J. Am. Chem. Soc.* **1957**, *79*, 1920.

4 Masamune, S.; Bates, G. S.; Corcoran, J. W. *Angew. Chem. Int. Ed. Engl.* **1977**, *16*, 585.

5 Westley, J. W. *Adv. Appl. Microbiol.* **1977**, *22*, 177.

6 Masamune, S.; Choy, W.; Petereson, J. S.; Sita, L. R. *Angew. Chem. Int. Ed. Engl.* **1985**, *24*, 1.

7 Verhé, R.; De Kimpe, N.; De Buyck, L.; Thierie, R.; Schamp, N. *Bull. Soc. Chim. Belg.* **1980**, *89*, 563.

8 Yoshida, Y.; Hayashi, R.; Sumihara, H.; Tanabe, Y. *Tetrahedron Lett.* **1997**, *38*, 8727.

9 Yoshida, Y.; Matsumoto, N.; Hamasaki, R.; Tanabe, Y. *Tetrahedron Lett.* **1999**, *40*, 4227.

10 Tanabe, Y.; Matsumoto, N.; Funakoshi, S.; Manta, N. *Synlett* **2001**, *12*, 1959.

11 Yachi, K.; Shinokubo, H.; Oshima, K. *J. Am. Chem. Soc.* **1999**, *121*, 9465.

12 Devant, R.; Braun, M. *Chem. Ber.* **1986**, *119*, 2191.

13 Yan, T.; Hung, A.; Lee, H.; Chang, C.; Liu, W. *J. Org. Chem.* **1995**, *60*, 3301.

14 Yan, T.; Hung, A.; Lee, H.; Liu, W.; Chang, C. *J. Chinese Chem. Soc.* **1995**, *42*, 691.

15 Palomo, C.; Oiarbide, M.; Gonzáles, A.; García, J.; Berrée, F.; Linden, A. *Tetrahedron Lett.* **1996**, *37*, 6931.

16 Phillips, A. J.; Guz, N. R. *Org. Lett.* **2002**, *4*, 2253.

17 González, Á.; Aiguadé, J.; Urpí, F.; Vilarrasa, J. *Tetrahedron Lett.* **1996**, *37*, 8949.

18 Duthaler, R. O.; Herold, P.; Lottenbach, W.; Oertle, K.; Riediker, M. *Angew. Chem. Int. Ed. Engl.* **1989**, *28*, 495.

19 Cambie, R. C.; Coddington, J. M.; Milbank, J. B. J.; Pausler, M. G.; Rustenhover, J. J.; Rutledge, P. S.; Shaw, G. L.; Sinkovich, P. I. *Australian J. Chem.* **1993**, *46*, 583.

20 Fringuelli, F.; Martinetti, E.; Piermatti, O.; Pizzo, F. *Gazz. Chim. Ital.* **1993**, *123*, 637.

21 Fringuelli, F.; Piermatte, O.; Pizzo, F.; Scappini, A. M. *Gazz. Chim. Ital.* **1995**, *125*, 195.

22 (a) Wang, Y.; Su, D.; Lin, C.; Tseng, H.; Li, C.; Yan, T. *Tetrahedron Lett.* **1999**, *40*, 3577. (b) Wang, Y.; Su, D.; Lin, C.; Tseng, H.; Li, C.; Yan, T. *J. Org. Chem.* **1999**, *64*, 6495.

23 Wang, Y.; Yan, T. *J. Org. Chem.* **2000**, *65*, 6752.

24 Cosp, A.; Romea, P.; Urpí, F.; Vilarrasa, J. *Tetrahedron Lett.* **2001**, *42*, 4629.
25 Ghosh, A. K.; Kim, J. Manuscript in preparation, Univ. of. Illinois–Chicago.
26 Reetz, M. T.; Peter, R. *Tetrahedron Lett.* **1981**, *22*, 4691.
27 Harrison, C. *Tetrahedron Lett.* **1987**, *28*, 4135.
28 Evans, D. A.; Rieger, D. L.; Bilodeau, M. T.; Urpí, F. *J. Am. Chem. Soc.* **1991**, *113*, 1047.
29 Annunziata, R.; Cinquini, M.; Cozzi, F.; Cozzi, P. G.; Coslandi, E. *Tetrahedron* **1991**, *47*, 7897.
30 Toru, T.; Hayakawa, T.; Nishi, T.; Watanabe, Y.; Ueno, Y. *Phosphorus, Sulphur, and Silicon* **1998**, *136–138*, 653.
31 Nakamura, S.; Hayakawa, T.; Nishi, T.; Watanabe, Y.; Toru, T. *Tetrahedron* **2001**, *57*, 6703.
32 Reetz, M. T.; Steinbach, R.; Kesseler, K. *Angew. Chem. Int. Ed. Engl.* **1982**, *21*, 864.
33 Crimmins, M. T.; King, B. W.; Tabet, E. A.; Chaudhary, K. *J. Org. Chem.* **2001**, *66*, 894.
34 Nerz-Stormes, M.; Thornton, E. R. *Tetrahedron Lett.* **1986**, *27*, 897.
35 Shirodkar, S.; Nerz-Stormes, M.; Thornton, E. R. *Tetrahedron Lett.* **1990**, *31*, 4699.
36 Nerz-Stormes, M.; Thornton, E. R. *J. Org. Chem.* **1991**, *56*, 2489.
37 Crimmins, M. T.; King, B. W.; Tabet, E. A. *J. Am. Chem. Soc.* **1997**, *119*, 7883.
38 Crimmins, M. T.; Chaudhary, K. *Org. Lett.* **2000**, *2*, 775.
39 Li, Z.; Wu, R.; Michalczyk, R.; Dunlap, R. B.; Odom, J. D.; Silks, L. A. III. *J. Am. Chem. Soc.* **2000**, *122*, 386.
40 Bonner, M. P.; Thornton, E. R. *J. Am. Chem. Soc.* **1991**, *113*, 1299.
41 Ahn, K. H.; Lee, S.; Lim, A. *J. Org. Chem.* **1992**, *57*, 5065.
42 Yan, T.; Lee, H.; Tan, C. *Tetrahedron Lett.* **1993**, *34*, 3559.
43 Yan, T.; Tan, C.; Lee, H.; Lo, H.; Huang, T. *J. Am. Chem. Soc.* **1993**, *115*, 2613.
44 Ghosh, A. K.; Fidanze, S.; Onishi, M.; Hussain, K. A. *Tetrahedron Lett.* **1997**, *38*, 7171.
45 Ghosh, A. K.; Kim, J. *Tetrahedron Lett.* **2001**, *42*, 1227.
46 Ghosh, A. K.; Kim, J. *Tetrahedron Lett.* **2002**, *43*, 5621.
47 Ghosh, A. K.; Liu, C.; Xu, X. Unpublished results, University of Illinois–Chicago.
48 Xiang, Y.; Olivier, E.; Ouimet, N. *Tetrahedron Lett.* **1992**, *33*, 457.
49 Ahn, K. H.; Yoo, D. J.; Kim, J. S. *Tetrahedron Lett.* **1992**, *22*, 6661.
50 Casper, D. M.; Burgeson, J. R.; Esken, J. M.; Ferrence, G. M.; Hitchcock, S. R. *Org. Lett.* **2002**, *4*, 3739.
51 Bold, G.; Duthaler, R. O.; Riediker, M. *Angew. Chem. Int. Ed. Engl.* **1989**, *28*, 497.
52 Duthaler, R. O.; Herold, P.; Wyler-Helfer, S.; Riediker, M. *Helv. Chim. Acta* **1990**, *73*, 659.
53 (a) Reetz, M. T.; Peter, R. *Tetrahedron Lett.* **1981**, *22*, 4691. (b) Harrison, C. R. *Tetrahedron Lett.* **1987**, *28*, 4135. (c) Panek, J. S.; Bula, O. A. *Tetrahedron Lett.* **1988**, *29*, 1661.

54 Mahrwald, R. *Org. Lett.* **2000**, *2*, 4011.
55 Siegel, C.; Thornton, E. R. *Tetrahedron Lett.* **1986**, *27*, 457.
56 Siegel, C.; Thornton, E. R. *J. Am. Chem. Soc.* **1989**, *111*, 5722.
57 Figueras, S.; Martín, R.; Romea, P.; Urpí, F.; Vilarrasa, J. *Tetrahedron Lett.* **1997**, *38*, 1637.
58 Evans, D. A.; Clark, J. S.; Metternich, R.; Novack, V. J.; Sheppard, G. S. *J. Am. Chem. Soc.* **1990**, *112*, 866.
59 Evans, D. A.; Dart, M. J.; Duffy, J. L.; Rieger, D. L. *J. Am. Chem. Soc.* **1995**, *117*, 9073.
60 Meyers, A. I.; Yamamoto, Y. *Tetrahedron* **1984**, *40*, 2309.
61 Murphy, P. J.; Procter, G.; Russell, A. T. *Tetrahedron Lett.* **1987**, *28*, 2037.
62 Annunziata, R.; Cinquini, M.; Cozzi, F.; Borgia, A. L. *J. Org. Chem.* **1992**, *57*, 6339.
63 Kazmaier, U.; Grandel, R. *Synlett* **1995**, 945.
64 Ghosh, A. K.; Onishi, M. *J. Am. Chem. Soc.* **1996**, *118*, 2527.
65 Ghosh, A. K.; Fidanze, S.; Onishi, M.; Hussain, K. A. *Tetrahedron Lett.* **1997**, *38*, 7171.
66 Fidanze, S. Ph.D. Thesis. University of Illinois–Chicago, **2000**.
67 Ghosh, A. K.; Kim, J. *Unpublished results.*
68 Ghosh, A. K.; Kim, J. *Org. Lett.* **2003**, *5*, 1063.
69 Crimmins, M. T.; McDougall, P. J. *Org. Lett.* **2003**, *5*, 591.
70 Duthaler, R. O.; Herlold, P.; Wyler-Helfer, S.; Riediker, M. *Helv. Chim. Acta* **1990**, *73*, 659.
71 Ghosh, A. K.; Fidanze, S. *Org. Lett.* **2000**, *2*, 2405.
72 (a) Evans, D. A.; Chapman, K. T.; Carriera, E. M. *J. Org. Chem.* **1988**, *53*, 3560. (b) Evans, D. A.; Chapman, K. T. *Tetrahedron Lett.* **1986**, *27*, 5939.
73 Hu, T.; Schaus, J. V.; Lam, K.; Palfreyman, M. G.; Wuonola, M.; Gustafson, G.; Pankek, J. S. *J. Org. Chem.* **1998**, *63*, 2401.
74 Crimmins, M. T.; King, B. W. *J. Am. Chem. Soc.* **1998**, *120*, 9084.
75 Ghosh, A. K.; Bischoff, A.; Cappiello, J. *Eur. J. Org. Chem.* **2003**, 821.
76 Ghosh, A. K.; Bischoff, A.; Cappiello, J. *Org. Lett.* **2001**, *3*, 2677.
77 (a) Dondoni, A.; Perrone, D.; Semola, T. *J. Org. Chem.* **1995**, *60*, 7927. (b) Dondoni, A.; Fantin, G.; Fogagnolo, M.; Medici, A.; Pedrini, P. *Synthesis* **1988**, 685. (c) Dondoni, A.; Fogagnolo, M.; Medici, A.; Pedrini, P. *Tetrahedron Lett.* **1985**, *26*, 5477.
78 Biswas, K.; Lin, H.; Njardarson, J. T.; Chappel, M. D.; Chou, T.; Guan, Y.; Tong, W. P.; He, Lifeng; Horwitz, S. B.; Danishefsky, S. J. *J. Am. Chem. Soc.* **2002**, *124*, 9825.
79 Chappell, M. D.; Stachel, S. J.; Lee, C. B.; Danishefsky, S. J. *Org. Lett.* **2000**, *2*, 1633.
80 (a) Lee, C. B.; Wu, Z.; Zhang, F.; Chappell, M. D.; Stachel, S. J.; Chou, T. C.; Guan, Y.; Danishefsky, S. J. *J. Am. Chem. Soc.* **2001**, *123*, 5249. (b) Wu, Z.; Zhang, F.; Danishefsky, S. J. *Angew. Chem. Int. Ed. Engl.* **2000**, *39*, 4505.
81 Ghosh, A. K.; Bischoff, A. *Org. Lett.* **2000**, *2*, 1753.
82 Ghosh, A. K.; Liu, C. *J. Am. Chem. Soc.* **2003**, *125*, 2374.

83 Ley, S. V.; Lygo, B.; Wonnacott, A. *Tetrahedron Lett.* **1989**, *26*, 535.
84 Hayward, C. M.; Yohannes, D.; Danishefsky, S. J. *J. Am. Chem. Soc.* **1993**, *115*, 9345.
85 Crimmins, M. T.; Katz, J. D.; Washburn, D. G.; Allwein, S. P.; McAtee, L. F. *J. Am. Chem. Soc.* **2002**, *124*, 5661.
86 Sheppeck, J. E. II; Liu, W.; Chamerlin, A. R. *J. Org. Chem.* **1997**, *62*, 387.
87 Chakraborty, T. K.; Jayaprakash, S.; Laxman, P. *Tetrahedron* **2001**, *57*, 9461.
88 Enders, D.; Geibel, G.; Osborne, S. *Chem. Eur. J.* **2000**, *6*, 1302.
89 (a) Enders, D. in *Asymmetric Synthesis, Vol 3* (J. D. Morrison, Ed.), Academic Press, Orlando, **1984**, 275. (b) Enders, D. *Chem. Scripta* **1985**, *65*, 139. (c) Enders, D.; Fey, P.; Kipphardt, H. *Org. Synth.* **1987**, *65*, 173, 183.
90 Paterson, I.; Perkins, M. V. *Tetrahedron*, **1996**, *52*, 1811.
91 Perkins, M. V.; Sampson, R. A. *Org. Lett.* **2001**, *3*, 123.
92 (a) Narasaka, K.; Pai, F. *Tetrahedron* **1984**, *40*, 2233. (b) Patterson, I.; Donghi, M.; Gerlach, K. *Angew. Chem. Int. Ed. Engl.* **2000**, *39*, 3315.
93 Stille, J. R.; Grubbs, R. H. *J. Am. Chem. Soc.* **1983**, *105*, 1664.
94 Maeda, K.; Shinokubo, H.; Oshima, K. *J. Org. Chem.* **1998**, *63*, 4558.
95 Mukaiyama, T.; Kagayama, A.; Igarashi, K.; Shiina, I. *Chem. Lett.* **1999**, 1157.

3
Boron and Silicon Enolates in Crossed Aldol Reaction

Teruaki Mukaiyama and Jun-ichi Matsuo

3.1
Introduction

Metal enolates play an important role in organic synthesis and metal enolate-mediated aldol type reactions, in particular, are very useful synthetic tools in stereoselective and asymmetric carbon–carbon bond formation. Generation and reactions of different metal enolates have been extensively studied and successful applications to the controlled formation of carbon–carbon bonds have been realized under mild conditions.

The aldol reaction has long been recognized as one of the most useful synthetic tools. Under classical aldol reaction conditions, in which basic media are usually employed, dimers, polymers, self-condensation products, or α,β-unsaturated carbonyl compounds are invariably formed as by-products. The lithium enolate-mediated aldol reaction is regarded as one useful synthetic means of solving these problems. Besides the well-studied aldol reaction based on lithium enolates, very versatile regio- and stereo-selective carbon–carbon bond forming aldol-type reactions have been established in our laboratory by use of boron enolates (1971), silicon enolates–Lewis acids (1973), and tin(II) enolates (1982). Here we describe the first two topics, boron and silicon enolate-mediated crossed aldol reactions, in sequence.

3.2
Crossed Aldol Reactions Using Boron Enolates

3.2.1
Discovery of Aldol Reaction Mediated by Boron Enolates

First, the background of how we first conceived the idea of using boron enolate (vinyloxyboranes) in aldol reactions is described. At the beginning of the 1970s, several reactions were being screened by utilizing characteristics of alkylthioboranes based on the concept of elements in combination, i.e. two elements in combination, create a novel reactivity different from when

Modern Aldol Reactions. Vol. 1: Enolates, Organocatalysis, Biocatalysis and Natural Product Synthesis.
Edited by Rainer Mahrwald

ISBN: 3-527-30714-1

$(CH_3)_2CO \xrightarrow{h\nu} H_2C{=}C{=}O$ (1) + $Bu_2B{-}SBu$ (2) $\longrightarrow \xrightarrow{H_2O}$ 4 (OH, O, SBu); 1 ⇢ 3 ($H_2C{=}C(SBu)_2$) crossed out

Scheme 3.1
Unexpected formation of β-hydroxy thioester **4** on reaction of ketene **1** with thioborane **2**.

they are used separately. When ketene **1** was mixed with two moles of butylthioborane **2**, which we assumed would result in the formation of ketene thioacetal **3**, *S*-butyl 3-hydroxy-3-methylbutanethiolate **4** was unexpectedly obtained (Scheme 3.1).

It was difficult at first to discover the mechanism of this reaction, but the product soon indicated the participation of acetone in this reaction. In the experiment, ketene **1** is generated by degradation of acetone under irradiation; a small amount of acetone is, therefore, introduced into the reaction mixture. Thus β-hydroxy thioester **4** is afforded by reaction of the three components acetone, ketene **1**, and alkylthioborane **2**. When a gaseous ketene **1**, free form acetone, is introduced into the mixture of alkylthioborane **2** and carbonyl compound, the expected β-hydroxy thioesters **5** are obtained in high yield (Eq. (1)) [1].

$$\text{RCHO} + \underset{\mathbf{1}}{H_2C{=}C{=}O} + \underset{\mathbf{2}}{Bu_2B{-}SBu} \longrightarrow \xrightarrow{H_2O} \underset{\mathbf{5}}{\text{RCH(OH)CH}_2\text{C(O)SBu}} \quad (1)$$

Investigation of this mechanism reveals that the key intermediate of this reaction is vinyloxyborane (boron enolate) **7** generated from ketene **1** and alkylthioborane **6** (Eq. (2)) [2]. Thus, our original study on organothioboranes led us, unexpectedly, to discover the widely utilized aldol reactions via boron enolates [3].

$$\underset{\mathbf{6}}{Bu_2B{-}SPh} + \underset{\mathbf{1}}{H_2C{=}C{=}O} \longrightarrow \left[H_2C{=}C(OBBu_2)(SPh) \right] \text{ vinyloxyborane } \mathbf{7} \xrightarrow{Me_2CO} [\text{cyclic transition state}] \longrightarrow (CH_3)_2C(OH)CH_2C(O)SPh \quad (2)$$

3.2.2
New Method for Direct Generation of Boron Enolates

Direct generation of boron enolates from parent carbonyl compounds had been desired to expand the synthetic utility of the boron enolate-mediated aldol reaction. Although several synthetic methods to generate vinyloxyboranes (boron enolates) were reported [3–5] no useful ones for direct generation of boron enolates from parent carbonyl compounds were known until 1976. After discovering the above-mentioned aldol reaction via boron enolates we had been exploring a useful method for direct generation of boron enolates from parent carbonyl compounds. It was then thought that increasing the Lewis acidity of boron by introducing an excellent leaving group on to boron would result in an increase in acidity of the carbonyl compounds by coordination of a carbonyl group to the boron compound; the corresponding boron enolate would then be formed by abstraction of the α-proton of the carbonyl compound with a weak base such as tertiary amine.

The trifluoromethanesulfonyloxy (triflate, TfO) group was chosen as the leaving group, and dibutylboryl triflate **8** was found to generate boron enolates **9** by the reaction with ketones in the presence of a weak base such as *N*-diisopropylethylamine or 2,6-lutidine (Eq. (3)) [6]. This is the first example of the use of a metal triflate in synthetic chemistry; a variety of metal triflates are now known to be versatile Lewis acids in organic synthesis. Subsequent addition of aldehydes afforded the corresponding aldols **10** in good yields. Thus, the crossed aldol reaction which starts from ketone and aldehyde is performed easily by applying dialkylboryl triflate under mild reaction conditions.

$$R^1C(O)CH_3 \xrightarrow[i\text{-}Pr_2NEt]{Bu_2BOTf\ \mathbf{8}} \left[R^1C(OBBu_2){=}CH_2 \right] (\mathbf{9}) \xrightarrow{R^2CHO} R^1C(O)CH_2CH(OH)R_2\ (\mathbf{10}) \quad (3)$$

After our first report, this aldol reaction has been investigated in detail by many research groups [3]. It is currently understood that the boron enolate-mediated aldol reaction proceeds via a more rigid chair-like six-membered transition state (**12** or **15**) than those of alkali metal enolates, because of a shorter bond length between boron and oxygen (Figure 3.1), that is, dialkylboron enolates have relatively short metal–ligand and metal–oxygen bonds, which are suited to maximizing 1,3-diaxial (R^3–L) interactions in the transition states. This facilitates the formation of more stable transition states (**12** and **15**), where R^3 occupies a pseudoequatorial position, when vinyloxyboranes (**11** and **14**) react with aldehydes to afford aldol adducts (**13** and **16**) stereoselectively. Therefore, aldol reactions via boron enolates give aldol adducts more stereoselectively than those via alkali metal enolates such as lithium enolates. This stereoselective aldol reaction is thus an outstanding method for the stereoselective synthesis of acyclic compounds.

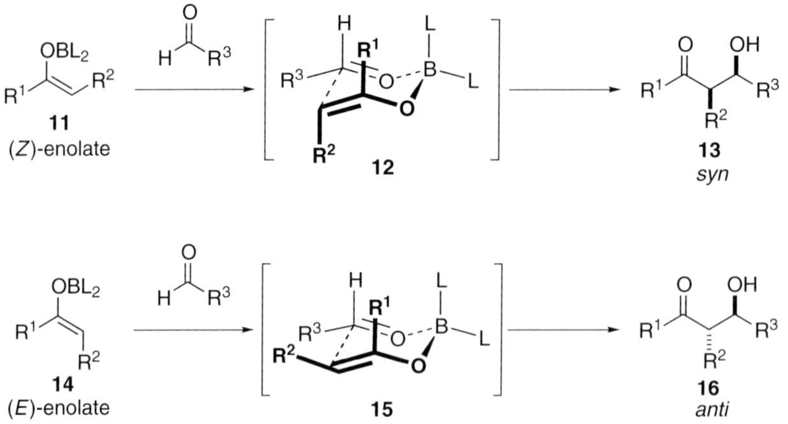

Fig. 3.1
Stereoselective aldol reaction of (*Z*) or (*E*) boron enolates and aldehydes.

3.2.3
Regioselectivity on Generation of Boron Enolates

Regioselective formation of boron enolates is conducted by using α-halo- [3–5], α-diazo- [3], or α,β-unsaturated ketones [3] but their direct generation from parent carbonyl compounds is a more important and synthetically useful method. The regioselectivity on generation of the boron enolate is controlled by the reaction conditions. The kinetic boron enolate of 2-pentanone **17** is formed by use of dibutylboryl triflate **8** and *N*-diisopropylethylamine at -78 °C in a short reaction time [6a], whereas the thermodynamic enolate **18** is predominantly generated by the use of 9-BBNOTf **19** and 2,6-lutidine at -78 °C in a long reaction time (Figure 3.2) [6b].

Subsequent aldol reactions of boron enolates with aldehydes proceed without loss of regiochemical integrity, but the reactions with ketones proceed slowly and the regiochemical integrity of an aldol product does not reflect that of an enolate.

Preparation of Dibutylboryl Triflate 8 (Eq. (4)) [6c, 7]

$$\mathrm{Bu_3B + TfOH \longrightarrow Bu_2BOTf}\ \mathbf{8} \qquad (4)$$

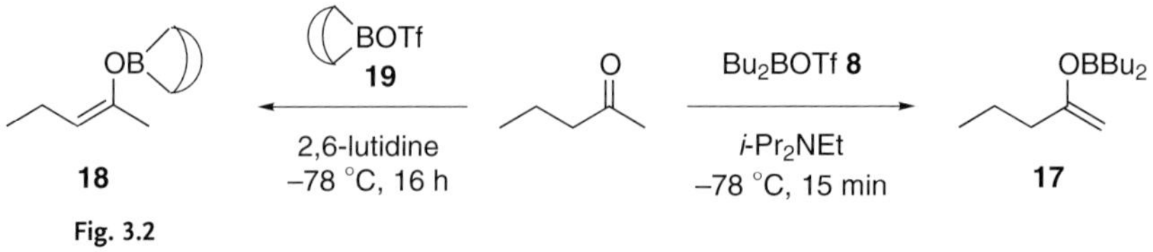

Fig. 3.2
Regioselective generation of boron enolates.

A small amount of trifluoromethanesulfonic acid (1.0 g) was added to tributylborane (15.16 g, 83.3 mmol) at room temperature under argon. The mixture was stirred and warmed to 50 °C until evolution of butane began (there is an induction period). After cooling of the mixture to 25 °C the remaining trifluoromethanesulfonic acid (11.51 g, total 83.3 mmol) was added dropwise at such a rate as to maintain a temperature between 25 and 50 °C. The mixture was then stirred for further 3 h at 25 °C. Distillation under reduced pressure gave pure dibutylboryl triflate **8** (19.15 g, 84%; bp 60 °C/2 mmHg).

Preparation of 9-BBN Triflate 19 (Eq. (5)) [6c]

$$\text{9-BBN-H} + \text{TfOH} \longrightarrow \text{9-BBN-OTf} \ (\mathbf{19}) \quad (5)$$

Trifluoromethanesulfonic acid (18.75 g, 125 mmol) was added to 9-BBN (15.33 g, 127 mmol) in hexane (100 mL) under argon. After overnight stirring the reaction mixture was concentrated and distilled in vacuo to afford 9-BBNOTf **19** (38 °C/0.03 mmHg, 28.84 g, 85%).

3.2.4 Stereoselective Formation of (*E*) or (*Z*) Boron Enolates

To obtain either *syn* or *anti* aldol adducts selectively it is important to generate boron enolates with the appropriate geometry (*E* or *Z*), that is, (*Z*) enolates **11** react with a variety of aldehydes to yield predominantly *syn* aldols **13**, whereas (*E*) enolates **14** react somewhat less stereoselectively to give *anti* aldol adducts **16** as the major products (Figure 3.1 and Table 3.1).

(*E*) Boron ketone enolates **20** are generated from a hindered dialkylboryl triflate (e.g. dicyclohexyl) and diisopropylethylamine at 0 °C whereas the (*Z*) isomer **21** is prepared by using a less hindered boryl triflate (e.g. dibutyl) at −78 °C (Figure 3.3) [8].

(*Z*) Ester boron enolates **22** [9] are selectively generated by using Bu_2BOTf and $i\text{-}Pr_2NEt$ when the methyl or ethyl esters are employed, whereas (*E*) ester boron enolates **23** [9] are also selectively generated by using $c\text{-}Hex_2BOTf$ and Et_3N when the *tert*-butyl ester is employed (Figure 3.4) [10].

Preparation of Dicyclohexylboryl Triflate (Eq. (6)) [11]

$$c\text{-}Hex_2BH + TfOH \longrightarrow c\text{-}Hex_2BOTf \quad (6)$$

A 250-mL round-bottomed flask capped with a rubber septum, containing a magnetic stirring bar and a connecting tube attached to a mercury bubbler was kept at 0 °C and charged with hexane (100 mL) and $c\text{-}Hex_2BH$ (26.7 g,

Tab. 3.1
Stereoselective formation of *syn* and *anti* aldols via (*Z*) and (*E*) boron enolates.

R^2–CH=C($OBBu_2$)–R^1 + PhCHO ⟶ *syn* aldol + *anti* aldol

Boron Enolate	*syn/anti*	Ref.	Boron Enolate[a]	*syn/anti*	Ref.
$OBBu_2$, Me; *Z*	>95:5	7a	$OBBu_2$, *t*-Bu; *Z*:*E* = >99:1	>97:3	8a
$OBBu_2$, Me; *E*	25:75	7a	$OBBu_2$, Ph; *Z*:*E* = 99:1	>97:3	8a
$OBBu_2$, Et; *Z*:*E* = >97:3	>97:3	8a	$OBBu_2$, S^tBu; *Z*:*E* = <5:95	10:90	8a
$OB(c\text{-}C_5H_9)_2$, *i*-Pr; *Z*:*E* = 19:81	18:82	8a	Bu_2BO, O, N, O (oxazolidinone); *Z*:*E* = >97:3	98:2	7c
$OB(c\text{-}C_5H_9)_2$, *c*-Hex; *Z*:*E* = 12:88	14:86	7b			

[a] The highest priority designation is assigned to the OBR_2 group with enolate substituents.

150 mmol). Trifluoromethanesulfonic acid (13.3 mL, 150 mmol) was added dropwise using a syringe with constant stirring. Hydrogen is rapidly evolved and should be safely vented. The stirring was continued at 0 °C for 2–3 h. All the suspended solid *c*-Hex_2BH dissolved and the homogeneous reaction mixture was left at 0 °C for 1–2 h without stirring. Two layers were obtained and the top layer was transferred into a dry 250-mL round-bottomed flask

(*E*)-**20** ($OB(c\text{-}Hex)_2$), R = *i*-Pr, *E*:*Z* = 81:19 ⟵ *c*-Hex_2BOTf, *i*-Pr_2NEt, 0 °C, 30 min — R–CO–Et — Bu_2BOTf, *i*-Pr_2NEt, –78 °C, 30 min ⟶ (*Z*)-**21** ($OBBu_2$), R = Et, *E*:*Z* = <3:97

Fig. 3.3
Stereoselective formation of (*Z*) or (*E*) boron enolates of ketones.

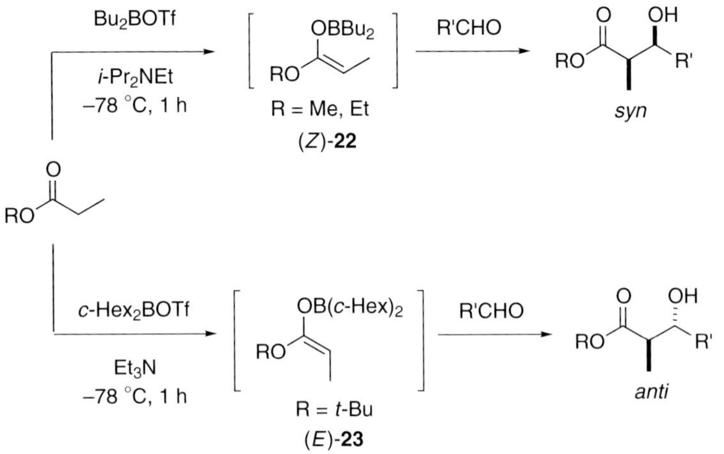

Fig. 3.4
Stereoselective formation of (*Z*) or (*E*) boron enolates of carboxylic esters.

leaving the small yellow layer (approx. 2 mL) behind. Solid *c*-Hex_2BOTf was obtained by removing the solvent using a water aspirator (15–20 mm). It was then recrystallized from hexane. Mp 88 °C, yield 80%. Stock solutions (1.00 M) in CCl_4 and in hexane were prepared and kept at 0 °C for enolboration.

Typical Experimental Procedure for Crossed Aldol Reaction via Boron Enolate (Eq. (7)) [6a]

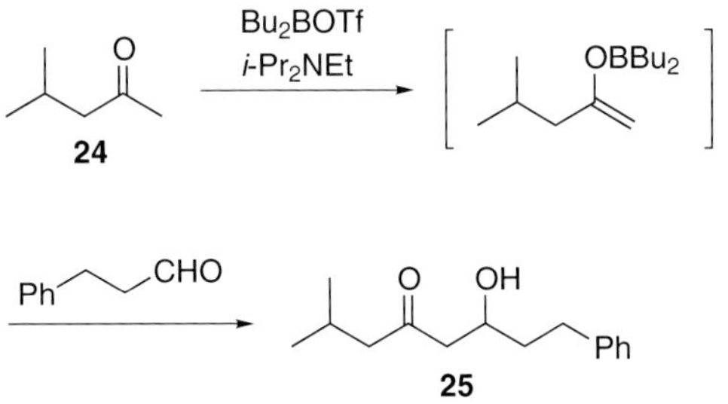

(7)

To a solution of dibutylboryl triflate (301 mg, 1.1 mmol) and diisopropylethylamine (142 mg, 1.1 mmol) in ether (1.5 mL) was added dropwise 2-methyl-4-pentanone **24** (100 mg, 1.0 mmol) in ether (1.5 mL) at −78 °C, under argon, with stirring. After stirring of the mixture for 30 min, 3-phenylpropanal (134 mg, 1.0 mmol) in ether (1.5 mL) was added at the same temperature. The reaction mixture was allowed to stand for 1 h, then added to pH 7 phosphate buffer at room temperature and extracted with ether. After removal of ether, the mixture was treated with 30% H_2O_2

(1 mL) in methanol (3 mL) for 2 h and H_2O was added. The mixture was concentrated to remove most of the methanol and extracted with ether. The organic layer was washed with 5% $NaHCO_3$ solution and brine, dried over Na_2SO_4, and concentrated. The crude oil was purified by preparative TLC to give 3-hydroxy-7-methyl-1-phenyl-5-octanone **25** (192 mg, 82%).

3.2.5
syn-Selective Asymmetric Boron Aldol Reactions

Conventional asymmetric aldol reactions have been performed by using chiral enolates and achiral carbonyl compounds. A chiral boron enolate generated from a chiral oxazolidone derivative (**26** and **28**), dialkylboron triflate, and diisopropylethylamine reacts stereoselectively with aldehydes to afford the corresponding *syn* aldol adducts (**27** and **29**) in good yields with excellent diastereoselectivity (Eqs. (8) and (9)) [12]. The opposite sense of asymmetric induction is achieved by changing the chiral auxiliary. Several other chiral auxiliaries have also been developed for highly diastereoselective synthesis of *syn* aldol adducts (Eqs. (10)–(13)) [13].

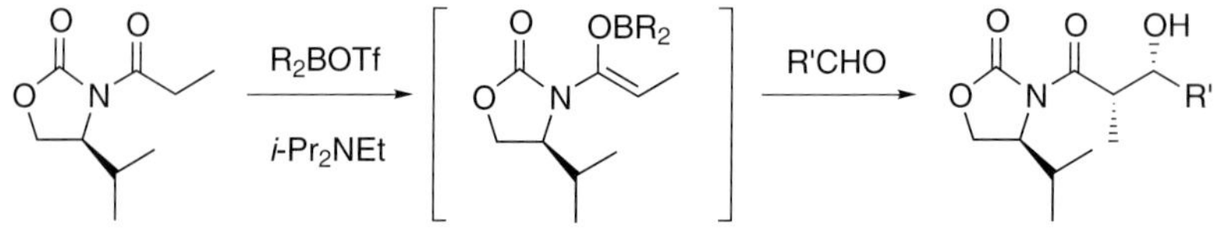

26 **27**

(8)

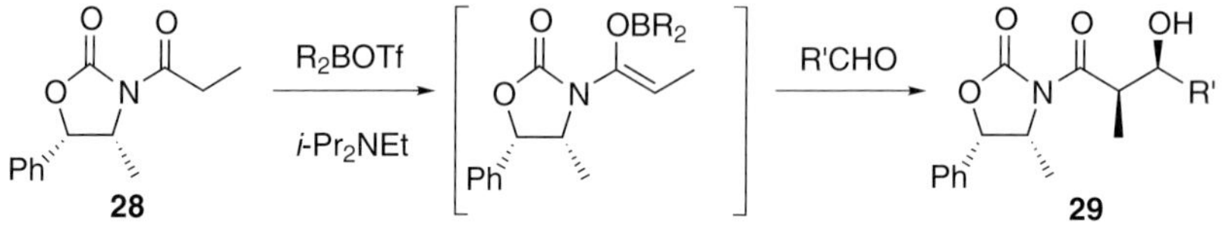

(9)

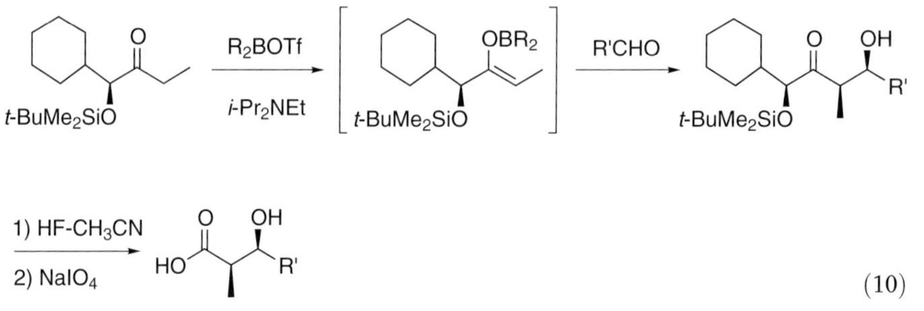

(10)

Me
Me
N
O
Ipc_2BOTf
i-Pr_2NEt
Me
Me
N
B
O
2
RCHO
1) H_2O_2
2) H_3O^+
3) CH_2N_2
O
OH
MeO
R
syn/*anti* = 10/90~5/95
77~85% ee (*anti*)

(11)

OMe
N
Ph
O
9-BBNOTf
i-Pr_2NEt
OMe
B
N
Ph
O
RCHO
1) H_2O_2
2) H_3O^+
3) CH_2N_2
O
OH
MeO
R
syn/*anti* = 97/3~98/2
40~60% ee (*syn*)

(12)

N
O
S
O_2
R_2BOTf, *i*-Pr_2NEt
CH_2Cl_2, -5°C
OBR_2
N
S
O_2
R'CHO
-78°C
O
OH
X_N
R'

(13)

3.2.6
anti-Selective Asymmetric Aldol Reaction

An *anti*-selective diastereoselective aldol reaction [14] has been performed by using enantiomerically pure carboxylic esters derived from (−)- or (+)-norephedrine **30** [15]. This method is applicable to a wide range of aldehydes with high selectivity (both *syn*/*anti* and diastereoselectivity of *anti* isomer). It is proposed that (*E*) boron enolates **31** are formed by this procedure and aldol reaction proceeds via the six-membered transition state (Eq. (14)). The aldol products **32** are converted to the corresponding alcohols ($LiAlH_4$, THF) or carboxylic acids (LiOH, THF–H_2O) without loss of stereochemical integrity.

Ph
O
Me
O
Bn
N
SO_2Mes
30
c-Hex_2BOTf
Et_3N
Ph
OB(*c*-$Hex)_2$
Me
O
Bn
N
SO_2Mes
31
RCHO
Ph
O
OH
Me
O
R
Bn
N
SO_2Mes
32

Mes:
Me
Me
Me

(14)

It is also reported that addition of Lewis acids to the reaction of chiral boron enolates and aldehydes changes *syn*-selectivity to *anti*-selectivity (Eqs. (15) and (16)) [16]. The change of stereoselection is rationalized by considering a Lewis acid-mediated open transition state (e.g. **33**).

Bu_2BOTf, *i*-Pr_2NEt; *i*-PrCHO, Et_2AlCl

95 : 5

33

(15)

1) Et_2BOTf, *i*-Pr_2EtN
2) RCHO
syn

1) Et_2BOTf, *i*-Pr_2EtN
2) RCHO, $TiCl_4$
anti

(16)

Thus, boron enolates prepared under mild conditions enable aldol-type reactions essentially under neutral conditions. Stereocontrolled synthesis of acyclic molecules has been achieved by employing boron enolate-mediated aldol reactions; this method has been extensively applied to the synthesis of natural products.

Typical Procedure for *syn*-Selective Asymmetric Boron Aldol Reaction (Eq. (17)) [13]

26 → (Bu_2BOTf, *i*-Pr_2NEt) → [OBR_2 enolate] → (*i*-PrCHO) → **34**

syn/*anti* = 497:1

(17)

To a 0.2–0.5 M solution of chiral imide **26** in CH_2Cl_2 under argon (0 °C) was added 1.1 equiv. dibutylboryl triflate followed by 1.2 equiv. diisopropylethylamine. After 30 min the reaction mixture was cooled (−78 °C) and 1.1 equiv. isobutyraldehyde was added and stirred for 0.5 h at −78 °C and then for 1.5 h at room temperature. The reaction was quenched with pH 7 phosphate buffer and the boron aldolate complex was oxidized with 30% hydrogen peroxide–methanol (0 °C, 1 h). The aldol product **34** was then isolated by ether extraction (*syn*/*anti* = 497:1).

Typical Procedure for *anti*-Selective Asymmetric Boron Aldol Reaction (Eq. (18)) [15c]

1) *c*-Hex_2BOTf, Et_3N
2) *i*-PrCHO

30 → **35** (18)

anti:*syn* = >98:2
96%de (*anti*)

To a solution of chiral ester **30** (4.80 g, 10 mmol) and triethylamine (3.40 mL, 24 mmol) in dichloromethane (50 mL) at −78 °C under nitrogen is added a solution of dicyclohexylboron triflate (1.0 M in hexane, 22 mL, 22 mmol) dropwise over 20 min. After the resulting solution has been stirred at −78 °C for 30 min, isobutyraldehyde (1.08 mL, 12 mmol) is added dropwise. The reaction mixture is stirred for 30 min at −78 °C then left to warm to room temperature over 1 h. The reaction is quenched by adding pH 7 buffer solution (40 mL), the mixture is diluted with methanol (200 mL), and 30% hydrogen peroxide (20 mL) is added. After the mixture has been stirred vigorously overnight, it is concentrated. Water is added to the residue, and the mixture is extracted with dichloromethane. The combined organic extracts are washed with water, dried over Na_2SO_4, filtered, and concentrated. Purification of the residue gives aldol adduct **35** (88%, *anti*/*syn* $\geq$ 24:1).

3.3 Crossed Aldol Reactions Using Silicon Enolates

3.3.1 Discovery of Silicon Enolate-mediated Crossed Aldol Reactions

The driving force of the above-mentioned aldol reaction with boron enolate is considered to be the interconversion of enol ketones (boron enolates) to their more stable ketones (β-boryloxy ketones) [17]. When the boron enolate-mediated aldol reaction was studied in our laboratory an investigation on the development of new reaction chemistry using titanium(IV) chloride was in progress [18]. A new and important idea was immediately came to mind that titanium(IV) chloride would effectively generate active electrophilic species as a result of its strong interaction with carbonyl compounds, and the complex thus formed would react easily even with relativly weaker carbon nucleophiles to form a new carbon–carbon bond. Use of a stable and isolable silyl enol ether [19] was suggested as a weak nucleophile and, just as expected, aldol reaction between silyl enol ether of acetophenone **36** and benzaldehyde in the presence of titanium(IV) chloride afforded the aldol product **37** in high yield (Eq. (19)) [20].

(19)

It is reported that enol ethers react with acetals or ketals, promoted by Lewis acids, to give aldol-type adducts: these reactions of alkyl enol ethers are, however, often accompanied by undesired side reactions [21]. Further, is difficult to perform crossed-aldol reactions selectively because conventional aldol reactions are conducted under equilibrium conditions using a basic or acidic catalyst in protic solvents [22]. Detailed studies of this new aldol reaction of silicon enolates, however, reveal a number of advantages over conventional methods.

First, it not only gives a variety of aldol adducts in high yields but also a regioselective aldol adduct when the silyl enol ether of an unsymmetrical ketone is used. That is, the aldol reaction proceeds with retention of the regiochemical integrity of the starting silyl enol ethers to afford the corresponding aldol regiospecifically. Starting silyl enol ethers can be conveniently prepared regioselectively under kinetically or thermodynamically controlled conditions.

Second, functional group selectivity is observed – i.e. reactions with aldehydes proceed at −78 °C whereas those with ketones proceed at elevated temperatures (ca. 0 °C). Chemoselectivity is observed with acceptors having two different kinds of carbonyl function, for example aldehyde and ketone or ester, in the same molecule. Treatment of phenylglyoxal with silyl enol ether **38** at −78 °C affords α-hydroxy-γ-diketone **39** (Eq. (20)) [20b]. The reaction of ketoesters **40** other than β-ketoesters with silyl enol ether **38** gives hydroxyketoesters **41** as sole products (Eq. (21)) [23].

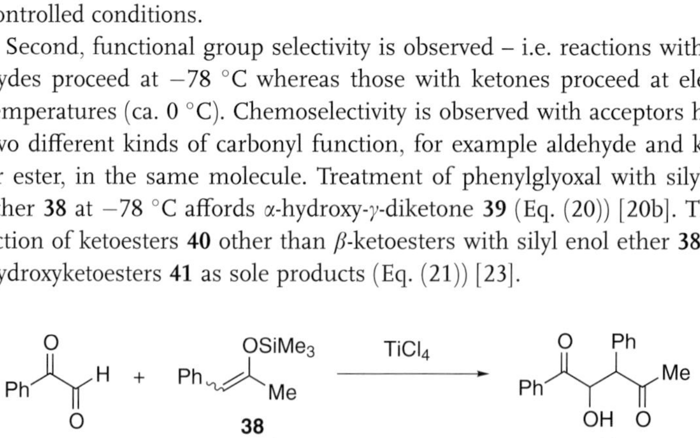

(20)

40 + 38 $\xrightarrow{TiCl_4}$ 41

n = 0, 2, 3
R = Me, Et

(21)

A directed aldol reaction between two ketones affords thermodynamically unfavorable aldols in high yields, because of stabilization of the aldol adducts by their intramolecular chelation with titanium **42** or by their conversion to silyl ethers **43** (Eq. (22)).

R^1COR^2 + $R^3R^4C{=}C(OSiMe_3)R^5$ $\xrightarrow{TiCl_4}$ [**42** or **43**] $\xrightarrow{H_2O}$ $R^1R^2C(OH)CR^3R^4COR^5$

(22)

Despite its remarkable power as a method for carbon–carbon bond formation, the level and sense of its stereoselectivity often vary. The *syn*/*anti* ratio for the aldol product is affected by the stereochemistry of the aldehyde and silyl enolate, and the character of the Lewis acid catalyst (Figure 3.5 and Table 3.2) [24]. The mechanistic basis for the aldol reaction has not yet been firmly established. Apart from a limited number of exceptions, however [25], the stereochemical observations have been rationalized by considering acyclic transition states (**44**–**47**) [26].

Typical Procedure for Titanium(IV) Chloride-catalyzed Aldol Reaction of Silicon Enolates (Eq. (23)) [20b]

48 + PhCHO $\xrightarrow{TiCl_4}$ **49**

(23)

A solution of 1-trimethylsilyloxy-1-cyclohexene **48** (426 mg, 2.5 mmol) in dichloromethane (10 mL) was added dropwise into a mixture of benzaldehyde (292 mg, 2.75 mmol) and $TiCl_4$ (550 mg, 2.75 mmol) in dichloro-

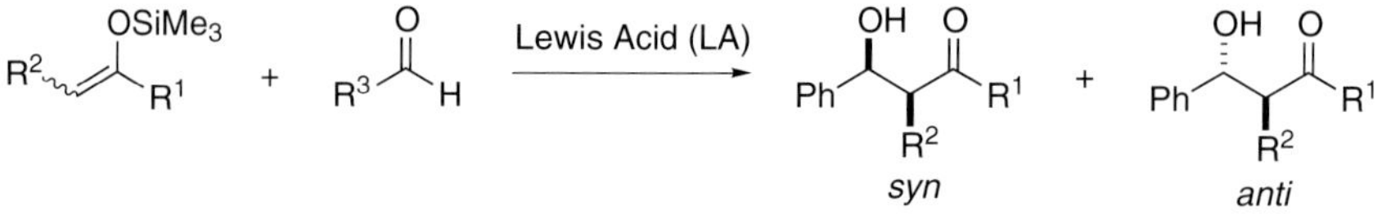

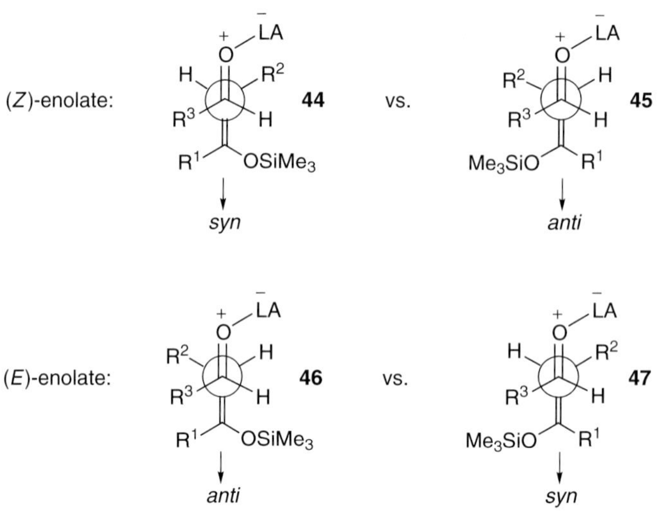

Fig. 3.5
Transition-state models for Lewis acid-catalyzed aldol reaction of silicon enolates.

methane (20 mL) under an argon atmosphere at −78 °C and the reaction mixture was stirred for 1 h. After hydrolysis at that temperature the resulting organic layer was extracted with ether, and the combined organic extract was washed with water, dried over Na_2SO_4, and concentrated. The residue was purified by chromatography to afford a diastereo mixture of aldol adducts **49** (92%, *syn*/*anti* = 75/25).

As an extension of this new procedure for carbon–carbon bond formation, the reaction between silyl enol ethers and acetals **50**, a typical protecting group of aldehydes, is performed to afford β-alkoxy carbonyl compound **51** in the presence of titanium(IV) chloride (Eq. (24)) [27]. A variety of substituted furans are readily prepared by application of the $TiCl_4$-promoted reaction of α-halo acetals **52** with silyl enol ethers (Eq. (25)) [27].

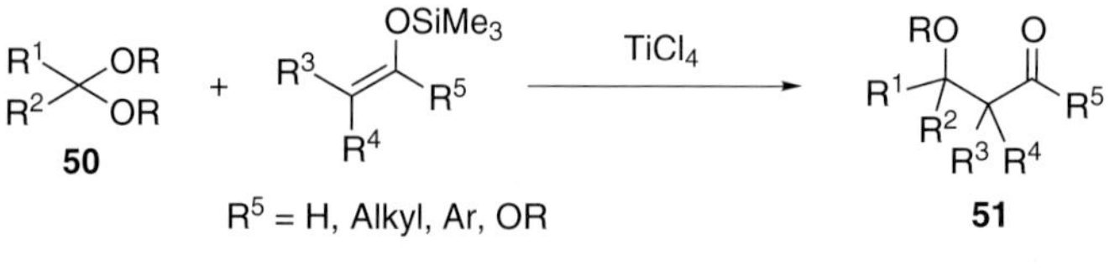

R^5 = H, Alkyl, Ar, OR (24)

Tab. 3.2
Stereochemical outcome of Lewis acid-catalyzed aldol reactions of a variety of silicon enolates.

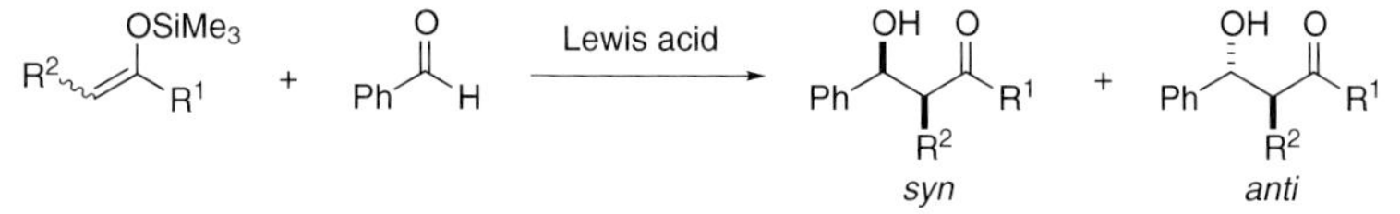

OSiMe$_3$, R^2, R^1 + PhCHO —Lewis acid→ syn + anti (Ph, OH, O, R^1, R^2)

Silicon Enolate[a]	Lewis acid	syn/anti	Ref.
OSiMe$_3$, OEt; Z:E = 100:0	$TiCl_4$	33:67	24a
OSiMe$_3$, OEt; Z:E = 15:85	$TiCl_4$	26:74	24a
t-Bu, OSiMe$_3$, OEt	$TiCl_4$	>92:8	24b
OSiMe$_3$, OEt, t-Bu	$TiCl_4$	>92:8	24b
OSiMe$_3$ (cyclohexenyl); Z:E = 0:100	$TiCl_4$	25:75	20b
OSiMe$_3$ (cyclopentenyl); Z:E = 0:100	$TiCl_4$	50:50	20b
OSiMe$_3$, Et; Z:E = 100:0	BF_3–OEt_2	60:40	24c
OSiMe$_3$, i-Pr; Z:E = 97:3	BF_3–OEt_2	56:44	24c
OSiMe$_3$, t-Bu; Z:E = 100:0	$TiCl_4$	4:96	24c
OSiMe$_3$, Ph; Z:E = 100:0	BF_3–OEt_2	47:53	24c
OSiMe$_3$, S^{t}Bu; Z:E = 10:90	BF_3–OEt_2	5:95	24d
OSiMe$_2$^{t}Bu, S^{t}Bu; Z:E = >95:5	BF_3–OEt_2	4:96	24d
OSiMe$_3$, OSiMe$_3$	$TiCl_4$	14:86	24b
t-Bu, OSiMe$_3$, OSiMe$_3$	$TiCl_4$	77:23	24b

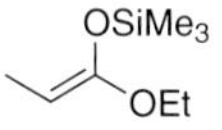

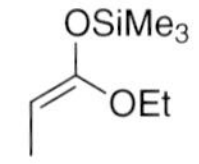

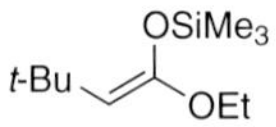

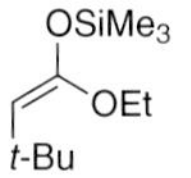

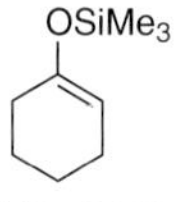

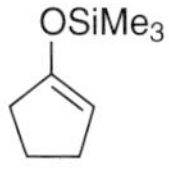

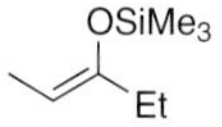

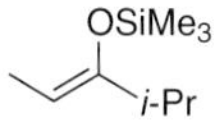

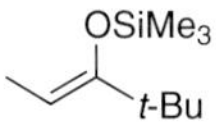

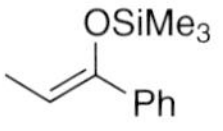

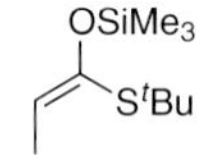

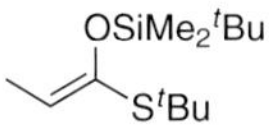

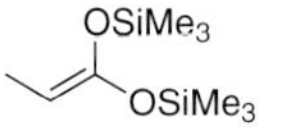

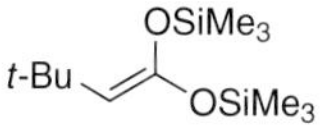

[a] The highest priority designation is assigned to the OSiR3 group with enolate substituents.

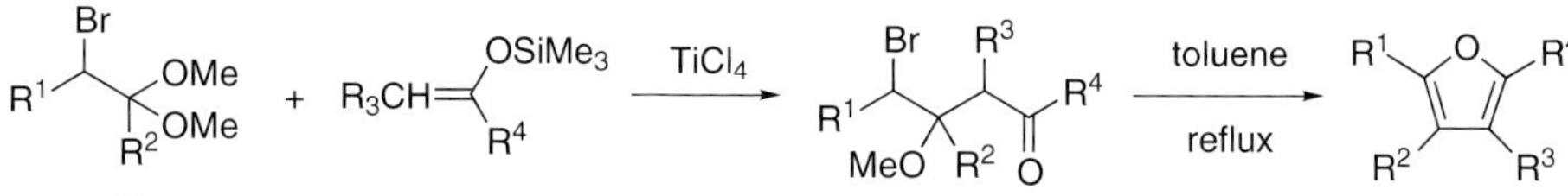

52 (Br, R^1, OMe, OMe, R^2) + R$_3$CH=C(OSiMe$_3$)R^4 —$TiCl_4$→ (Br, R^3, R^1, R^4, MeO, R^2, O) —toluene, reflux→ **53** (R^1, O, R^4, R^2, R^3)

(25)

OMe OMe 54 + OSiMe3 55 1) $TiCl_4$-$Ti(Oi\text{-}Pr)_4$ 2) H_2O OMe CHO 56 → → → OH Vitamin A

Scheme 3.2
Synthesis of vitamin A using aldol reaction of acetal **54** and silyl dienol ether **55**.

In the presence of titanium(IV) chloride, silyl dienol ether **55** derived from an α,β-unsaturated aldehyde reacts with acetal **54** selectively at the γ-position to give δ-alkoxy-α,β-unsaturated aldehydes **56**, albeit in low yields. Because titanium(IV) chloride is strongly acidic, polymerization of silyl dienol ether **55** proceeds. In these reactions addition of tetraisopropoxytitanium(IV) to titanium(IV) chloride increases the yield dramatically [28a] – vitamin A is successfully synthesized by utilizing this aldol reaction of silyl dienol ether **55** (Scheme 3.2) [28b].

When this reaction was further investigated using trimethylsilyltrifluoromethanesulfonate (Me_3SiOTf) as Lewis acid some interesting results were reported. Me_3SiOTf-mediated aldol reaction of silicon enolates and acetals tends to give *syn*-β-methoxy ketones as a major products, irrespective of the stereochemistry of the silicon enolate double bond, except for the (*Z*) silicon enolate of *tert*-butyl ethyl ketone **57** (Table 3.3) [29].

Because they are even more nucleophilic than silyl enol ethers, silyl ketene acetals **58**, derived from carboxylic esters, react with ketones and aldehydes in the presence of titanium(IV) chloride to give β-hydroxy esters **59** in high yields (Eq. (26)) [30, 31]. Although the Reformatsky reaction is well known as a good synthetic tool for synthesis of β-hydroxy esters, the titanium(IV) chloride-mediated reaction is a milder and more versatile method for synthesis of α-substituted β-hydroxy esters.

$$R^1C(O)R^2 + R^3R^4C{=}C(OSiMe_3)OR^5 \ (\mathbf{58}) \xrightarrow[2)\ H_2O]{1)\ TiCl_4} R^1R^2C(OH)\text{-}CR^3R^4\text{-}C(O)OR^5 \ (\mathbf{59}) \quad (26)$$

After the discovery of aldol reactions of silyl enolates with carbonyl compounds or acetals, as described above, silyl enolates become one of the most popular carbon nucleophiles in organic synthesis and are also employed for other reactions such as Michael reaction [32], Mannich reaction [33], etc.

Tab. 3.3
Me_3SiOTf-catalyzed aldol reaction of acetals and silicon enolates.

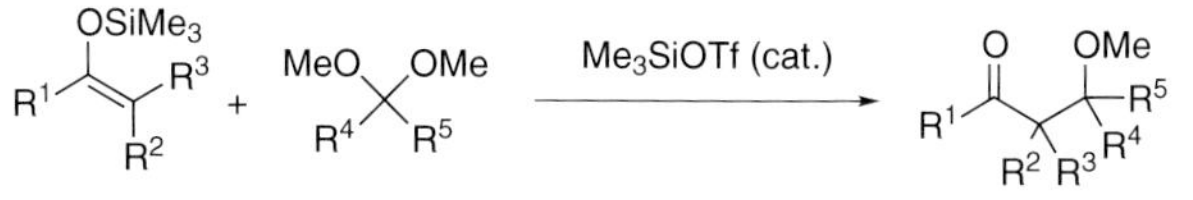

Entry	Silyl Enol Ether	Acetal	Product	syn/anti
1	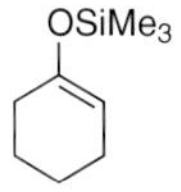	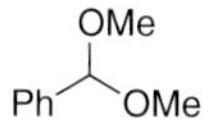	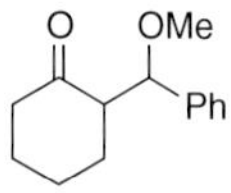 86%	92:8
2	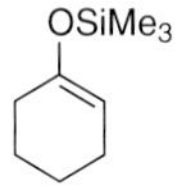	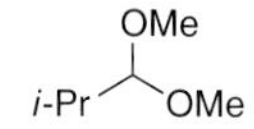	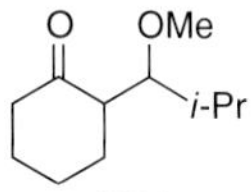 95%	86:14
3	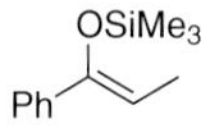	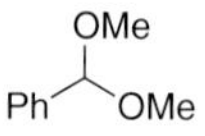	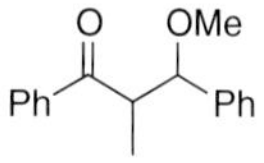 97%	84:16
4	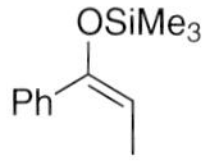	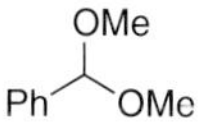	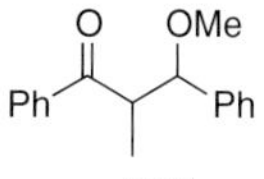 83%	71:29
5	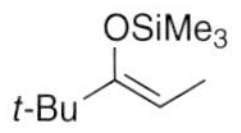 **57**	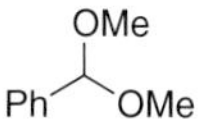	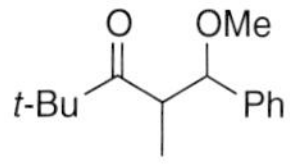 94%	5:95

[34]. Silyl enolates are superior to other metal enolates in isolation, regioselective formation, and unique reactivity under mild conditions.

3.3.2
Lewis Acid-catalyzed Aldol Reactions of Silicon Enolates

Initially, the titanium(IV) chloride-mediated aldol reaction of silyl enolates with aldehydes was investigated [20] and a catalytic amount of trityl salt **60** (e.g. trityl perchlorate) was found to promote the aldol reaction (Eq. (27)) [35]. Whereas the original reaction is performed by using a stoichiometric amount of titanium(IV) chloride, 5–10 mol% trityl salt is sufficient to drive the aldol reaction to completion. One interesting finding in this catalytic reaction is that the silicon enolate reacts with aldehydes to give the corresponding aldol adducts as their silyl ethers **61**.

$$R^1CHO + R^2\text{-}CH{=}C(OSiMe_3)R^3 \xrightarrow{\text{TrX } \mathbf{60} \text{ (cat.)}} R^1CH(OSiMe_3)CH(R^2)C(O)R^3 \; (\mathbf{61}) \quad (27)$$

TrX = $Ph_3C^+ X^-$
X = ClO_4, $SbCl_6$, OTf, PF_6, etc.

The aldol reaction of a variety of silyl enolates and acetals in the presence of a catalytic amount (1–10 mol%) of trityl perchlorate proceeded efficiently to afford β-methoxy ketone in high yield. In the presence of trityl tetrafluoroborate catalyst, the reaction of dithioacetal **62** with silyl enol ethers affords β-ethylthio ketones **63** (Eq. (28)) [36].

$$R^1R^2C(SEt)_2 \; (\mathbf{62}) + R^3CH{=}C(OSiMe_3)R^4 \xrightarrow{TrBF_4} R^1CH(R^2)\text{-}CH(SEt)\text{-}CH(R^3)C(O)R^4 \; (\mathbf{63}) \quad (28)$$

A specific combination of two weak acids, tin(II) chloride and chlorotrimethylsilane, is found to serve as an effective catalyst for the aldol reaction [37]. Neither chlorotrimethylsilane or tin(II) chloride has any accelerating effect at −78 °C even when more than one equivalent is added. In the combined presence of catalytic amounts of chlorotrimethylsilane and tin(II) chloride, however, the aldol reaction gives the desired product in more than 90% yield (Eq. (29)). It is supposed that the cationic silyl species generated by coordination of the chloride to the tin(II) atom catalyzes the aldol reaction. Because a similar cationic silicon Lewis acid, trimethylsilyl triflate, is known to catalyze aldol reactions of silicon enolates [38], and several other unique Lewis acid catalysts have also been reported (e.g. lanthanide catalysts [39]). Water-stable Lewis acids such as lanthanide triflates, which catalyze the aldol reaction in aqueous solvent or pure water, have received much attention in the development of economical and environmentally benign synthetic methods [40].

$$\text{1-(trimethylsilyloxy)cyclohexene} + PhCHO \xrightarrow[\text{2) } H_3O^+]{\text{1) } SnCl_2\text{-}Me_3SiCl \text{ (10 mol\%)}, \; -78\,^{\circ}C} \text{2-(hydroxy(phenyl)methyl)cyclohexanone} \quad (29)$$

Silicon enolate-mediated aldol reactions are performed by other activation methods. The aldol reaction is performed by using a catalytic amount of tetrabutylammonium fluoride (TBAF) or tris(diethylamino)sulfonium difluorotrimethylsiliconate (TASF) (Table 3.4) [41]. Aldehydes undergo this

Tab. 3.4
TBAF-catalyzed aldol reaction of silicon enolates [41d].

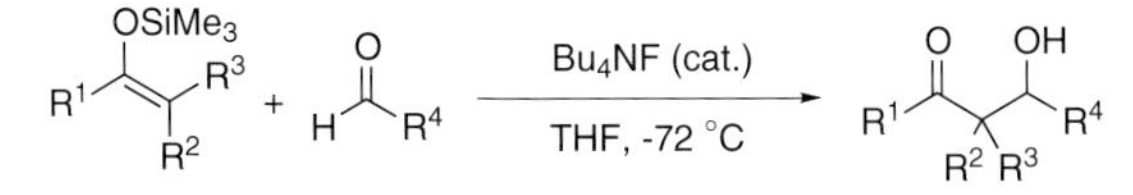

Entry	Silyl Enol Ether (Geometric Purity, %)	Aldehyde	Product (Diastereomer Ratio)
1	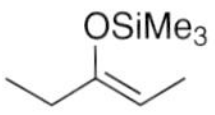 (97)	PhCHO	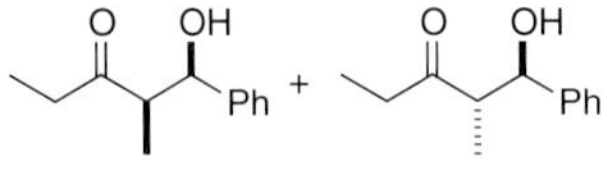 (86 : 14)
2	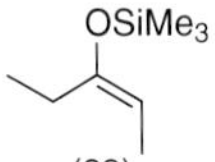 (89)	PhCHO	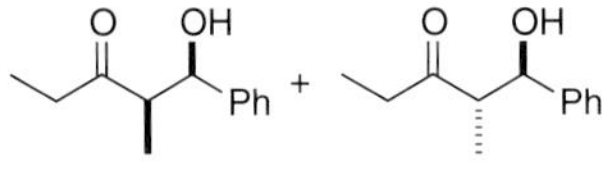 (44 : 56)
3	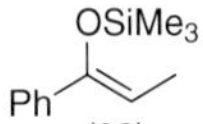 (93)	PhCHO	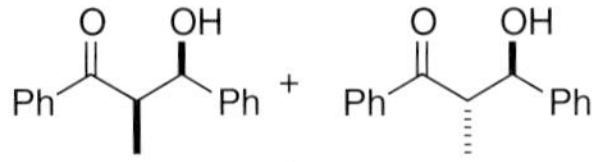 (93 : 7)
4	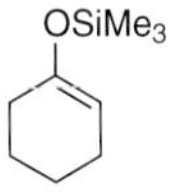	*i*-PrCHO	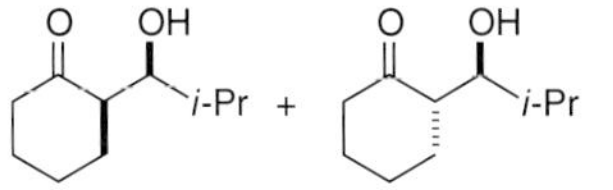 (97 : 3)
5	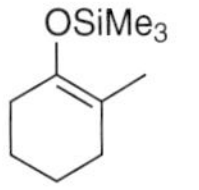	*n*-PrCHO	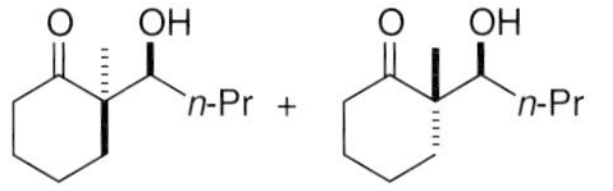 (86 : 14)
6	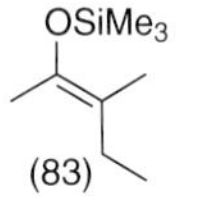 (83)	PhCHO	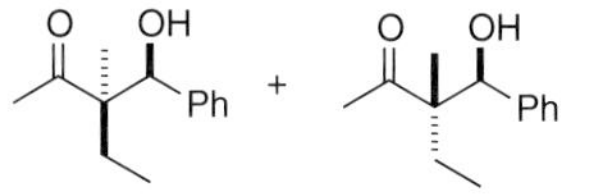 (29 : 71)
7	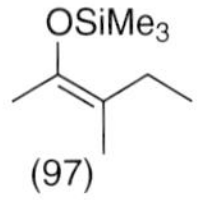 (97)	PhCHO	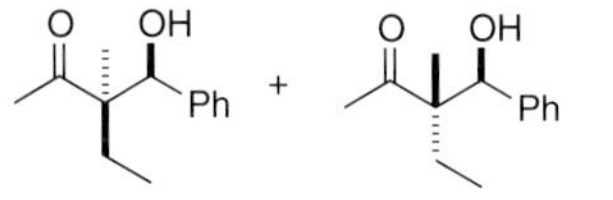 (64 : 36)
8	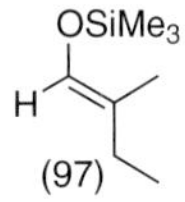 (97)	PhCHO	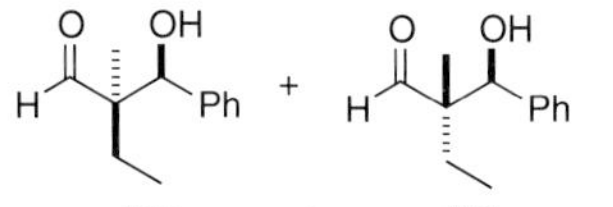 (35 : 65)

Scheme 3.3
Reaction mechanism for fluoride ion-catalyzed aldol reaction of silicon enolates.

type of aldol reaction quite readily (aromatic aldehydes gives aldol products more effectively than aliphatic aldehydes), whereas ordinary aliphatic and aromatic ketones form no aldol products. This reaction is considered to proceed through a catalytic cycle involving reversible steps (Scheme 3.3). A naked enolate **64** generated by the reaction of silicon enolate and fluoride ion reacts with aldehydes to aldol dianion **65**. Rapid chemical trapping of unstable aldol anions **65** with Me_3SiF to form **66** serves to drive this reaction in the forward direction. Often, especially with sterically demanding enolates, *syn* aldol adducts are obtained irrespective of enolate geometry; this is ascribed to an extended open transition state **67**. Recently, another open "skew" transition state **68** has been considered for less hindered enolates [42].

Preparation of Anhydrous Tetrabutylammonium Fluoride (TBAF) [41e]. Tetrabutylammonium fluoride, which is commercially available as its trihydrate, was dried over P_2O_5 (30–40 °C, 0.5 mmHg, overnight).

Typical Procedure for the TBAF-catalyzed Aldol Reaction (Eq. (30)) [41a]

$$\mathbf{48} + \text{PhCHO} \xrightarrow{\text{TBAF (10 mol\%)}} \mathbf{49} \qquad (30)$$

A mixture of the trimethylsilyl enolate **48** of cyclohexanone and benzaldehyde (1:1.1 mol ratio) was added in one portion to a solution of TBAF (10 mol%) in THF at −78 °C under argon. The resulting mixture was stirred at the same temperature for 3.5 h and poured into hexane. The hexane solution was washed with water, dried, and concentrated to give aldol silyl ether in 80% yield. The aldol silyl ether was dissolved in a 1:9 mixture of 1 M HCl and methanol and left to stand at room temperature for 5 min. The solution was diluted with 1:1 hexane–ether and treated with water. The organic layer was dried and concentrated to give the almost pure desired aldol adduct **49**.

Trimethylsilyl enolates are also activated by transmetalation to the corresponding metal enolates by use of MeLi [43], transition metal catalysts [44, 45], etc. [46].

3.3.3
Non-catalyzed Aldol Reactions of Silicon Enolates

Specially designed silicon enolates which readily form hypervalent silicates react with aldehydes in the absence of catalysts. These include: trichlorosilyl enolates **69** (Eq. (31)) [47], dimethyl(trifloxy)silyl enolate **70** (Eq. (32)) [48], enoxysilacyclobutanes **71** (Eq. (33)) [49], and dimethylsily enolates **72** (Eq. (34)) [50]. Owing to the electron-withdrawing group (**69** and **70**), the less sterically demanding group (**72**), or angle strain (**71**) on silicon, Lewis acidity of the silicon is increased. It is also reported that *O*-silyl enol derivatives of amides **73** (*O*-silyl ketene *N*,*O*-acetals) undergo non-catalyzed aldol reaction with aldehydes (Eq. (35)) [51].

$OSiCl_3$, OMe (**69**) + PhCHO → (CH_2Cl_2, 0 °C) Ph–CH(OH)–CH$_2$–C(O)OMe, 98% (31)

Ph–C(O)–CH$_2$CH$_3$ → ($Me_2Si(OTf)_2$, *i*-Pr_2NEt) [Ph–C($OSiMe_2(OTf)$)=CHCH$_3$ (**70**)] → (PhCHO, -78 °C) Ph–C(O)–CH(CH$_3$)–CH(OH)–Ph, 88%, *syn*/*anti* = 91:9 (32)

71 (R^3, Si, O, X, R^1, R^2) + R^4CHO → X, R^1 R^2, OH, R^4 (33)

X = OR, SR, NR_2

$OSiMe_2H$ (**72**) + PhCHO → (1) DMF, 50 °C, 48 h; 2) MeOH-HCl) 2-(hydroxy(phenyl)methyl)cyclohexanone, 79%, *syn*/*anti* = 58:42 (34)

OSiEt3 Me2N 73 + PhCHO —(CH_2Cl_2, -30 °C, 4 h)→ Me2N(C=O)CH(Me)CH(OSiEt3)Ph 80% *syn*/*anti* = 1:1.8 (35)

It is also known that non-catalyzed aldol reactions using silyl ketene acetals proceed at high temperature [52], or in H_2O [53], DMSO, DMF, and DME [54], or under high pressure [55].

3.3.4
Lewis Base-catalyzed Aldol Reactions of Trimethylsilyl Enolates

Since 1973, many Lewis acid-catalyzed aldol reactions have been studied, as described above. There have, however, been few examples of Lewis base-catalyzed aldol reaction using special silyl enolates, e.g. phosphoramide-catalyzed aldol reactions of trichlorosilyl enolates **69** (Eq. (36)) [47] and $CaCl_2$-catalyzed reactions of dimethylsilyl enolates **72** in aqueous DMF (Eq. (37)) [56].

OSiCl3, OMe **69** + *t*-BuCHO → *t*-Bu–CH(OH)–CH2–C(=O)OMe (36)

CD_2Cl_2: 50% conversion (120 min)
CD_2Cl_2+HMPA: 100% conversion (<3 min)

OSiMe2H **72** + PhCHO —(CaCl2 (cat.), DMF, 30 °C, 24 h)→ 2-(hydroxy(phenyl)methyl)cyclohexanone 90% *syn*/*anti* = 45:55 (37)

New challenges were then made to develop useful Lewis base-catalyzed aldol reactions of trimethylsilyl enolates, simple and the most popular silicon enolates. It has recently been found that aldol reactions of trimethylsilyl enolates with aldehydes proceed smoothly under the action of a catalytic amount of lithium diphenylamide or lithium 2-pyrrilidone in DMF or pyridine (Eq. (38)) [57]. This Lewis base-catalyzed aldol reaction of trimethylsilyl enolates [58] has an advantage over acid-catalyzed reactions in that aldol reaction of carbonyl compounds with highly-coordinative functional groups with Lewis acid catalysts are smoothly catalyzed by Lewis bases to afford the desired aldol adducts in high yields.

$$R^1CHO + R^2R^3C{=}C(OSiMe_3)R^4 \xrightarrow[\text{DMF or pyridine}]{\text{LiNPh}_2 \text{ or lithium 2-pyrrolidone (cat.)}} R^1CH(OH)CR^2R^3C(O)R^4$$

R^4 = R, OR, SR

(38)

Preparation of the 0.1 M Solution of Lithium Pyrrolidone in DMF. A solution of MeLi (1.14 M in ether, 0.52 mL, 0.59 mmol) was added to a solution of pyrrolidone (55.3 mg, 0.65 mmol) in THF at 0 °C. The solvents were removed in vacuo and DMF (5.9 mL) was added to the residue.

Typical Experimental Procedure for Lithium 2-Pyrrolidone-catalyzed Aldol Reaction (Eq. (39)) [57b,c]

$$4\text{-}Me_2NC_6H_4CHO + Me_2C{=}C(OSiMe_3)OMe\ (\mathbf{74}) \xrightarrow[\text{DMF, }-45\ ^\circ\text{C, 1 h}]{\text{lithium 2-pyrrolidone (10 mol\%)}} 4\text{-}Me_2NC_6H_4CH(OSiMe_3)CMe_2CO_2Me\ (\mathbf{75})\ 97\%$$

(39)

A solution of ketene silyl acetal **74** (146 mg, 0.84 mmol) in DMF (0.8 mL) was added to a solution of lithium pyrrolidone in DMF (0.1 M, 0.6 mL, 0.06 mmol) at −45 °C and a solution of aldehyde (89.5 mg, 0.60 mmol) in DMF (1.6 mL) was then added. The reaction mixture was stirred at the same temperature for 1 h and the reaction was then quenched by adding saturated NH_4Cl. The mixture was extracted with diethyl ether, and the combined organic extracts were washed with brine, dried over Na_2SO_4, filtered, and concentrated. The crude product was purified by preparative TLC (hexane–ethylacetate, 3:1) to give the silylated aldol adduct **75** (187 mg, 96%).

3.3.5
Diastereoselective Synthesis of Polyoxygenated Compounds

Stereoselective aldol reaction of the trimethylsilyl enolate of methyl 2-benzyloxyacetate **76** with enantiomerically pure trialkoxy aldehyde **77** is performed by using three equivalents of $MgBr_2{-}Et_2O$ as activator to afford an aldol adduct **78** in a high yield and excellent diastereoselectivity, whereas conventional Lewis acids such as $TiCl_4$ and $SnCl_4$ give the desired aldol

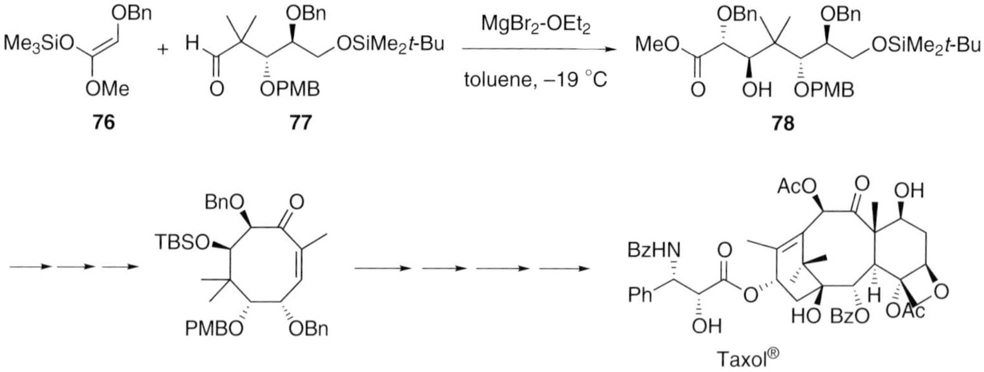

Scheme 3.4
Diastereoselective aldol reaction for preparing an acyclic polyoxy molecule **78**.

product **78** in low yields [59, 60]. This method has been used as a key step in the construction of the B-ring system of an antitumor agent, Taxol (Scheme 3.4) [61].

Diastereoselective Aldol Reaction Mediated by Magnesium Bromide–Diethyl Ether Complex (Eq. (40)) [61c]

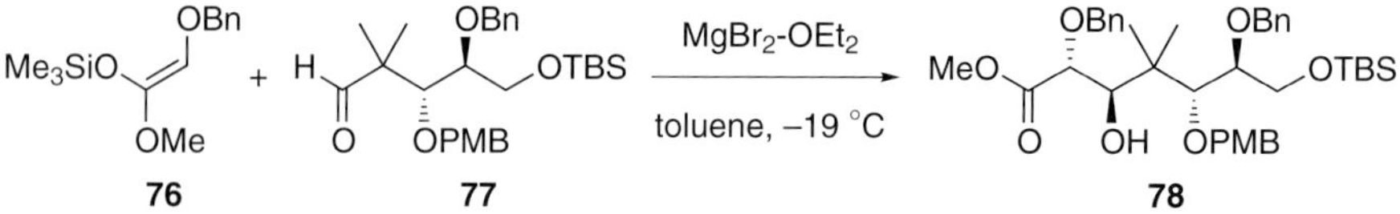

(40)

A solution of ketene trimethylsilyl acetal **76** (6.30 g, 24.9 mmol) in toluene (30 mL) and a solution of aldehyde **77** (8.10 g, 16.6 mmol) in toluene (30 mL) were successively added to a suspension of magnesium bromide–diethyl ether complex (12.9 g, 49.9 mmol) in toluene (100 mL) at −19 °C. The reaction mixture was stirred for 1 h at −19 °C then triethylamine (23 mL) and saturated aqueous $NaHCO_3$ were added. The mixture was extracted with diethyl ether and the combined organic extracts were washed with brine, dried over Na_2SO_4, filtered, and concentrated. The crude product was purified by column chromatography (silica gel, hexane–ethyl acetate, 9:1) to afford the aldol adduct **78** (9.60 g, 87%) as a colorless oil.

3.3.6 Asymmetric Aldol Reactions Using Chiral Tin(II) Lewis Acid Catalysts

The asymmetric aldol reaction is one of the most powerful tools for the construction of new carbon–carbon bonds by controlling the absolute configurations of newly formed chiral centers [62]. Among asymmetric aldol reactions that mediated by silicon enolate has been extensively studied by many research groups over the past two decades.

3.3.6.1 **Stoichiometric Enantioselective Aldol Reaction**

Many aldehydes react with the (*E*) silicon enolate [63] derived from propionic acid thioester **79**, to give *syn* aldol adducts in high yield and with perfect stereochemical control, by combined use of tin(II) triflate, chiral diamine **80**, and dibutyltin acetate (Eq. (41)) [64–66]

RCHO + **79** → $Sn(OTf)_2$ + **80** + *n*-$Bu_2Sn(OAc)_2$, CH_2Cl_2, −78 °C → *syn* + *anti*

70–96% yield
syn/*anti* = 100/0, >98% ee

81
Assumed three-component Promoter

(41)

The formation of an active complex **81** consisting of three components, tin(II) triflate, chiral diamine **80**, and dibutyltin acetate is assumed in these aldol reactions. The three-component complex would activate both aldehyde and silyl enolate (double activation), i.e. the chiral diamine-coordinated tin(II) triflate activates aldehyde while oxygen atoms of the acetoxy groups in dibutyltin acetate interact with the silicon atom of the silicon enolate. Because it has been found that the reaction does not proceed via tin(II) or tin(IV) enolates formed by silicon–metal exchange, silicon enolate is considered to attack the aldehydes directly [65]. The problem of this aldol reaction is that (*Z*) enolates [63] react with aldehydes more slowly, consequently affording the aldols in lower yield and with lower diastereo- and enantioselectivity.

Because optically active molecules containing 1,2-diol units are often observed in nature (e.g. carbohydrates, macrolides, polyethers), asymmetric aldol reaction of the silyl enolate of α-benzyloxythioacetate **82** with aldehydes has been investigated for simultaneous introduction of two vicinal hydroxy groups with stereoselective carbon–carbon bond-formation. It has, interestingly, been found that the *anti*-α,β-dihydroxy thioester derivatives **83** are obtained in high yields with excellent diastereo- and enantioselectivity by combined use of tin(II) triflate, chiral diamine **84**, and dibutyltin acetate (Eq. (42)) [67]. These results are unusual, because aldol reaction of simple silyl enolate **79** with aldehydes generally affords *syn* aldol adducts as mentioned above (Eq. (41)). Consideration of the transition states of these aldol reaction leads us to postulate the coordination of the oxygen atom of the silyl enolate **82** by the tin atom of tin(II) triflate, which is essential for *anti* selectivity.

anti-selective aldol reaction

RCHO + **82** (OSiMe$_3$, SEt, BnO) → [**84** (N-Et), Sn(OTf)$_2$, *n*-Bu$_2$Sn(OAc)$_2$, CH$_2$Cl$_2$, −78 °C] → *syn*-**83** (OH, O, R, SEt, OBn) + *anti*-**83** (OH, O, R, SEt, OBn)

syn/*anti* = 2/98–1/99
95–98% ee (*anti*)

(42)

To examine this hypothesis silicon enolate **85**, which has bulky *tert*-butyldimethylsilyl group, was prepared, to prevent coordination of the α oxygen atom to tin(II). As expected, *syn* aldol **86** is obtained in high stereoselectivity by reaction of the above-mentioned hindered silicon enolate **85**, tin(II) triflate, a chiral diamine **87**, and dibutyltin acetate (Eq. (43)) [68].

syn-selective aldol reaction

RCHO + **85** (OSiMe$_3$, SEt, TBSO) → [**87** (N-Pr), Sn(OTf)$_2$, *n*-Bu$_2$Sn(OAc)$_2$, CH$_2$Cl$_2$, −78 °C] → *syn*-**86** (OH, O, R, SEt, OTBS) + *anti*-**86** (OH, O, R, SEt, OTBS)

syn/*anti* = 88/12–97/3
82–94% ee (*syn*)

(43)

It is, therefore, possible to prepare *syn* and *anti* aldols selectively when the appropriate protecting group is chosen for the alkoxy part of the molecule. This method has been applied to the synthesis of several monosaccharides including branched, deoxy, and amino sugars [69]. One example is shown below (Scheme 3.5) [69c].

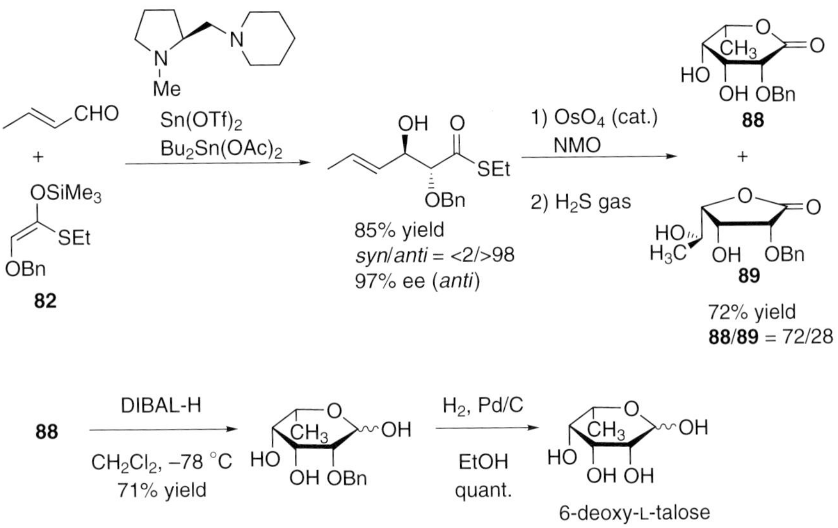

Scheme 3.5
Stereoselective synthesis of 6-deoxy-L-talose.

Typical Procedure for Stoichiometric Enantioselective Aldol Reaction Using a Chiral Tin(II) Catalyst System (Eq. (44)) [67]

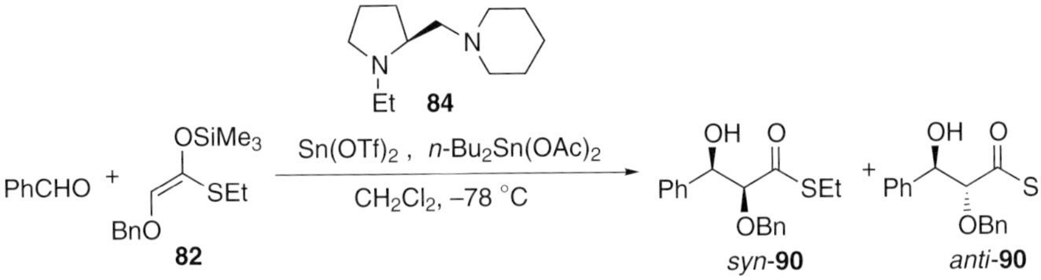

(44)

Dibutyltindiacetate (0.44 mmol) was added at room temperature to a solution of tin(II) triflate (0.4 mmol) and (*S*)-1-ethyl-2-[(piperidin-1-yl)methyl]-pyrrolidine **84** (0.48 mmol) in dichloromethane (1 mL). The mixture was stirred for 30 min then cooled to −78 °C. Dichloromethane solutions (0.5 mL) of the silyl enol ether of *S*-ethyl 2-benzyloxyethanethioate **82** (0.4 mmol) and benzaldehyde (0.27 mmol) were added successively. The reaction mixture was further stirred for 20 h then quenched with aqueous $NaHCO_3$. After the usual work up the desired aldol adduct **90** was obtained in 83% yield (*syn*/*anti* = 1:99, 96% ee (*anti*)).

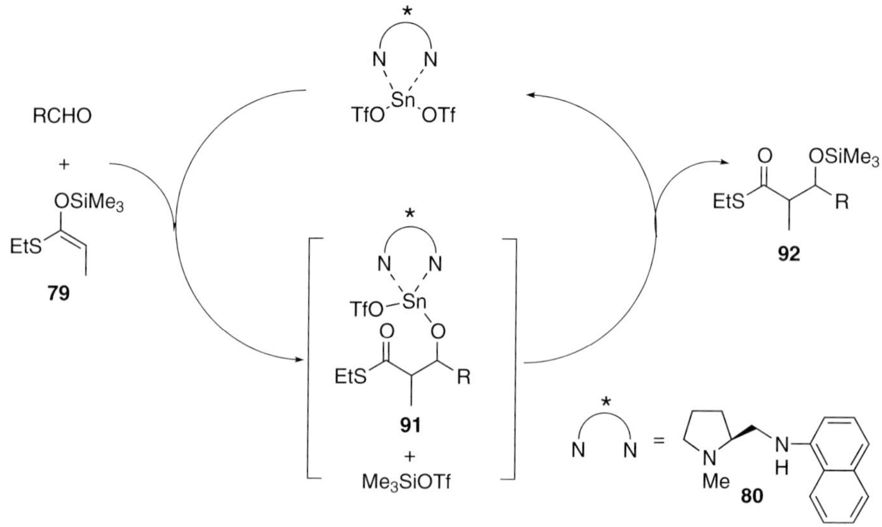

Fig. 3.6
Proposed catalytic cycle.

3.3.6.2 **Catalytic Enantioselective Aldol Reaction**

As described above, optically active aldol adducts are easily obtained by using a stoichiometric amount of chiral diamine, tin(II) triflate, and dibutyltin acetate. To perform the enantioselective aldol reaction by using a catalytic amount of the chiral catalyst, transmetalation of initially formed tin(II) alkoxide **91** to silyl alkoxide **92** with silyl triflate is an essential step (Figure 3.6). When the aldol reaction was conducted simply by reducing the amount of the chiral catalyst, aldol adducts were obtained with low stereoselectivity because Sn–Si exchange occurs slowly and undesired Me_3SiOTf-promoted aldol reaction affords racemic aldol adducts.

To keep the concentration of trimethylsilyl triflate as low as possible during the reaction a dichloromethane solution of silyl enolate and aldehyde was added slowly to the solution of the catalyst **80** (20 mol%), and the aldol product **92** was obtained in good yields with high enantioselectivity (Eq. (44)) [70]. Selectivity is improved by using propionitrile as solvent instead of dichloromethane. The rate of metal exchange between Sn–Si is faster in propionitrile than in dichloromethane [71].

After the first reports of the above-mentioned highly efficient catalytic enantioselective aldol reaction, some groups independently reported catalytic symmetric aldol reactions of silicon enolates with aldehydes using chiral boron [72], titanium [73], zirconium [74], and copper Lewis acids [75], or by transmetalation to chiral Pd(II) enolates [44]. Chiral phosphoramide-promoted aldol reactions of trichlorosilyl enol ethers have been reported as Lewis base-catalyzed asymmetric aldol reactions [76].

Typical Procedure for Catalytic Enantioselective Aldol Reaction Using a Chiral Tin(II) Catalyst System (Eq. (45)) [71]

(45)

(*S*)-1-Methyl-2-[(*N*-naphthylamino)methyl]pyrrolidine **80** (0.088 mmol) in propionitrile (1 mL) was added to a solution of tin(II) triflate (0.08 mmol, 20 mol%) in propionitrile (1 mL). The mixture was cooled to −78 °C and a mixture of silyl ketene acetal **79** (0.44 mmol) and an aldehyde (0.4 mmol) was then added slowly over 3 h. The mixture was further stirred for 2 h, then quenched with saturated aqueous $NaHCO_3$. After the usual work up the aldol-type adduct was isolated as the corresponding trimethylsilyl ether **92**.

References

1 Mukaiyama, T.; Inomata, K.; *Bull. Chem. Soc. Jpn.* **1971**, *44*, 3215. The mechanism shown in this paper was corrected as mentioned in reference 2.

2 (a) Mukaiyama, T.; Inomata, K.; Muraki, M.; *J. Am. Chem. Soc.* **1973**, *95*, 967. (b) Inomata, K.; Muraki, M.; Mukaiyama, T.; *Bull. Chem. Soc. Jpn.* **1973**, *46*, 1807.

3 Review: (a) Mukaiyama, T.; *Org. React.* **1982**, *28*, 203. (b) Cowden, C. J.; Paterson, I.; *Org. React.* **1997**, *51*, 1. (c) Kim, B. M.; Williams, S. F.; Masamune, S.; in *Comprehensive Organic Synthesis*, (B. M. Trost, Ed.), Pergamon Press, London, **1991**. (d) Evans, D. A.; Nelson, J. V.; Taber, T. R.; *Topics Stereochem.* **1982**, *13*, 1.

4 From α-iodoketones and Et_3B: Aoki, Y.; Oshima, K.; Utimoto, K.; *Chem. Lett.* **1995**, 463.

5 From α-iodoketones with 9-BBN and 2,6-lutidine; (a) Mukaiyama, T.; Imachi, S.; Yamane, K.; Mizuta, M.; *Chem. Lett.* **2002**, 698. (b) Mukaiyama, T.; Takuwa, T.; Yamane, K.; Imachi, S.; *Bull. Chem. Soc. Jpn.* **2003**, *76*, 813. Interestingly, some boron enolates were isolated by distillation. See: Hoffmann, R. W.; Ditrich, K.; Froech, S.; *Tetrahedron* **1985**, *41*, 5517, and 5b.

6 (a) Mukaiyama, T.; Inoue, T.; *Chem. Lett.* **1976**, 559. (b) Inoue, T.; Uchimaru, T.; Mukaiyama, T.; *Chem. Lett.* **1977**,

153. (c) Inoue, T.; Mukaiyama, T.; *Bull. Chem. Soc. Jpn.* **1980**, *53*, 174.

7 (a) Masamune, S.; Mori, S.; Van Horn, D. E.; Brooks, D. W.; *Tetrahedron Lett.* **1979**, 1665. (b) Van Horn, D. E.; Masamune, S.; *Tetrahedron Lett.* **1979**, 2229. (c) Evans, D. A.; Cherpeck, R. E.; Tanis, S.; unpublished results.

8 (a) Evans, D. A.; Nelson, J. V.; Vogel, E.; Taber, T. R.; *J. Am. Chem. Soc.* **1981**, *103*, 3099. (b) Evans, D. A.; Vogel, E.; Nelson, J. V.; *J. Am. Chem. Soc.* **1979**, *101*, 6120. (c) Hirama, M.; Masamune, S.; *Tetrahedron Lett.* **1979**, 2225. (d) Hirama, M.; Garvey, D. S.; Lu, L. D.-L.; Masamune, S.; *Tetrahedron Lett.* **1979**, 3937. The degree of (*E*),(*Z*)-selection depends on the structure of carbonyl compounds to some extent.

9 The highest priority designation is assigned to the OBR_2 group concerning enolate substituents.

10 (a) Abiko, A.; Liu, J.-F.; Masamune, S.; *J. Org. Chem.* **1996**, *61*, 2590. (b) Abiko, A.; *J. Synth. Org. Chem. Jpn.* **2003**, *61*, 26. (c) Abiko, A.; *Org. Synth.* **2002**, *79*, 116.

11 (a) Brown, H. C.; Ganesan, K.; Dhar, R.; *J. Org. Chem.* **1993**, *58*, 147. (b) Abiko, A.; *Org. Synth.* **2002**, *79*, 103.

12 Evans, D. A.; Bartroli. J.; Shih, T. L.; *J. Am. Chem. Soc.* **1981**, *103*, 2127.

13 (a) Masamune, S.; Choy, W.; Kerdesky, F. A. J.; Imperiali, B.; *J. Am. Chem. Soc.* **1981**, *103*, 1566. (b) Meyers, A. I.; Yamamoto, Y.; *J. Am. Chem. Soc.* **1981**, *103*, 4278. (c) Meyers, A. I.; Yamamoto, Y.; *Tetrahedron* **1984**, *40*, 2309. (d) Oppolzer, W.; *Pure Appl. Chem.* **1988**, *60*, 39. (e) Oppolzer, W.; Blagg, J.; Rodriguez, I.; *J. Am. Chem. Soc.* **1990**, *112*, 2767.

14 (a) Meyers, A. I.; Yamamoto, Y.; *Tetrahedron* **1984**, *40*, 2309. (b) Masamune, S.; Sato, T.; Kim, B.-M.; Wollman, T. A.; *J. Am. Chem. Soc.* **1986**, *108*, 8279. (c) Reetz, M. T.; Rivadeneira, E.; Niemeyer, C.; *Tetrahedron Lett.* **1990**, *27*, 3863. (d) Corey, E. J.; Kim, S. S.; *J. Am. Chem. Soc.* **1990**, *112*, 4976. (e) Gennari, C.; Hewkin, C. T.; Molinari, F.; Bernardi, A.; Comotti, A.; Goodman, J. M.; Paterson, I.; *J. Org. Chem.* **1992**, *57*, 5173. (f) Gennari, C.; Moresca, D.; Vieth, S.; Vulpeti, A.; *Angew. Chem., Int. Ed. Engl.* **1993**, *32*, 1618.

15 (a) Inoue, T.; Liu, J.-F.; Buske, D. C.; Abiko, A.; *J. Org. Chem.* **2002**, *67*, 5250. (b) Abiko, A.; Liu, J.-F.; Masamune, S.; *J. Am. Chem. Soc.* **1997**, *119*, 2586, and references are cited therein. (c) Abiko, A.; *Org, Synth.* **2002**, *79*, 116. (d) Abiko, A.; *J. Synth. Org. Chem. Jpn.* **2003**, *61*, 24.

16 (a) Walker, M. A.; Heathcock, C. H.; *J. Org. Chem.* **1991**, *56*, 5747. (b) Oppolzer, W.; Lienard, P.; *Tetrahedron Lett.* **1993**, *34*, 4321.

17 These phenomena had already been experienced in the former research of developing new reactions such as new-type 1,4-addition polymerization. (a) Mukaiyama, T.; Fujisawa, T.; Hyugaji, T.; *Bull. Chem. Soc. Jpn.* **1962**, *35*, 687. (b) Mukaiyama, T.; Fujisawa, T.; Nohira, H.; Hyugaji, T.; *J. Org. Chem.* **1962**, *27*, 3337. (c) Mukaiyama, T.; Fujisawa, T.; *Macromol. Synth.* **1972**, *4*, 105. (d) Fujisawa, T.; Tamura, Y.; Mukaiyama, T.; *Bull. Chem. Soc. Jpn.* **1964**, *37*, 793. (e)

Nohira, H.; Nishikawa, Y.; Mukaiyama, T.; *Bull. Chem. Soc. Jpn.* **1964**, *37*, 797. (f) Mukaiyama, T.; Sato, K.; *Bull. Chem. Soc. Jpn.* **1963**, *36*, 99.
18 Mukaiyama, T.; *Angew. Chem. Int. Ed.* **1977**, *16*, 817.
19 Brownbridge, P.; *Synthesis* **1983**, *1*, 85.
20 (a) Mukaiyama, T.; Narasaka, K.; Banno, K.; *Chem. Lett.* **1973**, 1011. (b) Mukaiyama, T.; Banno, K.; Narasaka, K.; *J. Am. Chem. Soc.* **1974**, *96*, 7503.
21 (a) Isler, O.; Schudel, P.; *Adv. Org. Chem.* **1963**, *14*, 115. (b) Effenberger, F.; *Angew. Chem. Int. Ed. Engl.* **1969**, *8*, 295.
22 House, H. O.; *Modern Synthetic Reactions*, 2nd edn, W. A. Benjamin, Menlo Park, **1972**.
23 (a) Banno, K.; Mukaiyama, T.; *Chem. Lett.* **1975**, 741. (b) Banno, K.; Mukaiyama, T.; *Bull. Chem. Soc. Jpn.* **1976**, *49*, 2284.
24 (a) Chan, T. H.; Aida, T.; Lau, P. W. K.; Gorys, V.; Harpp, D. N.; *Tetrahedron Lett.* **1979**, 4029. (b) Dubois, J.-E.; Axiotis, G.; Bertounesque, E.; *Tetrahedron Lett.* **1984**, *25*, 4655. (c) Heathcock, C. H.; Hug, K. T.; Flippin, L. A.; *Tetrahedron Lett.* **1984**, *25*, 5973. (d) Gennari, C.; Beretta, M. G.; Bernardi, A.; Moro, G.; Scolastico, C.; Todeschini, R.; *Tetrahedron* **1986**, *42*, 893.
25 (a) Myers, A. G.; Widdowson, K. L.; *J. Am. Chem. Soc.* **1990**, *112*, 9672. (b) Myers, A. G.; Widdowson, K. L.; Kukkola, P. J.; *J. Am. Chem. Soc.* **1992**, *114*, 2765.
26 Denmark, S. E.; Lee, W.; *J. Org. Chem.* **1994**, *59*, 707.
27 Mukaiyama, T.; Hayashi, M.; *Chem. Lett.* **1974**, 15.
28 (a) Mukaiyama, T.; Ishida, A.; *Chem. Lett.* **1975**, 319. (b) Mukaiyama, T.; Ishida, A.; *Chem. Lett.* **1975**, 1201.
29 Murata, S.; Suzuki, M.; Noyori, R.; *J. Am. Chem. Soc.* **1980**, *102*, 3248.
30 Saigo, K.; Osaki, M.; Mukaiyama, T.; *Chem, Lett.* **1975**, 989.
31 Saigo, K.; Osaki, M.; Mukaiyama, T.; *Chem, Lett.* **1976**, 769.
32 (a) Narasaka, K.; Soai, K.; Mukaiyama, T.; *Chem. Lett.* **1974**, 1223. (b) Narasaka, K.; Soai, K.; Aikawa, Y.; Mukaiyama, T.; *Bull. Chem. Soc. Jpn.* **1976**, *49*, 779. (c) Saigo, K.; Osaki, M.; Mukaiyama, T.; *Chem. Lett.* **1976**, 163. (d) Danishefsky, S.; Vaughan, K.; Gadwood, R. C.; Tsuzuki, K.; *J. Am. Chem. Soc.* **1980**, *102*, 4262. (e) Jung, M. E.; Pan, Y.-G.; *Tetrahedron Lett.* **1980**, *21*, 3127. (f) Jung, M. E.; McCombs, C. A.; Takeda, Y.; Pan, Y.-G.; *J. Am. Chem. Soc.* **1981**, *103*, 6677. (g) Heathcock, C. H.; Norman, M. H.; Uehling, D. E.; *J. Am. Chem. Soc.* **1985**, *107*, 2797.
33 (a) Ishitani, H.; Ueno, M.; Kobayashi, S.; *J. Am. Chem. Soc.* **1997**, *119*, 7153. (b) Kobayashi, S.; Ishitani, H.; Ueno, M.; *J. Am. Chem. Soc.* **1998**, *120*, 431. (c) Ishitani, H.; Ueno, M.; Kobayashi, S.; *J. Am. Chem. Soc.* **2000**, *122*, 8180, and references are cited therein.
34 [4+2]cycloaddition reactions, alkylations, acylations, hydroborations, oxidative processes, etc.
35 (a) Mukaiyama, T.; Kobayashi, S.; Murakami, M.; *Chem. Lett.* **1985**, 447. (b) Kobayashi, S.; Murakami, M.; Mukaiyama, T.; *Chem. Lett.* **1985**, 1535.

36 Ohshima, M.; Murakami, M.; Mukaiyama, Y.; *Chem. Lett.* **1985**, 1871.

37 Iwasawa, N.; Mukaiyama, T.; *Chem. Lett.* **1987**, 463.

38 (a) Murata, S.; Suzuki, M.; Noyori, R.; *Tetrahedron* **1988**, *44*, 4259. (b) Muraki, C.; Hashizume, S.; Nagami, K.; Hanaoka, M.; *Chem. Pharm. Bull.* **1990**, *38*, 1509.

39 (a) Vougioukas, A. E.; Kagan, H. B.; *Tetrahedron Lett.* **1987**, *28*, 5513. (b) Gong, L.; Streitwieser, A.; *J. Org. Chem.* **1990**, *55*, 6235. (c) Kobayashi, S.; Hachiya, I.; Takahori, T.; *Synthesis* **1993**, 371. (d) Kobayashi, S.; Hachiya, I.; Ishitani, H.; Araki, M.; *Synlett* **1993**, 472. (e) Kobayashi, S.; Hachiya, I.; *J. Org. Chem.* **1994**, *59*, 3590.

40 (a) Kobayashi, S.; Nagayama, S.; Busujima, T.; *J. Am. Chem. Soc.* **1998**, *120*, 8287. (b) Kobayashi, S.; Hachiya, I.; *J. Org. Chem.* **1994**, *59*, 3590. (c) Loh, T.-P.; Chua, G.-L.; Vittal, J. J.; Wong, M.-W.; *Chem. Commun.* **1998**, 861. (d) Manabe, K.; Kobayashi, S.; *Synlett* **1999**, 547.

41 (a) Noyori, R.; Yokoyama, K.; Sakata, J.; Kuwajima, I.; Nakamura, E.; *J. Am. Chem. Soc.* **1977**, *99*, 1265. (b) Noyori, R.; Nishida, I.; Sakata, J.; *J. Am. Chem. Soc.* **1981**, *103*, 2106. (c) Kuwajima, I.; Nakamura, E.; *Acc. Chem. Res.* **1985**, *18*, 181. (d) Nakamura, E.; Yamago, S.; Machii, D.; Kuwajima, I.; *Tetrahedron Lett.* **1988**, *29*, 2207. (e) Yamago, S.; Machii, D.; Nakamura, E.; *J. Org. Chem.* **1991**, *56*, 2098.

42 (a) Nakamura, E.; Yamago, S.; Machii, D.; Kuwajima, I.; *Tetrahedron Lett.* **1988**, *29*, 2207. (b) Yamago, S.; Machii, D.; Nakamura, E.; *J. Org. Chem.* **1991**, *56*, 2098.

43 Stork, G.; Hudrlik, P. F.; *J. Am. Chem. Soc.* **1968**, *90*, 4462.

44 (a) Sodeoka, M.; Ohrai, K.; Shibasaki, M.; *J. Org. Chem.* **1995**, *60*, 2648. (b) Sodeoka, M.; Tokunoh, R.; Miyazaki, F.; Hagiwara, E.; Shibasaki, M.; *Synlett* **1997**, 463.

45 (a) Krüger, J.; Carreira, E. M.; *J. Am. Chem. Soc.* **1998**, *120*, 837. (b) Fujimura, O.; *J. Am. Chem. Soc.* **1998**, *120*, 10032.

46 Transmetalation to boron enolates: Wada, M.; *Chem. Lett.* **1981**, 153. Also see ref. 41c.

47 Denmark, S. E.; Stavenger, R. A.; *Acc. Chem. Res.* **2000**, *33*, 432.

48 Kobayashi, S.; Nishido, K.; *J. Org. Chem.* **1993**, *58*, 2647.

49 (a) Myers, A. G.; Kephart, S. E.; Chen, H.; *J. Am. Chem. Soc.* **1992**, *114*, 7922. (b) Denmark, S. E.; Griedel, B. D.; Coe, D. M.; Schnute, M. E.; *J. Am. Chem. Soc.* **1994**, *116*, 7026.

50 Miura, K.; Sato, H.; Tamaki, K.; Ito, H.; Hosomi, A.; *Tetrahedron Lett.* **1998**, *39*, 2585.

51 Myers, A. G.; Widdowson, K. L.; *J. Am. Chem. Soc.* **1990**, *112*, 9672.

52 (a) Creger, P. L.; *Tetrahedron Lett.* **1972**, 79. (b) Kita, Y.; Tamura, O.; Itou, F.; Yasuda, H.; Kishino, H.; Ke, Y. Y.; Tamura, Y.; *J. Org. Chem.* **1988**, *53*, 554.

53 (a) Loh, T.-P.; Feng, L.-C.; Wei, L.-L.; *Tetrahedron*, **2000**, *56*, 7309. (b) Lubineau, A.; *J. Org. Chem.* **1986**, *51*, 2142.

54 Génisson, Y.; Gorrichon, L.; *Tetrahedron Lett.* **2000**, *41*, 4881.

55 Yamamoto, Y.; Maruyama, K.; *J. Am. Chem. Soc.* **1983**, *105*, 6963.

56 MIURA, K.; NAKAGAWA, T.; HOSOMI, A.; *J. Am. Chem. Soc.* **2002**, *124*, 536.

57 (a) FUJISAWA, H.; MUKAIYAMA, T.; *Chem. Lett.* **2002**, 182. (b) FUJISAWA, H.; MUKAIYAMA, T.; *Chem. Lett.* **2002**, 858. (c) MUKAIYAMA, T.; FUJISAWA, H.; NAKAGAWA, T.; *Helv. Chim. Acta* **2002**, *85*, 4518.

58 As another example, 20 mol% of tris(2,4,6-trimethoxyphenyl)phosphine catalyzed the aldol reaction of trimethylsilyl ketene acetals. See: MATSUKAWA, S.; OKANO, N.; IMAMOTO, T.; *Tetrahedron Lett.* **2000**, *41*, 103.

59 FUJISAWA, H.; SASAKI, Y.; MUKAIYAMA, T.; *Chem. Lett.* **2001**, 190.

60 Magnesium halide-mediated aldol reactions: (a) TAKAI, K.; HEATHCOCK, C. H.; *J. Org. Chem.* **1985**, *50*, 3247. (b) UENISHI, J.; TOMOZANE, H.; YAMATO, M.; *Tetrahedron Lett.* **1985**, *26*, 3467. (c) BERNARDI, A.; CARDANI, S.; COLONBO, L.; POLI, G.; SCHIMPERNA, G.; SCOLASTICO, C.; *J. Org. Chem.* **1987**, *52*, 888. (d) COREY, E. J.; LI, W.; REICHARD, G. A.; *J. Am. Chem. Soc.* **1998**, *120*, 2330.

61 (a) MUKAIYAMA, T.; SHIINA, I.; IWADARE, H.; SAKOH, H.; TANI, Y.; HASAGAWA, M.; SAITOU, K.; *Proc. Jpn. Acad.* **1997**, *73B*, 95. (b) MUKAIYAMA, T.; SHIINA, I.; IWADARE, H.; SAITOH, M.; NISHIMURA, T.; OHKAWA, N.; SAKOH, H.; NISHIMURA, K.; TANI, Y.; HASEGAWA, M.; YAMADA, K.; SAITOH, K.; *Chem. Eur. J.* **1999**, *5*, 121. (c) SHIINA, I.; SHIBATA, J.; IBUKA, R.; IMAI, Y.; MUKAIYAMA, T.; *Bull. Chem. Soc. Jpn.* **2001**, *74*, 113.

62 Review: (a) CARREIRA, E. M.; in *Comprehensive Asymmetric Catalysis*, (JACOBSEN, E. N.; PFALTZ, A.; YAMAMOTO, H.; eds), Springer, Heidelberg, **1999**, Vol. *3*, p. 998. (b) MAHRWALD, R.; *Chem. Rev.* **1999**, *99*, 1095. (c) GRÖGER, H.; VOGL, E. M.; SHIBASAKI, M.; *Chem. Eur. J.* **1998**, *4*, 1137. (d) NELSON, S. G.; *Tetrahedron: Asymmetry* **1998**, *9*, 357. (e) BACH, T.; *Angew. Chem. Int. Ed. Engl.* **1994**, *33*, 417.

63 The highest priority is assigned to $OSiR_3$ group concerning enolate substituents.

64 KOBAYASHI, S.; MUKAIYAMA, T.; *Chem. Lett.* **1989**, 297.

65 KOBAYASHI, S.; UCHIRO, H.; FUJISHITA, Y.; SHIINA, I.; MUKAIYAMA, T.; *J. Am. Chem. Soc.* **1991**, *113*, 4247.

66 MUKAIYAMA, T.; KOBAYASHI, S.; *J. Organomet. Chem.* **1990**, *382*, 39.

67 MUKAIYAMA, T.; UCHIRO, H.; SHIINA, I.; KOBAYASHI, S.; *Chem. Lett.* **1990**, 1019.

68 MUKAIYAMA, T.; SHIINA, I.; KOBAYASHI, S.; *Chem Lett.* **1991**, 1902.

69 (a) KOBAYSHI, S.; ONOZAWA, S.; MUKAIYAMA, T.; *Chem. Lett.* **1992**, 2419. (b) MUKAIYAMA, T.; ANAN, H.; SHIINA, I.; KOBAYAHI, S.; *Bull. Soc. Chim. Fr.* **1993**, *130*, 388. (c) MUKAIYAMA, T.; SHIINA, I.; KOBAYASHI, S.; *Chem, Lett.* **1990**, 2201.

70 MUKAIYAMA, T.; KOBAYASHI, S.; UCHIRO, H.; SHIINA, I.; *Chem. Lett.* **1990**, 129.

71 KOBAYASHI, S.; FUJISHITA, Y.; MUKAIYAMA, T.; *Chem. Lett.* **1990**, 1455.

72 (a) FURUTA, K.; MARUYAMA, T.; YAMAMOTO, H.; *J. Am. Chem.*

Soc. **1991**, *113*, 1041. (b) Parmee, E. R.; Tempkin, O.; Masamune, S.; Abiko, A.; *J. Am. Chem. Soc.* **1991**, *113*, 9365. (c) Furuta, K.; Maruyama, T.; Yamamoto, H.; *Synlett* **1991**, 439. (d) Kiyooka, S.; Kaneko, Y.; Kume, K.; *Tetrahedron Lett.* **1992**, *33*, 4927.

73 (a) Mikami, K.; Matsukawa, M.; *J. Am. Chem. Soc.* **1994**, *116*, 4077. (b) Mikami, K.; Matsukawa, M.; *J. Am. Chem. Soc.* **1993**, *115*, 7039. (c) Mikami, K.; *Pure Appl. Chem.* **1996**, *68*, 639. (d) Keck, G. E.; Krishnamurthy, D.; *J. Am. Chem. Soc.* **1995**, *117*, 2363. (e) Carreira, E. M.; Singer, R. A.; Lee, W.; *J. Am. Chem. Soc.* **1994**, *116*, 8837. (f) Singer, R. A.; Carreira, E. M.; *J. Am. Chem. Soc.* **1995**, *117*, 12360. (g) Carreira, E. M.; Singer, R. A.; Lee, W.; *J. Am. Chem. Soc.* **1995**, *117*, 3649.

74 Ishitani, H.; Yamashita, H.; Shimizu, H.; Kobayashi, S.; *J. Am. Chem. Soc.* **2000**, *122*, 5403.

75 Evans, D. A.; Murry, J. A.; Kozlowski, M. C.; *J. Am. Chem. Soc.* **1996**, *118*, 5814. (b) Evans, D. A.; Kozlowski, M. C.; Tedrow, J. S.; *Tetrahedron Lett.* **1997**, *42*, 7481. (c) Evans, D. A.; Kozlowski, M. C.; Burgey, C. S.; MacMillan, D. W. C.; *J. Am. Chem. Soc.* **1997**, *119*, 7893. (d) Evans, D. A.; MacMillan, D. W. C.; Campos, K. R.; *J. Am. Chem. Soc.* **1997**, *119*, 10859.

76 Denmark, S. E.; Winter, S. B. D.; Su, X.; Wong, K.-T.; *J. A. Chem. Soc.* **1996**, *118*, 7404. (b) Denmark, S. E.; Wong, K.-T.; Stavenger, R. A.; *J. Am. Chem. Soc.* **1997**, *119*, 2333.

4
Amine-catalyzed Aldol Reactions

Benjamin List

4.1
Introduction

Aldolizations can be catalyzed by both Lewis and Brønsted acids and bases [1]. This catalytic diversity is possible because aldol reactions combine a nucleophilic addition, which is acid-catalyzed, with an enolization, which is catalyzed by both acids and bases (Scheme 4.1).

Nature's aldolases use combinations of acids and bases in their active sites to accomplish direct asymmetric aldolization of unmodified carbonyl compounds. Aldolases are distinguished by their enolization mode – Class I aldolases use the Lewis base catalysis of a primary amino group and Class II aldolases use the Lewis acid catalysis of a Zinc(II) cofactor. To accomplish enolization under essentially neutral, aqueous conditions, these enzymes decrease the pK_a of the carbonyl donor (typically a ketone) by converting it into a cationic species, either an iminium ion (**5**) or an oxonium ion (**8**). A relatively weak Brønsted base co-catalyst then generates the nucleophilic species, an enamine- (**6**) or a zinc enolate (**9**), via deprotonation (Scheme 4.2).

Although chemists also use acids and bases to catalyze aldolizations, the aldolase-like direct catalytic asymmetric aldol reaction remained an illusive challenge for a long time. Indirect aldol reactions utilizing preformed enolate equivalents provided the only viable strategy for catalytic asymmetric aldol synthesis [2–10]. Among the first purely chemical direct asymmetric aldol catalysts were bifunctional Lewis acid-Brønsted base metal complexes that resemble class II aldolases [11–17]. Amino acids that mimic the enamine catalysis of class I aldolases have also been studied. Initially, these efforts have concentrated on proline-catalyzed enantiogroup differentiating intramolecular aldolizations [18–28]. More recently the first direct amine-catalyzed asymmetric intermolecular aldol reactions have been described [29–33]. This chapter reviews both asymmetric and non-asymmetric aminocatalytic intra- and intermolecular aldolizations. Not included here are

Modern Aldol Reactions. Vol. 1: Enolates, Organocatalysis, Biocatalysis and Natural Product Synthesis.
Edited by Rainer Mahrwald

ISBN: 3-527-30714-1

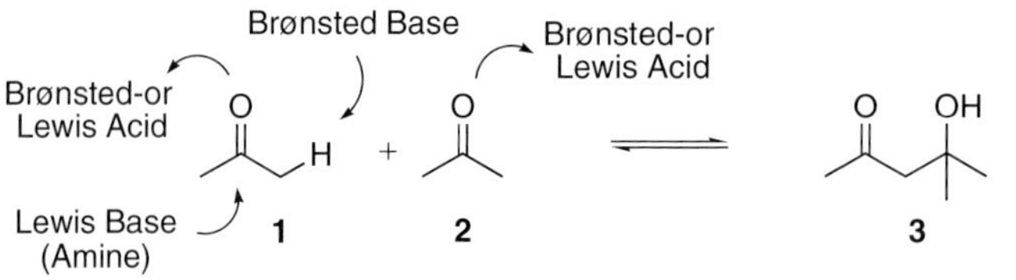

Scheme 4.1
Catalysis of the aldol reaction.

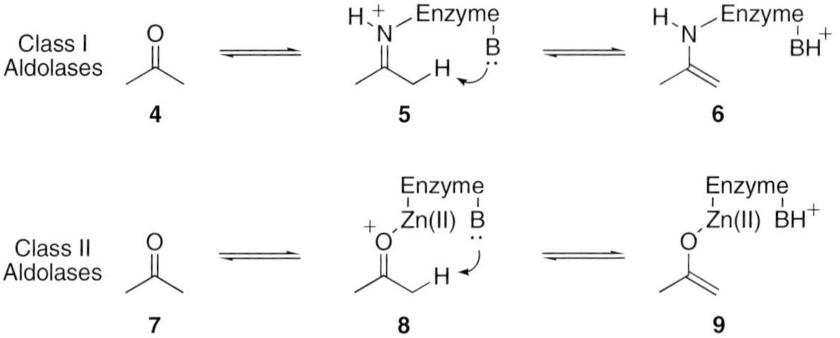

Scheme 4.2
Two enzymatic strategies for enolization of carbonyl compounds.

other interesting organocatalysts of the aldol reaction, for example simple tertiary amine Brønsted bases and phase-transfer type quaternary ammonium salts [34–36]. Mechanistically related class I aldolases and catalytic antibodies have been fundamentally important in the development of the reactions described below and are discussed in another chapter of this book.

4.2
Aminocatalysis of the Aldol Reaction

Although aminocatalysis of the aldol reaction via enamine intermediates is an important enzymatic strategy and several bioorganic studies of the subject have appeared, applications in preparative organic synthesis, particularly in intermolecular aldol addition reactions, have been published only sporadically. Despite the often-used Mukaiyama-aldol reaction of enol ethers and Stork's well-developed enamine chemistry [37, 38], aldolizations of *preformed* enamines are rare. One report describes Lewis acid-catalyzed aldolizations of preformed enamines with aldehydes that furnish aldol addition products [39]. Aldol *condensation* reactions of preformed enamines with aldehydes have also been described [40]. Only enamine-catalytic aldolizations, which are primary and secondary amine-catalyzed aldol reactions, will be discussed in this chapter, however.

2 H–C(=O)–CH3 **10** —(amine, aq. buffer)→ **11** + **12** (1)

13 —(amine, aq. buffer)→ 2 **14** (2)

14 + **15** —(amine, aq. buffer)→ **16** (3)

Scheme 4.3
Some amine-catalyzed aldolizations in aqueous buffers.

4.2.1 Intermolecular Aldolizations

Class I aldolase-like catalysis of the intermolecular aldol reaction with amines and amino acids in *aqueous solution* has been studied sporadically throughout the last century. Fischer and Marschall showed in 1931 that alanine and a few primary and secondary amines in neutral, buffered aqueous solutions catalyze the self-aldolization of acetaldehyde to give "aldol" (**11**) and crotonaldehyde (**12**) (Scheme 4.3, Eq. (1)) [41]. In 1941 Langenbeck et al. found that secondary amino acids such as sarcosine also catalyze this reaction [42]. Independently, Westheimer et al. and other groups showed that amines, amino acids, and certain diamines catalyze the retro-aldolization of "diacetone alcohol" (**13**) and other aldols (Scheme 4.3, Eq. (2)) [43–47]. More recently Reymond et al. [48] studied the aqueous amine catalysis of cross-aldolizations of acetone with aliphatic aldehydes furnishing aldols **16** (Scheme 4.3, Eq. (3)) and obtained direct kinetic evidence for the involvement of enamine intermediates.

It is believed that amine-catalyzed aldolizations, in a manner similar to class I aldolase-catalyzed reactions, proceed via a catalytic cycle that involves formation of (**A**) carbinolamine-, (**B**) iminium ion-, and (**C**) enamine intermediates (Scheme 4.4). Either the enamine generation or the subsequent C–C-bond-forming step (**D**) are rate limiting. Water addition to give a new carbinolamine (**E**) and its fragmentation (**F**) close the catalytic cycle.

Houk and Bahmanyar have used density functional theory calculations to investigate transition states of the C–C-bond-forming step in amine-catalyzed aldolizations (Scheme 4.5) [49, 50]. Accordingly, the cyclic transition state of the primary amine-catalyzed aldol reaction has a half-chair conformation featuring N–H–O hydrogen-bonding for charge stabilization (**A**). In secondary amine-catalyzed aldol reactions the calculated zwitterionic transition state (**B**) is somewhat higher in energy and leads to an oxetane

Scheme 4.4
The enamine catalytic cycle for aldolizations.

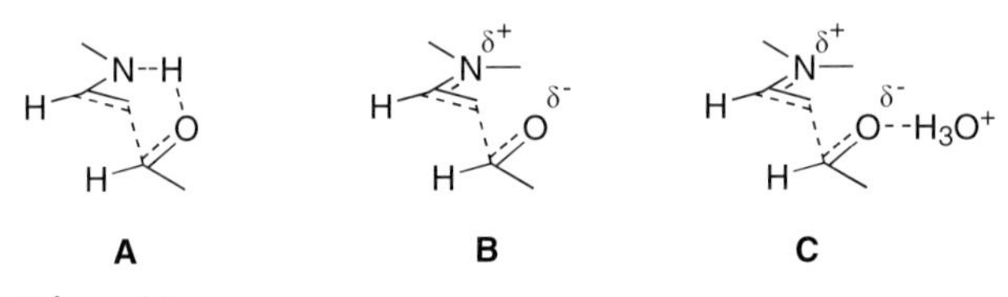

Scheme 4.5
DFT calculation-based transition states for primary (**A**) and secondary (**B**, **C**) amine-catalyzed aldolizations.

intermediate in an overall [2 + 2] cycloaddition reaction. When an external proton source was included (**C**), the reaction was found to have no barrier and led directly to the iminium ion intermediate (Scheme 4.5).

4.2.1.1 Aldehyde Donors

Despite these mechanistic and theoretical studies, intermolecular amine-catalyzed aldolizations have only rarely been used on a preparative scale. A few noteworthy exceptions in which aldehydes are used as donors are shown in Scheme 4.6 [51–55]. These reactions are often performed neat or in the presence of small amounts of an organic solvent. The catalyst usually used is either a primary or secondary amine, a combination of an amine with a carboxylic acid, or simply an amino acid. These catalyst systems have previously been used in the Knoevenagel condensation and it is apparent that synthetic amine-catalyzed aldolizations originate from Knoevenagel's chemistry [56].

A Mannich-type condensation mechanism involving an iminium ion electrophile similar to the aminocatalytic Knoevenagel reaction has recently been proposed for the amine-catalyzed self-aldolization of propionaldehyde (Eq. (6)) [55]. Although this mechanism is not unreasonable it should be

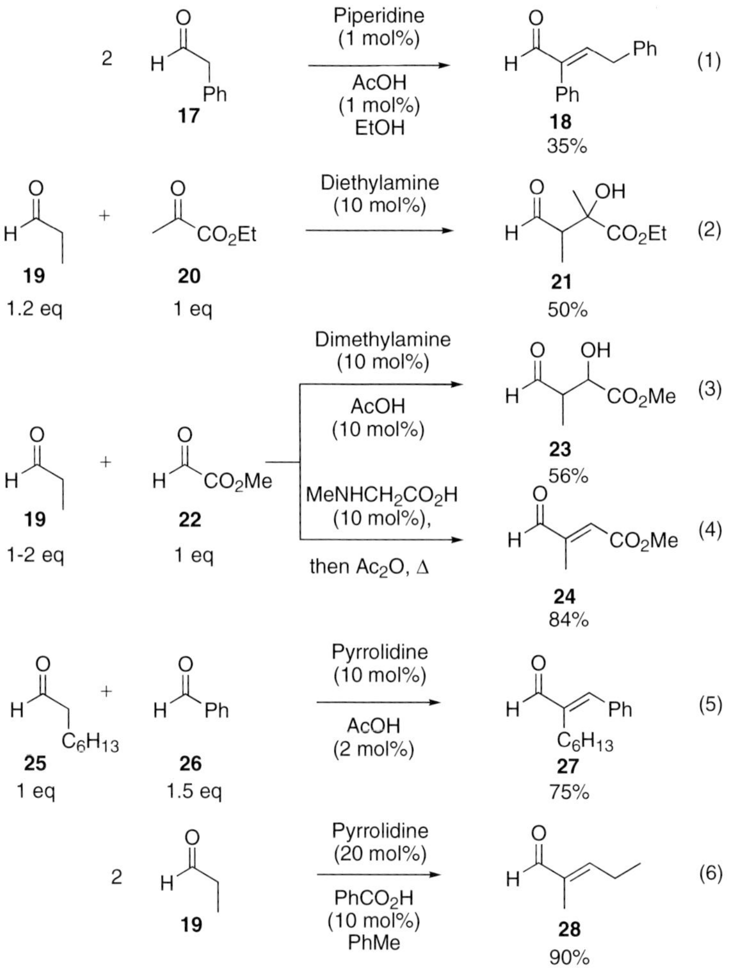

Scheme 4.6
Preparative amine-catalyzed intermolecular aldolizations with aldehyde donors.

noted that aldol addition products are often isolated in such reactions (e.g. Eqs. (2)–(4)). Obviously, a Mannich-type mechanism has to be excluded in such cases. In general both, the iminium ion Mannich-type mechanism and the aldolase-type enamine mechanism are plausible and seem to be energetically comparable. Which of these two mechanisms actually operates in a given reaction depends on the donor component: In Knoevenagel-type reactions where easily enolized low-pK_a malonate-type donors are used, primary and secondary amine catalysis most probably involves activation of the carbonyl acceptor as an iminium ion. In amine-catalyzed aldol reactions of aldehyde and ketone donors, which have higher pK_a (i.e. acetone p$K_a = 20$, propionaldehyde p$K_a = 17$), activation of the donor as an enamine predominates.

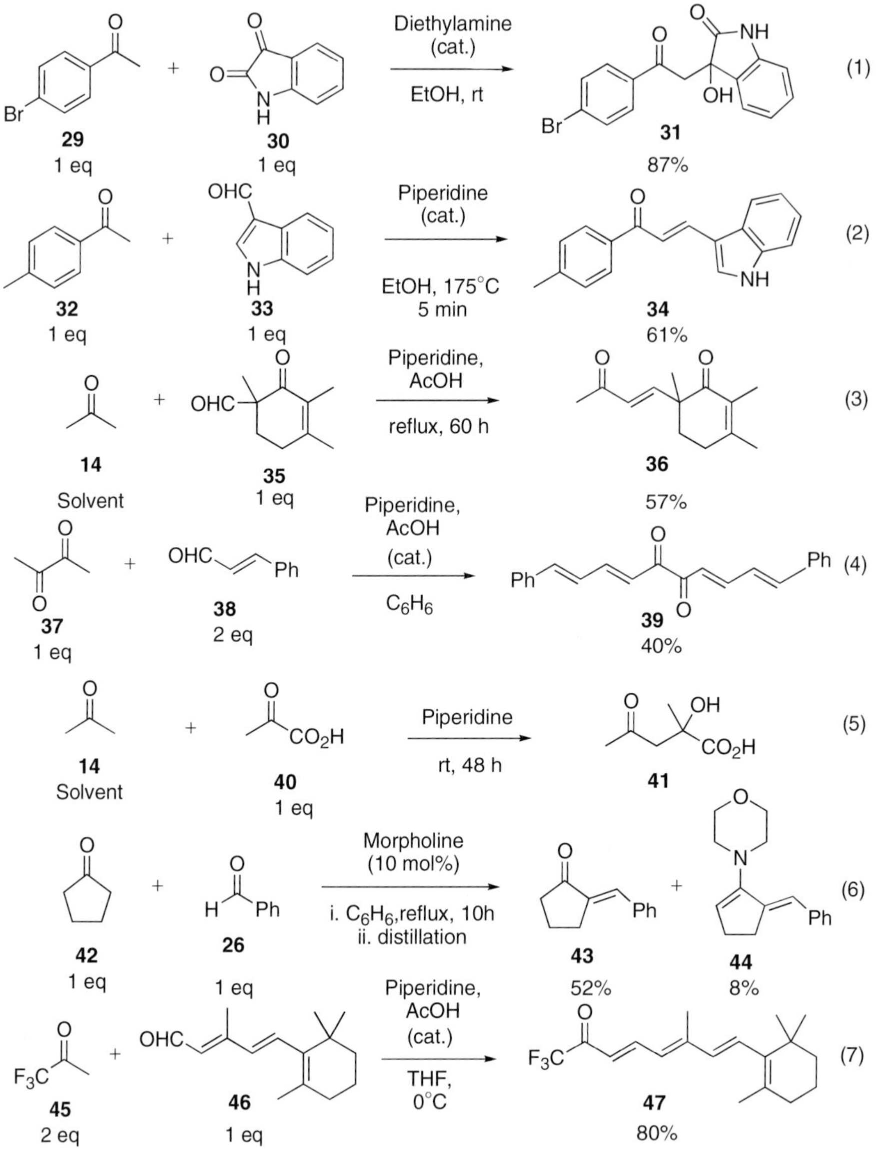

Scheme 4.7
Some secondary amine-catalyzed intermolecular aldolizations with ketone donors.

4.2.1.2 **Ketone Donors**

Ketone donors have only rarely been used in amine-catalyzed aldolizations. Selected examples that have been published over the last 70 years are shown in Scheme 4.7 [57–68].

The aminocatalysts used are the same that are also used when aldehyde-donors are employed. Aldehydes and ketones can be used as acceptors and

Enolendo-Aldolization (1)

48 → 49

Enolexo-Aldolization (2)

50 → 51

Scheme 4.8
The two modes of intramolecular aldolization.

aldol addition and condensation products have been isolated. Apparently, α-unbranched aliphatic aldehydes have never been used as acceptors in non-asymmetric reactions with ketone donors.

Morpholine-catalyzed Aldolization of Cyclopentanone with Benzaldehyde to Give Enone 43 [40]. Cyclopentanone (37 g, 0.5 mol), morpholine (4.4 g, 0.05 mol, 10 mol%), and benzaldehyde (53 g, 0.5 mol) were heated under reflux in dry benzene (100 mL) for 10 h using a Dean–Stark trap. The resulting crude product was distilled to furnish enone **43** (45 g, 52%, $Bp._{0.05}$ = 115–120 °C).

4.2.2
Intramolecular Aldolizations

There are two types of intramolecular aldol reaction, enolendo and enolexo aldolizations (Scheme 4.8, Eqs. (1) and (2), respectively). Both types can be catalyzed by amines and several examples have been published.

4.2.2.1 Enolexo Aldolizations

One of the earliest aminocatalytic intramolecular aldolizations was applied in Woodward's 1952 total synthesis of steroids [69]. A 5-enolexo aldol condensation established the D-ring of the steroid skeleton. Hexanedial **52** on treatment with a catalytic amount of piperidinium acetate gave cyclopentenal **53** (Scheme 4.9).

52 → **53**: Piperidine (cat.), AcOH (cat.), C_6H_6, 1h, 60 °C; 66%

Scheme 4.9
An intramolecular amine-catalyzed aldolization in Woodward's total synthesis of steroids.

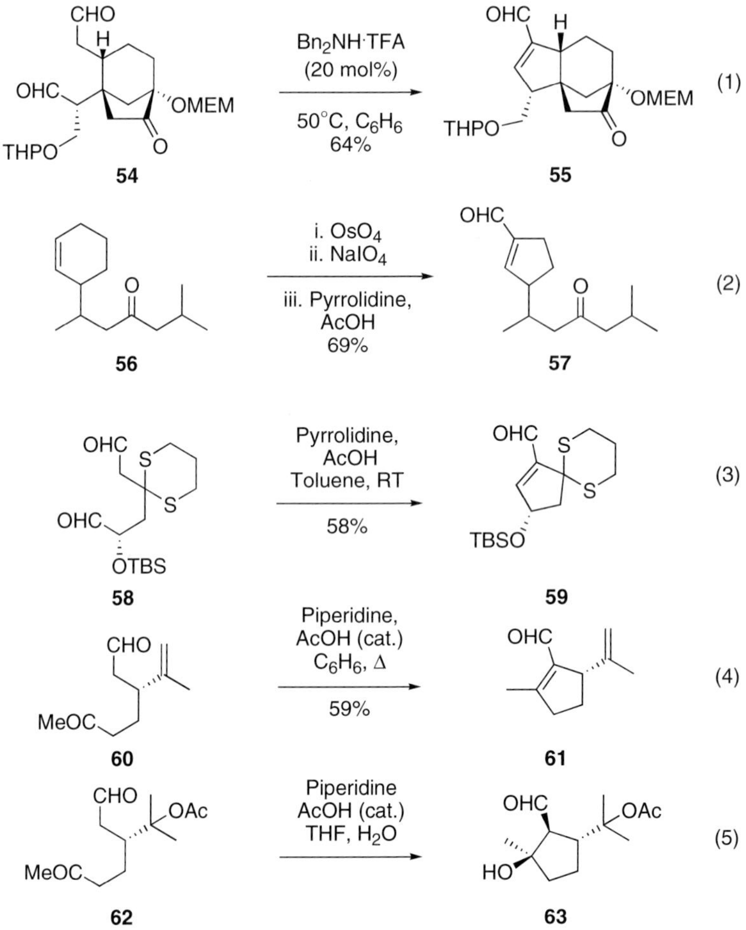

Scheme 4.10
Amine-catalyzed 5-enolexo aldolizations.

Since then amine-catalyzed 5-enolexo aldolizations have been applied in a variety of other synthetic contexts (Scheme 4.10) [70–75]. Typically, they constitute part of a strategy for the ring contraction of cyclohexenes to cyclopentenes via oxidative cleavage and aldolization (Eq. (2)).

As in intermolecular aldol reactions, amine-catalyzed 5-enolexo aldolizations can either lead to aldol condensation (Eqs. (1)–(4)) or addition products (Eq. (5)). These reactions usually involve an aldehyde group as the aldol donor; the corresponding amine-catalyzed 5-enolexo aldolizations of ketone donors have so far not been realized. The acceptor carbonyl group can either be an aldehyde or a ketone.

The regioselectivity of aminocatalytic aldolizations of ketoaldehydes such as **60** and **62** contrasts with that of the corresponding Brønsted base-

Piperidine, AcOH(cat.) C_6H_6, Δ

aq. KOH, Et_2O, rt

64 65 (1) 66 (2)

Scheme 4.11
Contrasting regioselectivity of amine- and base-catalyzed 5-enolexo aldolizations.

catalyzed processes. Similarly, ketoaldehyde **64** on treatment with piperidinium acetate in benzene gave only enal **65**, whereas treatment with aq. KOH in ether furnished mainly its regioisomer **66** (Scheme 4.11) [76].

The regioselectivity of the amine-catalyzed process might reflect a high aldehyde enamine (**67**) to ketone enamine (**68**) ratio, because of steric hindrance ($A^{1,3}$-strain) in the ketone enamine (Scheme 4.12). In the base-

Scheme 4.12
Rationalization of the contrasting regioselectivity of amine- and base-catalyzed aldolizations.

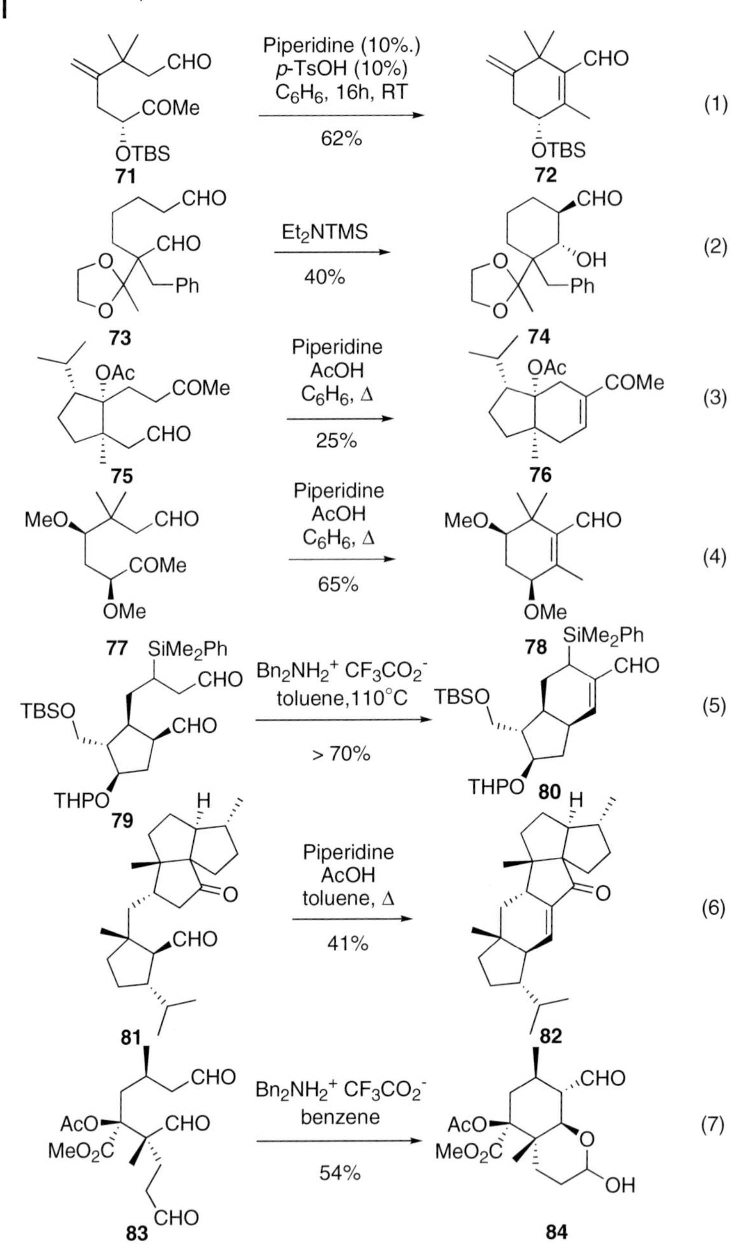

Scheme 4.13
Amine-catalyzed 6-enolexo aldolizations.

Scheme 4.14
Amine-catalyzed 6-enolendo aldolizations in the synthesis of Hagemann's ester (**89**) and aldol **91**.

catalyzed reaction the more stable and sterically relatively unhindered ketone enolate (**69**) leading to the opposite regioisomer dominates and, in addition, its aldolization might be faster than that of the less nucleophilic aldehyde enolate (**70**) (Scheme 4.12).

6-Enolexo aldolizations giving aldol addition or condensation products have also been described, examples are shown in Scheme 4.13 [77–82].

Typically, aldehydes function as the aldol donor in these reactions but ketones can also be used (Eq. (6)). Both ketones and aldehydes can play the role of the acceptor carbonyl group.

4.2.2.2 Enolendo Aldolizations

More than the corresponding enolexo-aldolizations, amine-catalyzed 6-enol*endo* aldol reactions have frequently been used synthetically. An early example is the synthesis of Hageman's ester (**89**) from formaldehyde and acetoacetate via piperidine-catalyzed Knoevenagel–Michael–aldol tandem-reaction followed by decarboethoxylation (Scheme 4.14, Eq. (1)) [83]. Related product **91** was obtained when thiophene-carbaldehyde **90** was used (Eq. (2)) [84].

Piperidine-Catalyzed Synthesis of Aldol 91. A mixture of thiophene-2-carbaldehyde (**90**, 8 g, 71 mmol) and ethyl acetoacetate (**85**, 20 g, 154 mmol, 2.1 equiv.) was cooled to 0 °C and treated with five drops of piperidine under stirring. After 48 h at room temperature a solid was formed, isolated by filtration, and recrystallized twice from diethyl ether to furnish aldol **91** (fine colorless needles Mp. = 106 °C) in almost quantitative yield.

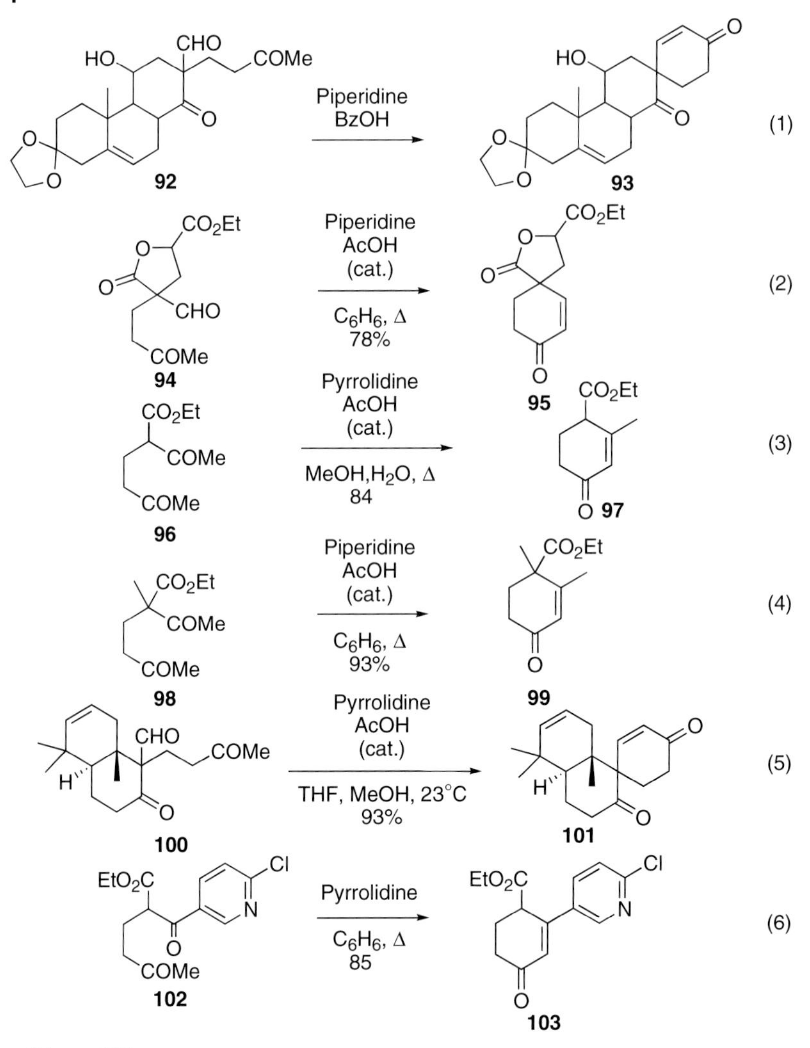

Scheme 4.15
Selected amine-catalyzed 6-enolendo aldolizations.

Other examples of amine-catalyzed 6-enolendo aldolizations are shown in Scheme 4.15 [85–90].

In enolendo aldolizations the aldol donor is always a ketone whereas both aldehyde and ketone carbonyl groups can serve as the acceptor.

Probably the most important aminocatalytic 6-enolendo aldolizations are used in the syntheses of bicyclic ketones **106** and **109** (the Wieland–Miescher ketone) from triketones **104** and **108** via pyrrolidine-catalyzed cycloaldolization (Scheme 4.16) [91–93]. Enones **106** and **109** are important intermediates in natural product synthesis, particularly as AB- and CD-fragments in the synthesis of steroids. It has been convincingly demon-

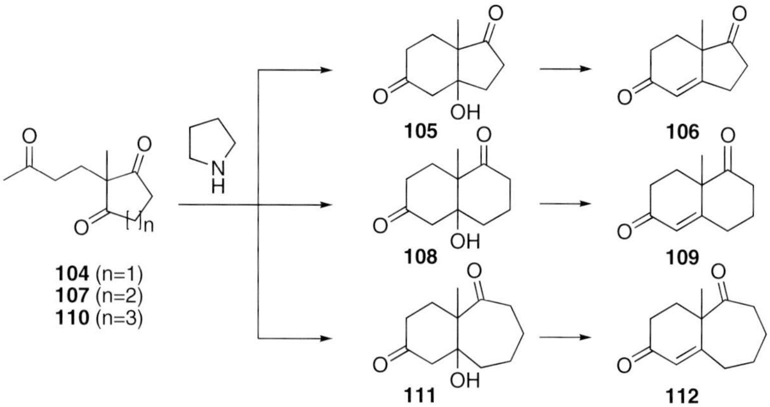

Scheme 4.16
Pyrrolidine-catalyzed 6-enolendo aldolizations.

strated by Spencer et al. that these reactions involve enamine intermediates [94–97].

4.3
Asymmetric Aminocatalysis of the Aldol Reaction

The first example of an aminocatalytic asymmetric aldol reaction was the Hajos–Parrish–Eder–Sauer–Wiechert cyclization, a proline-catalyzed enantiogroup-differentiating 6-enolendo aldolization of some di- and triketones [18–21, 27, 28]. Discovered in the early 1970s this reaction was the first example of a highly enantioselective organocatalytic process. Although the significance of the Hajos–Parrish–Eder–Sauer–Wiechert reaction was recognized almost from the beginning, it was another 30 years until the first proline-catalyzed intermolecular aldol reactions were described [29–33].

4.3.1
Intramolecular Aldolizations

4.3.1.1 Enolendo Aldolizations

Hajos and Parrish at Hoffmann La Roche discovered that proline-catalyzed intramolecular aldol reactions of triketones such as **104** and **107** furnish aldols **105** and **108** in good yields and with high enantioselectivity (Scheme 4.17). Acid-catalyzed dehydration of the aldol addition products then gave condensation products **106** and **109** (Eqs. (1) and (2)). Independently, Eder, Sauer, and Wiechert at Schering AG in Germany directly isolated the aldol condensation products when the same cyclizations were conducted in the presence of proline (10–200 mol%) and an acid co-catalyst (Eqs. (3) and (4)).

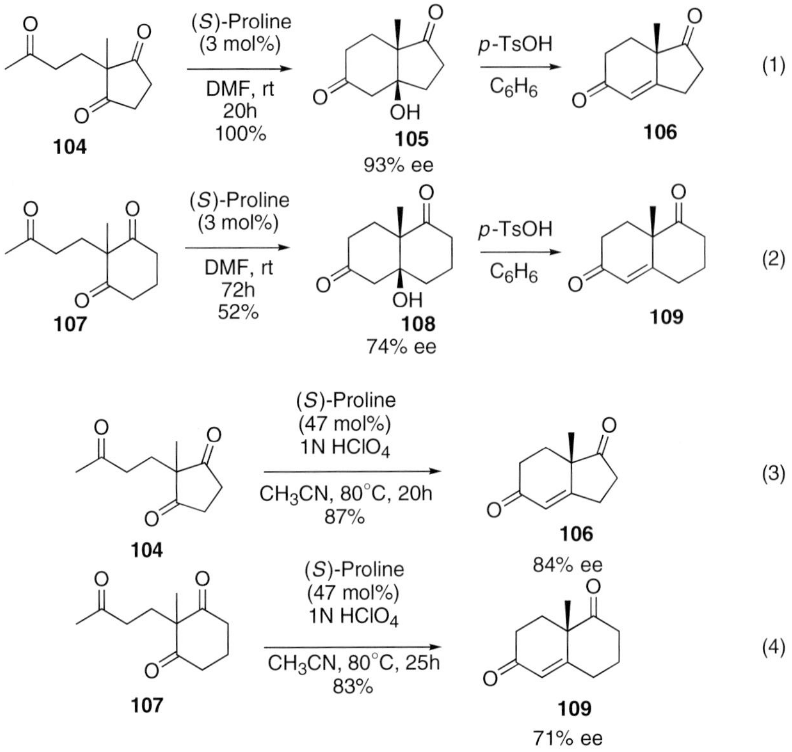

Scheme 4.17
Hajos–Parrish–Eder–Sauer–Wiechert reactions.

A proline-catalyzed cyclization of ketone **107** to give enone **109** directly without added acid was described by Fürst et al. [28].

Proline-Catalyzed Cyclization of Triketone 107 to Give the Wieland–Miescher Ketone (109) [28]. Crude oily trione **107** (ca. 2 mol) in DMSO (2 L) was treated with (*S*)-proline (11.5 g, 0.1 mol, 5 mol%) and the resulting mixture was stirred under argon at 18 °C for 5 days. The solvent was removed at 70 °C (0.1 torr) and the resulting dark green oil (420 g) was adsorbed on a column charged with silica (3 kg) Elution with 17% ethyl acetate in hexanes afforded 276 g product which was further purified by distillation (b.p. 108–112 °C/0.25 torr, bath temp. 150 °C) to give the enantiomerically enriched ketone **109** (255 g, 1.43 mol, 72%) in 70% ee ($[\alpha]_D = +70°$).

Proline-catalyzed enolendo aldolizations have been applied to a number of substrates, most often in steroid synthesis. Selected products from such Hajos–Parrish–Eder–Sauer–Wiechert reactions are shown in Scheme 4.18 [98–104].

Although several different catalysts have been studied in such enolendo aldolizations, proline has typically been preferred. It can, however, be advan-

113 89%, 90% ee

114 71%, >99% ee

115 76%, 81% ee

116 80%, 27% ee

117 35% (ee n.d.)

118 47%, 16% ee

119 60%, 57% ee

120 64%, 90% ee

121 49%, 34% ee

122 64%, 90% ee

Scheme 4.18
Selected products from proline-catalyzed 6-enolendo aldolizations.

123 → **124**

(S)-Proline, 1N $HClO_4$, CH_3CN, 80°C, 10d, 67% — 27% ee (1)

(S)-Phenylalanine, 1N $HClO_4$, CH_3CN, 80°C, 40h, 82% — 86% ee (2)

125 → **126**

(S)-Proline, DMSO, 65°C, 5d — 32% ee (3)

(S)-Phenylalanine, 1N $HClO_4$, CH_3CN, 80°C, 40h — 95% ee (4)

Scheme 4.19
Compared with proline, phenylalanine is a superior catalyst for some intramolecular aldolizations.

tageous to use primary amino acid catalysts such as phenylalanine, particularly when non-methyl ketones are employed. For example, Danishefsky et al. found that the proline-catalyzed cyclization of triketone **123** furnished product **124** in 27% ee whereas 86% ee was obtained when phenylalanine was used as the catalyst (Scheme 4.19, Eqs. (1) and (2)) [105]. Agami et al.

R	ee %
Ph	47
n-C_5H_{11}	20
Me	42
i-Pr	8
t-Bu	0

Scheme 4.20
Enantiogroup-differentiating aldol cyclodehydrations of 4-substituted 2,6-heptanedions **127**.

Scheme 4.21
Mechanistic models proposed for the Hajos–Parrish–Eder–Sauer–Wiechert reaction.

made similar observations in the cyclization of ketone **125** (Eqs. (3) and (4)) [106].

Attempts have been made to expand the scope of the Hajos–Parrish–Eder–Sauer–Wiechert reaction to an enantiogroup differentiating aldolization of acyclic 4-substituted 2,6-heptandiones **127** to give cyclohexenones **128** (Scheme 4.20) [107, 108]. The yield and enantioselectivity of this reaction are relatively modest compared with the parent cyclo-aldolization.

Several different mechanisms have been proposed for the Hajos–Parrish–Eder–Sauer–Wiechert reaction (Scheme 4.21). These include the original Hajos model (**A**), which assumes activation of the acceptor carbonyl as a carbinolamine [21]; the Agami model (**B**), in which a side-chain enamine reacts with a ring carbonyl group, mediated by a second proline molecule [109–114]; Swaminathan's model (**C**), proposing a heterogeneous mechanism [115]; and the Houk model (**D**), in which the side-chain enamine reacts with the ring acceptor carbonyl group, under concomitant activation via hydrogen-bonding to proline's carboxylic acid group [25, 116, 117].

Model D has recently been supported experimentally [118]. It was shown that, in contrast to earlier reports [119], there are no non-linear (Scheme

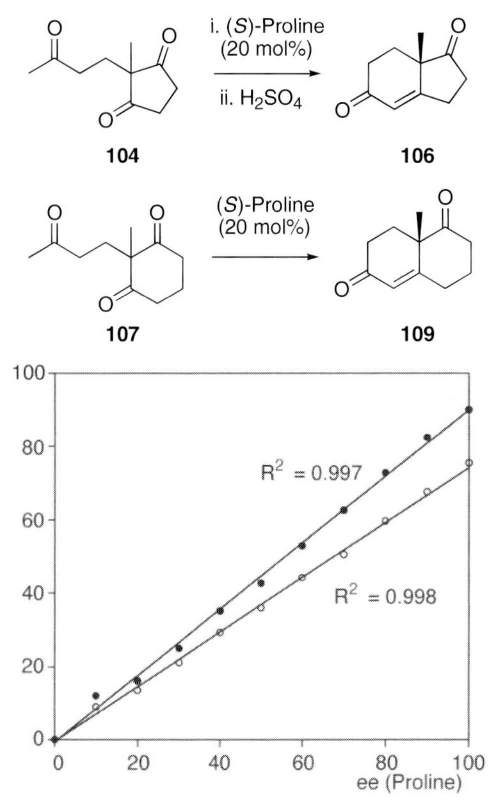

Scheme 4.22
No non-linear effects in proline-catalyzed Hajos–Parrish–Eder–Sauer–Wiechert reactions.

4.22) or dilution effects in the asymmetric catalysis, and that proline-catalyzed aldolizations are first order in proline.

4.3.1.2 Enolexo Aldolizations

Although amine-catalyzed enolexo aldolizations (Section 4.2.2.1) are relatively common and catalytic asymmetric enolendo aldolizations have been known for three decades (Section 4.3.1.1), the first catalytic asymmetric enolexo aldolizations were developed only very recently. It was discovered that a variety of achiral heptanedials (**129**) and 7-oxoheptanal on treatment with a catalytic amount of (*S*)-proline furnished *anti* aldols **130** with excellent enantioselectivity (Scheme 4.23) [120–122].

Whereas non-asymmetric amine-catalyzed enolexo-aldolizations often give

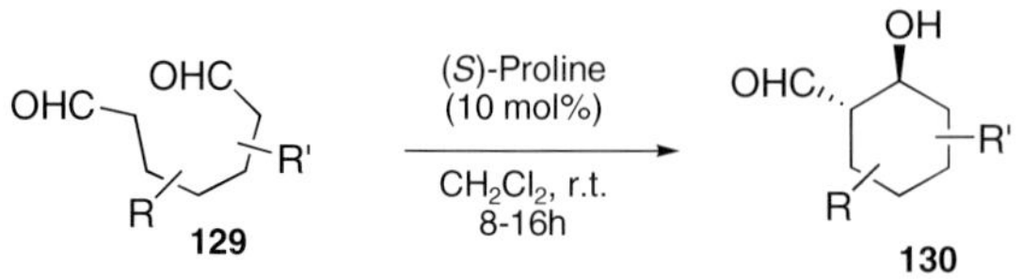

129 → **130**: (*S*)-Proline (10 mol%), CH_2Cl_2, r.t., 8-16h

Dicarbonyl	Yield %	Products	ee%	dr
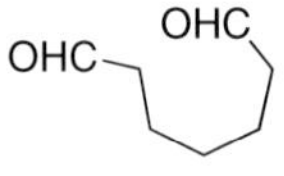 **129a**	95	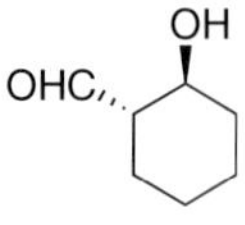 **130a**	99	10:1
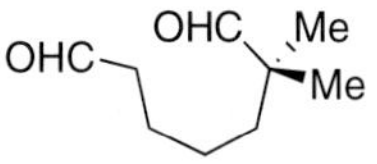 **129b**	74	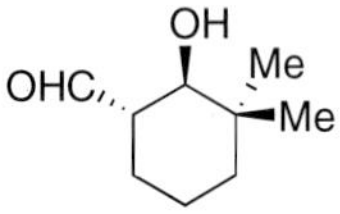 **130b**	98	>20:1
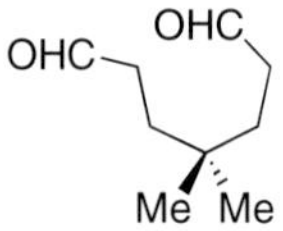 **129c**	75	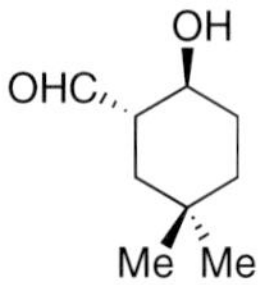 **130c**	97	>20:1
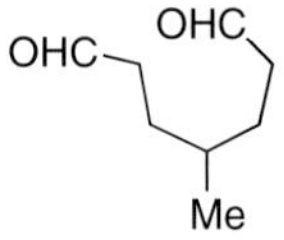 **129d**	76	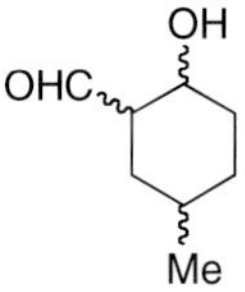 **130d**	75 : 89 : >95 : 8	22 : 5 : 5 : 1
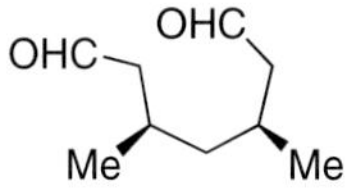 *meso*-**129e**	88	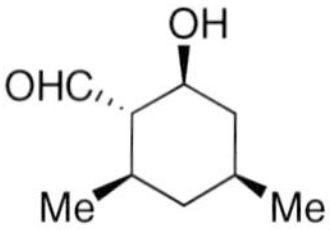 **130e**	99	1:1
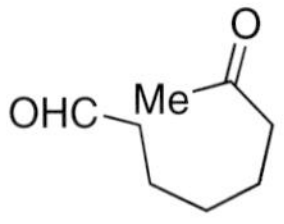 **129f**	92	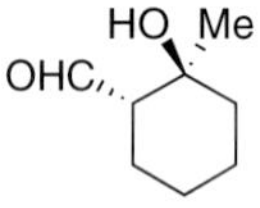 **130f**	99	2:1

Scheme 4.23
Proline-catalyzed enantioselective 6-enolexo-aldolizations.

aldol condensation products (Section 4.2.2.1), the corresponding proline-catalyzed process selectively provides the aldol addition products. The corresponding 5-enolexo aldolizations are much less stereoselective. For example, treating hexanedial (**131**) with a catalytic amount of (*S*)-proline fur-

OHC⌃⌃⌃CHO **131** —(S)-Proline (10%), CH_3CN, r.t., 7h; 85%→ **132a** (79% ee) + **132b** (37% ee), 2 : 1

Scheme 4.24
A proline-catalyzed 5-enolexo-aldolization.

nished aldol **132** with only modest diastereo- and enantioselectivity (Scheme 4.24).

It can be expected that proline-catalyzed enolexo-aldolizations will find utility in natural-product synthesis and that new catalysts will eventually be discovered that lead to improved selectivity in the potentially very useful 5-enolexo variant.

4.3.2 Intermolecular Aldolizations

4.3.2.1 Ketone Donors

The first amine-catalyzed, asymmetric intermolecular aldol reactions were developed by List et al. in 2000 [29–33]. Initially it was found that excess acetone in DMSO containing sub-stoichiometric amounts of (*S*)-proline reacted with some aromatic aldehydes and isobutyraldehyde to give the corresponding acetone aldols (**134**) with good yields and enantioselectivity (Scheme 4.25). Particularly high ee were achieved with α-branched aldehydes. Similarly to the intramolecular enolendo variant, the only side-product in proline-catalyzed intermolecular aldol reactions are the condensation products (Scheme 4.25).

Proline-catalyzed Aldolization of Acetone with Isobutyraldehyde to Give Aldol 134f [29]. A mixture of (*S*)-proline (2.3 g, 0.02 mol, 20 mol%) and isobutyraldehyde (7.2 g, 0.1 mol) was stirred in acetone (40 mL) and dry DMSO (160 mL) at room temperature. The mixture became completely homogenous within ca. 1 h. After 36 h at room temperature (*S*)-proline (ca. 2.0 g, 87%) precipitated and was recovered for reuse via filtration, washing with ethyl acetate, and drying. The combined organic solvents were washed with half-concentrated ammonium chloride solution and the aqueous layers were back-extracted with ethyl acetate. The organic layers were dried ($MgSO_4$), filtered, and concentrated. Column chromatography on SiO_2 (20% ethyl acetate in hexanes) furnished (*R*)-4-hydroxy-5-methyl-hexan-2-one (**134f**, 12.0 g, 0.92 mol, 92%) as a colorless liquid. The ee was determined to be 96% by chiral stationary phase HPLC analysis (Chiralpak AS, 2% iPrOH in hexanes, 1 mL min^{-1}, $t_{(R)} = 18.0$ min, $t_{(S)} = 20.1$ min).

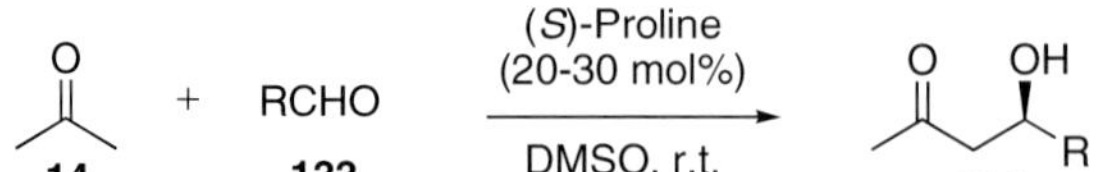

14 (excess) + RCHO 133 → (S)-Proline (20-30 mol%), DMSO, r.t. → 134

Aldehyde	Yield %	Product	ee%[a]
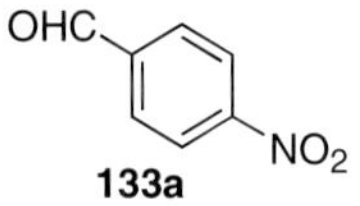 133a	68	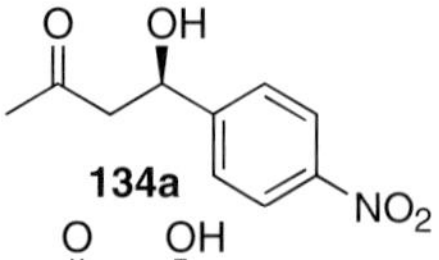 134a	76
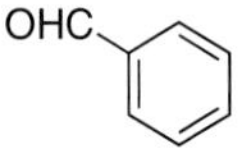 133b	62	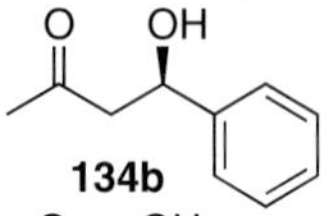 134b	72[b]
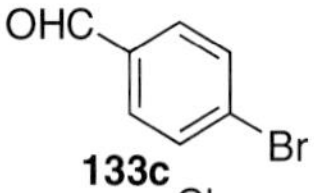 133c	74	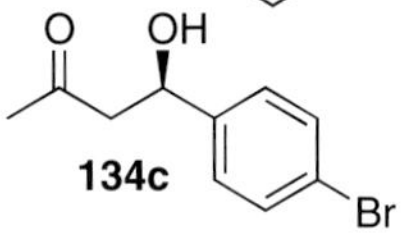 134c	65
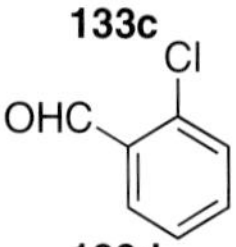 133d	94	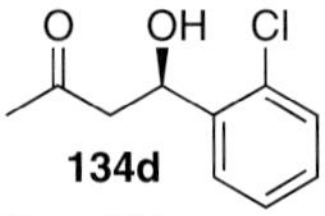 134d	69
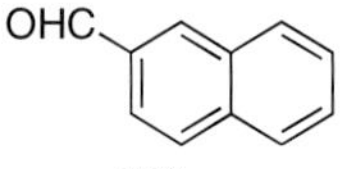 133e	54	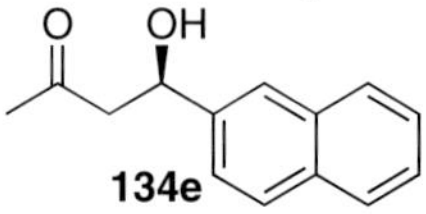 134e	77
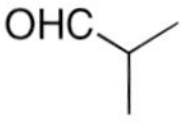 133f	97	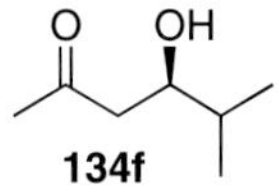 134f	96
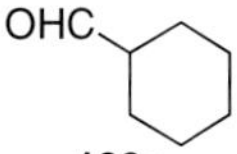 133g	63	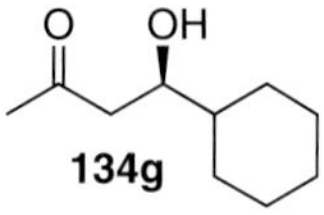 134g	84
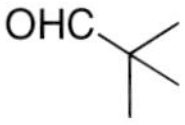 133h	81	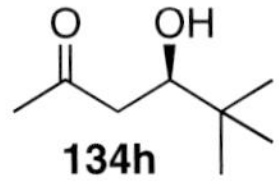 134h	>99
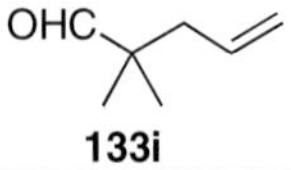 133i	85	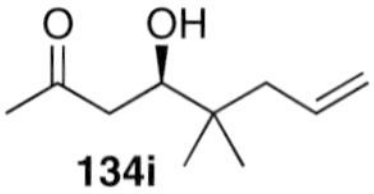 134i	>99

a)The enantiomeric excess was typically determined using chiral stationary phase HPLC-analysis (Chiralpak AS, not AD as originally reported). b) The originally reported ee of **134b** (60%) has been corrected.

Scheme 4.25
Proline-catalyzed asymmetric intermolecular aldolizations of acetone with aromatic- and α-branched aldehydes.

14 (excess) + RCH_2CHO **135** → [(*S*)-Proline (10-20 mol%), Acetone or $CHCl_3$, r.t.] → **136**

Aldehyde	Yield %	Product	ee%
135a	31	**136a**	67
135b	35	**136b**	73
135c	34	**136c**	72
135d	34	**136d**	73
135e	22	**136e**	36

Scheme 4.26
Proline-catalyzed aldolizations of acetone with α-unbranched aldehydes.

α-Unbranched aldehydes (**135**) proved to be extremely challenging substrates under the original conditions. Self-aldolization of the aldehyde was a major side reaction. It was found later that by using acetone or acetone–$CHCl_3$ mixtures instead of the commonly used DMSO as solvents, and 10–20 mol% proline as catalyst, the cross-aldol products (**136**) could be isolated in modest yields and with acceptable enantioselectivity (Scheme 4.26) [123].

The new reaction conditions effectively suppressed aldehyde self-aldolization. The main side product was now the corresponding acetone cross aldol condensation product, typically formed in comparable yields with the desired aldol addition product.

Proline-catalyzed Aldolization of Acetone with *n*-Hexanal to Give Aldol 136b [31]. *n*-Hexanal (2.40 mL, 20 mmol) and (*S*)-proline (230 mg, 2 mmol, 10 mol%) were stirred in 100 mL dry acetone for 168 h. Silica gel (ca. 5 g) was added and the mixture was evaporated. The residue was poured on to a preloaded silica gel column and chromatographed with hexanes–ethyl acetate (4:1) to give (*E*)-non-3-en-2-one (1.37 g, 49%) and aldol **136b** (1.11 g, 35%) in 73% ee (AS, 2% 2-propanol–hexanes, 1 mL min^{-1}, $t_{(R)} = 13.3$ min,

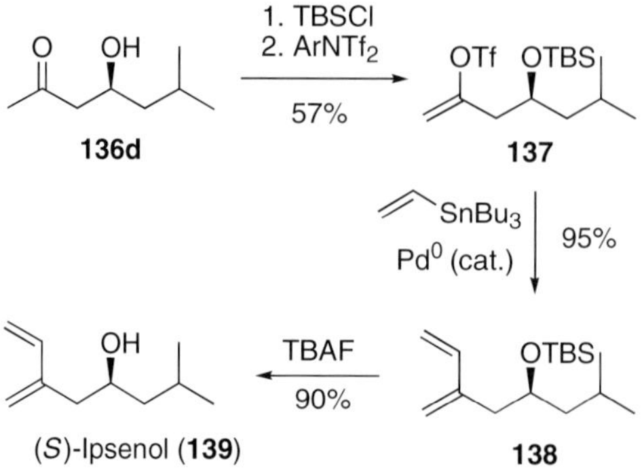

Scheme 4.27
Asymmetric synthesis of (*S*)-ipsenol.

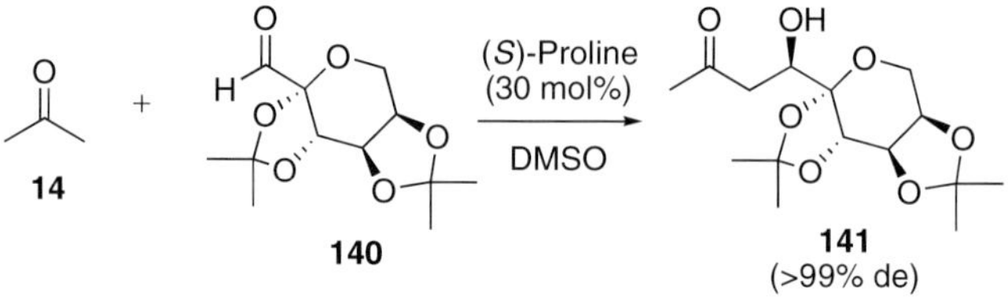

Scheme 4.28
Application of a proline-catalyzed intermolecular aldolization in the synthesis of carbohydrate derivatives.

$t_{(S)} = 15.7$ min). Almost identical results were obtained using an aqueous work-up (phosphate buffered saline–ethyl acetate).

At the time these relatively modest results represented the state-of-the-art in direct catalytic asymmetric aldolizations with α-unbranched aldehyde acceptors. Neither Shibasaki's nor Trost's bimetallic catalysts, the only alternative catalysts available, gave superior results. Moreover, even non-asymmetric amine-catalyzed cross aldolizations with α-unbranched acceptors are still unknown. That the practicality of the process can compensate for the modest yield and enantioselectivity was illustrated by a straightforward synthesis of the natural pheromone (*S*)-ipsenol (**139**) from aldol **136d**, featuring a high-yielding Stille coupling (Scheme 4.27).

A highly stereoselective proline-catalyzed acetone aldol reaction has recently been used in the synthesis of sugar derivatives such as **141** (Scheme 4.28) [124].

Proline-catalyzed aldolizations with ketones other than acetone have also been described. Because the ketone component is typically used in large excess, one is limited to readily available and inexpensive smaller ketones such as butanone, cyclopentanone, cyclohexanone, and hydroxyacetone. Depending on the aldehyde component excellent enantio- and

142 + R'CHO **143** —(*S*)-Proline (10-30%)→ **144-150**

Products		Yield %	dr (*anti*/*syn*)
144a 85% ee	**144b** 76% ee	85	1:1
145a 86% ee	**145b** 89% ee	41	7:1
146a 97% ee	**146b** -	68	>20:1
147a 95% ee	**147b** 20% ee	77	3:1
148a >99% ee	**148b** -	62	>20:1
149a 97% ee	**149b** 84% ee	38	1.7:1
150a 67% ee	**150b** 32% ee	95	1.5:1

Scheme 4.29
Examples of proline-catalyzed intermolecular aldolization with ketone donors other than acetone.

(*anti*)-diastereoselectivity can be achieved in such reactions (Scheme 4.29) [30, 31, 125].

Interestingly, the stereoselectivity of reactions of cyclohexanone with isobutyraldehyde and benzaldehyde were first predicted by using density functional theory calculations on models based on Houk's calculated transition state of the Hajos–Parrish–Eder–Sauer–Wiechert reaction [125]. The transition states of inter- and intramolecular aldol reactions are almost superimposable and readily explain the observed enantiofacial selectivity. Relative transition state energies were then used to predict the diastereo- and enantioselectivity of the proline-catalyzed reactions of cyclohexanone with isobutyraldehyde and benzaldehyde. The predictions are compared with the experimental results in Scheme 4.30. The good agreement clearly validates the theoretical studies, and provides support for the proposed mechanism. Additional density functional theory calculation also support a similar mechanism [126, 127].

The exceptionally high stereoselectivity found in proline-catalyzed aldol reactions involving hydroxyacetone has recently been used in a total synthesis of brassinolide (**156**), a steroidal plant-growth regulator [128]. Readily available aldehyde **151** underwent proline-catalyzed aldolization with hy-

(*S*)-Proline
DMSO

R = *i*-Pr	**98.5% (>99%)**	**<1% (<1%)**	**<1% (<1%)**	**1.5 (<1%)**
R = Ph	**46.5% (54.5%)**	**44% (44.5%)**	**6% (0.1%)**	**3.5% (0.8%)**

Experiment (Density functional theory calculation)

Scheme 4.30
Theoretical prediction and experimental verification of diastereo- and enantioselectivity of proline-catalyzed aldolizations.

Proline (cat.)
DMSO, rt, 12h
84%,

151 152 153 154

known

Brassinolide (**156**)

Scheme 4.31
Formal total synthesis of brasinolide featuring a proline-catalyzed aldol reaction.

droxyacetone to give diol **153** in good yield and diastereoselectivity. This material was converted to derivative **154**, a known precursor of brassinolide (Scheme 4.31).

Other catalysts besides proline have also been investigated. A particularly large amount of data has been collected for the aldol reaction of acetone with *p*-nitrobenzaldehyde (Scheme 4.32). Simple primary α-amino acids and acyclic *N*-methylated α-amino acids are not catalytically active under standard reaction conditions. Of the simple cyclic amino acids studied, azetidine, pyrrolidine, and piperidine 2-carboxylate, proline is clearly the best catalyst. α-, α′-, and, in particular, *N*-methylation reduce the efficiency and whereas substitution of the 3- and 4-positions are tolerated without dramatic effects. Proline amide is essentially catalytically inactive under the standard reaction conditions (DMSO, room temperature, 2 h) but after three days, the aldol could be isolated in good yields, albeit with very low enantioselectivity. Clearly, the carboxylic acid plays an important role in the catalysis and in determining enantioselectivity. That the enantioselectivity can be improved was shown with penicillamine derivative **168**, proline derived diamine salt **169**, and amide **170** [129–132].

By screening a wide variety of diamines Yamamoto et al. showed that diamine salt **169** catalyzes the aldol reaction of several different ketones with *p*-nitrobenzaldehyde to give aldols in good yields and selectivities (Scheme 4.33) [130, 131].

Comparing results from reaction of cyclopentanone and cyclohexanone with *p*-nitrobenzaldhyde catalyzed by proline, **168**, or diamine, **169**, illus-

Catalyst		Conditions	Yield %	*ee* %
	157	30 mol% Cat., 2h, rt DMSO/Acetone (4:1)	<10	n.d.
	158	30 mol% Cat., 2h, rt DMSO/Acetone (4:1)	<10	n.d.
	159	30 mol% Cat., 2h, rt DMSO/Acetone (4:1)	<10	n.d.
	160	30 mol% Cat., 2h, rt DMSO/Acetone (4:1)	55	40
	161	30 mol% Cat., 2h, rt DMSO/Acetone (4:1)	68	76
	162	30 mol% Cat., 2h, rt DMSO/Acetone (4:1)	<10	n.d.
	163	30 mol% Cat., 2h, rt DMSO/Acetone (4:1)	26	61
	164	30 mol% Cat., 2h, rt DMSO/Acetone (4:1)	<10	n.d.
		Same conditions, 72h	80	20
	165	30 mol% Cat., 2h, rt DMSO/Acetone (4:1)	85	78
	166	30 mol% Cat., 2h, rt DMSO/Acetone (4:1)	67	73
	167	30 mol% Cat., 24h, rt DMSO/Acetone (4:1)	<10	n.d.
	168	30 mol% Cat., 24h, rt DMSO/Acetone (4:1)	81	85
	169	3 mol% Cat., 2h, Acetone, 30°C	60	88
	170	20 mol% Cat., 24h, Acetone, -25°C	66	93

Scheme 4.32
Aminocatalysts studied for reaction of acetone with *p*-nitrobenzaldehyde.

Solvent + **133a** —[**169** (10%)]→ **171-173**

Products	Yield %	dr (*anti*/*syn*)
134a 88% ee	60[a]	-
171a 84% ee; **171b** 5% ee	88	0.75
172a 96% ee; **172b** 61% ee	97	2.85
173a 84% ee; **173b** 16% ee	81	1.17

a) Only 3 mol% of the catalyst was used.

Scheme 4.33
Diamine salt **169** catalyzes asymmetric aldol reactions of various ketones.

trates that it can be advantageous to use catalysts other than proline for selected reactions (Scheme 4.34).

The best enantioselectivity in aminocatalytic intermolecular acetone aldolizations has recently been achieved by Tang et al. [132]. Study of selected proline-derived amides led this group to identify hydroxy amide **170** as a superb catalyst for reactions of acetone (solvent) with a variety of aldehydes (Scheme 4.35). Although the yields are typically slightly better when proline is used as the catalyst, enantioselectivity is invariably higher when amide **170** is used.

The mechanism of catalysis by amide **170** has not yet been elucidated. Density functional theory calculations support activation of the aldehyde

Products	Yield %	*dr* (*anti*/*syn*)	*ee* % (*anti*)	*ee* % (*syn*)	Conditions
171	73	1.7	69	63	A
	63	1.6	63	60	B
	88	0.8	85	5	C
172	65	1.7	89	67	A
	65	1.7	90	69	B
	97	2.8	96	61	C

A: 161, 20 mol%, rt, Ketone/DMSO (1:4) Solvent
B: 168, 20 mol%, rt, Ketone/DMSO (1:4) Solvent
C: 169, 10 mol%, 30°C, Ketone Solvent

Scheme 4.34
Comparing yields, enantioselectivity, and diastereoselectivity of different amino catalysts in the aldolization of cyclic ketones with *p*-nitrobenzaldehyde.

with two simultaneous hydrogen bonds from the amide-NH and the alcohol-OH in the transition state (**A**, Scheme 4.36). An alternative explanation involves a protonated oxazoline which might be formed in situ via dehydration of the hydroxy amide [133]. The corresponding transition state (**B**) might then be very similar to that proposed for the proline- and diamine salt-catalyzed reactions.

Very recently *N*-terminal prolyl peptides have been suggested as another attractive class of aldol catalyst. Tang et al. [132] mention in a footnote that the dipeptide Pro–Thr–OMe catalyzes the reaction of acetone with *p*-nitrobenzaldehyde to give the corresponding aldol in 69% ee. Reymond et al. have studied a peptide library and also found that several *N*-terminal prolyl peptides catalyze the same asymmetric aldolization [124]. Independently, Martin and List showed that *N*-terminal prolyl peptides catalyze the direct asymmetric aldol reaction of acetone with *p*-nitrobenzaldehyde and the asymmetric Michael reaction between acetone and *β*-nitrostyrene (Scheme 4.37) [135]. Mechanisms have so not yet been proposed for these reactions.

14 Solvent + RCHO **133** → (**170**, 20 mol%; -25°C, 1-2 d) → **134**

Aldehyde	Yield %	Product	ee%
133a	66 (68)	**134a**	93 (76)
133b	51 (62)	**134b**	83 (72)
133c	77 (74)	**134c**	90 (65)
133d	83 (94)	**134d**	85 (69)
133e	93 (54)	**134e**	84 (77)
133f	43 (97)	**134f**	98 (96)
133g	77 (63)	**134g**	98 (84)
133h	51 (81)	**134h**	>99 (>99)
135a	12 (31)	**136a**	86 (67)

Scheme 4.35
Hydroxyamide **170**-catalyzed aldolizations of acetone with different aldehydes. Numbers in parentheses refer to the corresponding proline-catalyzed reactions.

A B

Scheme 4.36
Mechanistic considerations of the hydroxyamide-catalyzed aldolization.

14 + 133a —Peptide→ 134a

Catalyst	Yield %	*ee* %	Ref.
H-Pro-Thr-OMe	n.r.	69	132
H-Pro-Gly-OH	99	46	134
H-Pro-Glu-Leu-Phe-OH	96	66	134
H-Pro-Aib-Glu-Phe-OH	94	37	134
H-Pro-Asp-Leu-Phe-OH	95	50	134
H-Pro-Aib-Asp-Phe-OH	97	12	134
H-Pro-Ala-OH	90	70	135
H-Pro-Trp-OH	77	65	135
H-Pro-Asp-OH	75	74	135
H-Pro-Glu-OH	72	68	135
H-Pro-Val-OH	89	70	135
H-Pro-Arg-OH	91	31	135
H-Pro-Ser-OH	87	77	135
H-Pro-LysHCl-OH	62	66	135
H-Pro-Gly-Gly-OH	68	53	135
H-Pro-His-Ala-OH	85	56	135

Scheme 4.37
N-terminal prolyl-peptides catalyze enantioselective aldolizations.

(S)-Proline (30 mol%), [bmim][PF6], 25h, rt
14 + 133b → 134b (1)
58%, 71% ee
(62%, 72% ee in DMSO)

(S)-Proline (30 mol%), [bmim][PF6], 24h, rt
176 + 177 → 178 (2)
59%, 76% ee

(S)-Proline (50 mol%), [emim][OTf], 48h, rt
176 + 177 → 178 + 179 (3)
29%, > 98% ee
68%, 88% ee, dr = 99:1

[bmim][PF6] (174)
[emim][OTf] (175)

Scheme 4.38
Proline-catalyzed aldolizations in ionic liquids.

In addition to studying alternative catalysts, researchers have also investigated new reaction conditions for proline-catalyzed aldolizations. It has, for example, been found that the commonly used DMSO could be replaced with the room temperature ionic liquids [bmim][PF_6] (**174**) or [emim][OTf] (**175**) with comparable yields and selectivity (Scheme 4.38) [136–138].

The effect of solvent structure on the regioselectivity can, apparently, be quite strong as is illustrated by the reaction of butanone with *p*-trifluorobenzaldehyde (Scheme 4.38, Eqs. (2) and (3)).

Proline-catalyzed aldolizations can also be conducted under aqueous conditions, although rates and enantioselectivity generally decrease with increasing water concentration [48, 139]. An interesting observation has been made by Peng et al., who showed that a combination of proline with sodium dodecyl sulfonate can be used as a catalyst system, indicative of catalysis in aqueous micelles (Scheme 4.39) [140]. The enantioselectivity of the reaction studied has not been reported.

Proline has also been attached to a polymer support. Takemoto et al. described the first example of this approach [141]. Hydroxyproline was linked

14 + 133a —Catalyst, H_2O→ 134a

Catalyst	Conditions	Yield %
Proline (40 mol%)	5d, rt	15
SDS (20%	5d, rt	12
Proline (40 mol%) + SDS (20%)	1d, rt	87

SO_3Na

SDS =Sodium dodecyl sulfonate

Scheme 4.39
Proline-catalyzed aldolizations in inverse micelles.

180 —1. NaH; 2. H^+; 3. OH^-→ 181

104 —181, 34%→ 105 (18% ee)

Scheme 4.40
The first polymer-supported proline-catalyst.

to a polystyrene resin and the resulting material (**181**) was shown to catalyze the Hajos–Parrish–Eder–Sauer–Wiechert reaction, albeit in low yield and with low enantioselectivity (Scheme 4.40).

Proline has also recently been attached to a poly(ethylene glycol) support. The resulting catalyst (**183**) effectively catalyzes intermolecular aldolizations, a Robinson annexation, and Mannich reactions. It can be used under

MeO-(CH_2CH_2O)$_n$-CH_2CH_2-O ... OH
n = 110 **182**

MeO-(CH_2CH_2O)$_n$-CH_2CH_2-O ... CO_2H
183

14 + OHC **133f** → (**183**, DMSO, 130 h) **134f** 81%, >98% ee

+ → (**183**, DMSO, 130 h) **109** 55%, 75% ee

Scheme 4.41
Poly(ethylene glycol)-supported proline-catalysis.

homogenous conditions, then precipitated by solvent change and reused (Scheme 4.41) [142, 143].

4.3.2.2 Aldehyde Donors

In addition to ketones, aldehydes can also be used as aldol donors in proline-catalyzed reactions [144]. Barbas et al. found that treating acetaldehyde solutions with proline provided aldehyde **185**, an aldol trimer of acetaldehyde, in 84% ee and 4% yield (Scheme 4.42, Eq. (1)) [145, 146]. As shown by Jørgensen et al., other simple α-unbranched aldehydes can also be used as donors in proline-catalyzed cross aldolization with activated non-enolizable ketone acceptors to give aldols **188** in high enantioselectivity and yield (Scheme 4.42, Eq. (2)) [147].

Finally, Northrup and MacMillan found that proline to also catalyzed cross aldolizations of two different aldehydes under carefully developed conditions using syringe pump techniques [148]. These reactions furnish *anti* aldols **191** in excellent enantioselectivity and good yield and diastereoselectivity (Scheme 4.43).

Proline-catalyzed Aldolization of Propionaldehyde with Isobutyraldehyde to Give Aldol 191e [148]. A solution of freshly distilled propionaldehyde (1.81 mL, 25 mmol) in DMF (12.5 mL) pre-cooled to 4 °C was added slowly over

3 (184) → (S)-Proline (cat.), THF, 4°C → 185, 4% yield, ee = 84% (1)

186 (1 eq) + 187 (1 eq) → (S)-Proline (50 mol%.), CH_2Cl_2 → 188 (2)

R		Yield %	ee %
Me	(**188a**)	90	90
Et	(**188b**)	91	85
i-Pr	(**188c**)	88	85
$CH_2CH{=}CH_2$	(**188d**)	94	88
n-C_6H_{13}	(**188e**)	91	84
Ph	(**188f**)	97	0

Scheme 4.42
Proline catalyzed aldolreactions with aldehyde donors.

the course of 20 h to a stirred suspension of isobutyraldehyde (4.54 mL, 50 mmol), (*S*)-proline (288 mg, 2.5 mmol, 10 mol%), and DMF (12.5 mL) at 4 °C. After 30 h the resulting solution was diluted with diethyl ether and washed successively with water and brine. The combined aqueous layers were back-extracted with three portions of dichloromethane. The organic layers were combined, dried over anhydrous $MgSO_4$, and concentrated in vacuo. Flash chromatography (20:7 pentane–diethyl ether) afforded aldol **191e** as a clear colorless oil (2.65 g, 20.6 mmol, 82%) in >99% ee and 96:4 *anti:syn*. $[\alpha]_D = -17.9°$ ($c = 1$, $CHCl_3$). The product ratios were determined by GLC analysis of the acetal derived from 2,2-dimethylpropane-1,3-diol on a Bodman Chiraldex β-DM column (110 °C isothermal, 23 psig); (2*S*,3*S*) *anti* isomer $t_r = 31.8$ min, (2*R*,3*R*) *anti* isomer $t_r = 33.9$ min, (2*R*,3*S*) and (2*S*,3*R*) *syn* isomers $t_r = 29.4$, 29.8 min.

The discovery of the direct enantioselective cross-aldolization of aldehydes, which not too long ago would have probably been regarded as impossible by many, can be viewed as a major breakthrough in the aldol field.

(*S*)-Proline (10 mol%), 4°C, DMF

Product	%Yield	*dr*	%*ee*
191a	80	4:1	99
191b	88	3:1	97
191c	87	14:1	99
191d	81	3:1	99
191e	82	24:1	>99
191f	80	24:1	98
191g	75	19:1	91

Scheme 4.43
Direct asymmetric cross-aldolization of aldehydes.

Finally, aldols are made in the most atom economical and practical way without use of stoichiometric bases, auxiliaries, or other reagents.

Clearly, aminocatalytic aldolizations have great potential in academic and industrial synthesis and nothing other than frequent use and further development of these methods can be expected.

References

1 C. H. Heathcock, In *Comprehensive Organic Synthesis*; B. M. Trost, I. Fleming, C. H. Heathcock, Eds.; Pergamon: Oxford, **1991**; Vol. *2*, pp 133.
2 J. S. Johnson, D. A. Evans, *Acc. Chem. Res.* **2000**, *33*, 325.
3 T. D. Machajewski, C.-H. Wong, *Angew. Chem., Int. Ed.* **2000**, *39*, 1352.
4 B. Alcaide, P. Almendros, *Eur. J. Org. Chem.* **2002**, 1595.
5 S. E. Denmark, R. A. Stavenger, *Acc. Chem. Res.* **2000**, *33*, 432.
6 S. G. Nelson, *Tetrahedron: Asymmetry* **1998**, *9*, 357.
7 M. Shibasaki, H. Sasai, T. Arai, T. Iida, *Pure Appl. Chem.* **1998**, *70*, 1027.
8 E. M. Carreira, R. A. Singer, *Drug Discovery Today* **1996**, *1*, 145.
9 I. Peterson, *Pure Appl. Chem.* **1992**, *64*, 1821.
10 S. Masamune, W. Choy, J. S. Peterson, L. R. Sita, *Angew. Chem., Int. Ed. Engl.* **1985**, *24*, 1.
11 Y. M. A. Yamada, N. Yoshikawa, H. Sasai, M. Shibasaki, *Angew. Chem., Int. Ed. Engl.* **1997**, *36*, 1871.
12 N. Yoshikawa, Y. M. A. Yamada, J. Das, H. Sasai, M. Shibasaki, *J. Am. Chem. Soc.* **1999**, *121*, 4168.
13 N. Kumagai, S. Matsunaga, N. Yoshikawa, T. Ohshima, M. Shibasaki, *Org. Lett.* **2001**, *3*, 1539.
14 B. M. Trost, H. Ito, *J. Am. Chem. Soc.* **2000**, *122*, 12003.
15 B. M. Trost, E. R. Silcoff, H. Ito, *Org. Lett.* **2001**, *3*, 2497.
16 M. Nakagawa, H. Nakao, K.-I. Watanabe, *Chem. Lett.* **1985**, 391.
17 Y. Yamada, K. I. Watanabe, H. Yasuda, *Utsonomiya Daigaku Kyoikugakubu Kiyo, Dai-2-bu* **1989**, *39*, 25. Some of these results could not be reproduced, see: M. A. Calter, R. K. Orr, *Tetrahedron Lett.* **2003**, *44*, 5699.
18 Z. G. Hajos, D. R. Parrish, German Patent DE 2102623, 1971.
19 U. Eder, R. Wiechert, G. Sauer, German Patent DE 2014757, 1971.
20 U. Eder, G. Sauer, R. Wiechert, *Angew. Chem., Int. Ed. Engl.* **1971**, *10*, 496.
21 Z. G. Hajos, D. R. Parrish, *J. Org. Chem.* **1974**, *39*, 1615.
22 N. Cohen, *Acc. Chem. Res.* **1976**, *9*, 512–41.
23 K. Drauz, A. Kleemann, J. Martens, *Angew. Chem., Int. Ed. Engl.* **1982**, *21*, 584–608.
24 J. Martens, *Chem. Ztg* **1986**, *110*, 169.
25 M. E. Jung, *Tetrahedron* **1976**, *32*, 3.
26 E. R. Jarvo, S. J. Miller, *Tetrahedron* **2002**, *58*, 2481–2495.
27 N. Harada, T. Sugioka, H. Uda, T. Kuriki, *Synthesis* **1990**, 53.
28 J. Gutzwiller, P. Buchschacher, A. Fürst, *Synthesis* **1977**, 167.
29 B. List, R. A. Lerner, C. F. Barbas, III, *J. Am. Chem. Soc.* **2000**, *122*, 2395.
30 W. Notz, B. List, *J. Am. Chem. Soc.* **2000**, *122*, 7386.

31 B. List, P. Pojarliev, C. Castello, *Org. Lett.* **2001**, *3*, 573.
32 B. List, *Tetrahedron* **2002**, *58*, 5572.
33 B. List, *Synlett* **2001**, 1675.
34 S. Hosokawa, K. Sekiguchi, M. Enemoto, S. Kobayashi, *Tetrahedron Lett.* **2000**, *41*, 6429.
35 V. Pistara, P. L. Barili, G. Catelani, A. Corsaro, F. D'Andrea, S. Fisichella, *Tetrahedron Lett.* **2000**, *41*, 3253.
36 T. Ooi, M. Taniguchi, M. Kameda, K. Maruoka, *Angew. Chem. Int. Ed.* **2002**, *41*, 4542.
37 G. Storck, A. Brizzolara, H. Landesman, J. Szmuszkovicz, R. Terrell, *J. Am. Chem. Soc.* **1963**, *85*, 207.
38 T. Mukaiyama, K. Banno, K. Narasaka, *J. Am. Chem. Soc.* **1974**, *96*, 7503.
39 O. Takazawa, K. Kogami, K. Hayashi, *Bull. Chem. Soc. Jpn.* **1985**, *58*, 2427–2428.
40 See for example: L. Birkhofer, S. M. Kim, H. D. Engels, *Chem. Ber.* **1962**, *95*, 1495.
41 F.G. Fischer, A. Marschall, *Ber.* **1931**, *64*, 2825.
42 W. Langenbeck, G. Borth, *Ber.* **1942**, *75*, 951.
43 F. H. Westheimer, H. Cohen, *J. Am. Chem. Soc.* **1938**, *60*, 90.
44 R. P. Koob, J. G. Miller, A. R. Day, *J. Am. Chem. Soc.* **1951**, *73*, 5775.
45 J. C. J. Speck, A. A. Forist, *J. Am. Chem. Soc.* **1957**, *79*, 4659.
46 J. Hine, W. H. Sachs, *J. Org. Chem.* **1974**, *39*, 1937.
47 R. M. Pollack, S. Ritterstein, *J. Am. Chem. Soc.* **1972**, *94*, 5064.
48 J.-L. Reymond, Y. Chen, *J. Org. Chem.* **1995**, *60*, 6970.
49 S. Bahmanyar, K. N. Houk, *J. Am. Chem. Soc.* **2001**, *123*, 11273.
50 Also see: A. Sevin, J. Maddaluno, C. Agami, *J. Org. Chem.* **1987**, *52*, 5611.
51 W. Treibs, K. Krumbholz, *Chem. Ber.* **1952**, *85*, 1116.
52 J. Schreiber, C.-G. Wermuth, *Bull. Soc. Chim. Fr.* **1965**, *8*, 2242.
53 F. Merger, J. Frank, Eur. Pat. Appl. 0 421 271 A2, 1991.
54 A. Kuiterman, H. Dielemanns, R. Green, A. M. C. F. Castelijns, US patent 5,728,892, 1998.
55 T. Ishikwa, E. Uedo, S. Okada, S. Saito, *Synlett* **1999**, *4*, 450.
56 For an excellent review, see: Tietze, L. F., In: *Comprehensive Organic Synthesis*, Vol. 2; Trost, B. M., Ed.; Pergamon Press: New York, **1991**, 341–394.
57 H. G. Lindwall, J. S. Maclennan, *J. Am. Chem. Soc.* **1932**, *54*, 4739.
58 G. Kobayashi, S. Furukawa, *Chem. Pharm. Bull.* **1964**, *12*, 1129.
59 R. B. Van Order, H. G. Lindwall, *J. Org. Chem.* **1945**, *10*, 128.
60 R. C. Blume, H. G. Lindwall, *J. Org. Chem.* **1946**, *22*, 185.
61 A. L. Wilds, C. Djerassi, *J. Am. Chem. Soc.* **1946**, *68*, 1715.
62 H. Schlenk, *Chem. Ber.* **1948**, *81*, 175.
63 E. C. Constable, P. Harverson, D. R. Smith, L. A. Whall, *Tetrahedron* **1994**, *26*, 7799.

64 E. Buchta, G. Satzinger, *Chem. Ber.* **1959**, *92*, 449.

65 D. Mead, R. Loh, A. E. Asato, R. S. H. Liu, *Tetrahedron Lett.* **1985**, *26*, 2873.

66 A. H. Lewin, M. R. Nilsson, J. P. Burgess, F. I. Carroll, *Org. Prep. Proced. Int.* **1995**, *27*, 621.

67 A. Daut, Ö. Sezer, O. Anac, *Rev. Roum. Chim.* **1997**, *42*, 1045.

68 H. Bayer, C. Batzl, R. W. Hartmann, A. Mannschreck, *J. Med. Chem.* **1991**, *34*, 2685.

69 Woodward, R. B., Sondheimer, F., Taub, D., Heusler, K., MacLamore, W. M. *J. Am. Chem. Soc.* **1952**, *74*, 4223.

70 E. J. Corey, R. L. Danheiser, *J. Am. Chem. Soc.* **1978**, *100*, 8031.

71 B. B. Snider, K. Yang, *J. Org. Chem.* **1990**, *55*, 4392.

72 X.-X. Shi, Q.-Q. Wu, X. Lu, *Tetrahedron: Asymmetry* **2002**, *13*, 461.

73 A. Srikrishna, N. C. Babu, *Tetrahedron Lett.* **2001**, *42*, 4913.

74 B. Traber, H. Pfander, *Tetrahedron Lett.* **2000**, *41*, 7197.

75 W. R. Roush, D. A. Barda, *Tetrahedron Lett.* **1997**, *38*, 8785.

76 R. Lalande, J. Moulines, J. Duboudin, *C. R. Bull. Soc. Chim. Fr.* **1962**, 1087.

77 C.-Q. Wei, X.-R. Jiang, Y. Ding, *Tetrahedron* **1998**, *54*, 12623.

78 H. Hagiwara, H. Ono, N. Komatsubara, T. Hoshi, T. Suzuki, M. Ando, *Tetrahedron Lett.* **1999**, *40*, 6627.

79 M. Allard, J. Levisalles, H. Rudler, *Bull. Soc. Chim. Fr.* **1968**, 303.

80 Y. Ding, X.-R. Jiang, *J. Chem. Soc., Chem. Commun.* **1995**, *16*, 1693.

81 A. Takahashi, M. Shibasaki, *Tetrahedron Lett.* **1987**, *28*, 1893.

82 J. Wright, G. J. Drtina, R. A. Roberts, L. A. Paquette, *J. Am. Chem. Soc.* **1988**, *110*, 5806.

83 P. Wieland, K. Miescher, *Helv. Chim. Acta* **1950**, *33*, 2215.

84 N. P. Buu-Hoi, N. Hoan, D. Lavit, *J. Chem. Soc.* **1950**, 230.

85 A. Wettstein, K. Heusler, H. Ueberwasser, P. Wieland, *Helv. Chim. Acta* **1957**, *40*, 323.

86 E. Plieninger, G. Ege, R. Fischer, W. Hoffmann, *Chem. Ber.* **1961**, *94*, 2106.

87 A. L. Begbie, B. T. Golding, *J. Chem. Soc. Perkin I* **1972**, 602.

88 W. Kreiser, P. Below, *Tetrahedron Lett.* **1981**, *22*, 429.

89 E. J. Corey, M. A. Tius, J. Das, *J. Am. Chem. Soc.* **1980**, *102*, 7612.

90 B. Roy, H. Watanabe, T. Kitahara, *Heterocycles* **2001**, *55*, 861.

91 S. Swaminathan, M. S. Newman, *Tetrahedron* **1958**, *2*, 88.

92 S. Ramachandran, M. S. Newman, *Org. Synth.* **1961**, *41*, 38.

93 R. Selvarajan, J. P. John, K. V. Narayanan, S. Swaminathan, *Tetrahedron* **1966**, *22*, 949.

94 T. A. Spencer, H. S. Neel, T. W. Flechtner, R. A. Zayle, *Tetrahedron Lett.* **1965**, *43*, 3889.

95 T. A. Spencer, K. K. Schmiegel, *Chem. Ind.* **1963**, 1765.

96 R. D. Roberts, H. E. Ferran, Jr., M. J. Gula, T. A. Spencer, *J. Am. Chem. Soc.* **1980**, *102*, 7054.

97 T. A. Spencer, *Bioorganic Chemistry, Vol. 1*, Academic Press, New York San Francisco London, 1977.

98 S. Takano, C. Kasahara, K. Ogasawara, *J. Chem. Soc. Chem. Comm.* **1981**, 635.
99 J. C. Blazejewski, *J. Fluorine Chemistry* **1990**, *46*, 515.
100 S. Terashima, S. Sato, K. Koga, *Tetrahedron Lett.* **1979**, *36*, 3469.
101 A. Przézdziecka, W. Stepanenko, J. Wicha, *Tetrahedron: Asymmetry* **1999**, *10*, 1589.
102 S. Kwiatkowski, A. Syed, C. P. Brock, D. S. Watt, *Synthesis* **1989**, 818.
103 D. Rajagopal, R. Narayanan, S. Swaminathan, *Tetrahedron Lett.* **2001**, *42*, 4887.
104 N. Ramamurthi, S. Swaminathan, *Indian J. Chem. Sect. B* **1990**, *29*, 401.
105 S. Danishefsky, P. Cain, *J. Am. Chem. Soc.* **1976**, *98*, 4975.
106 C. Agami, F. Meynier, C. Puchot, *Tetrahedron* **1984**, *40*, 1031.
107 C. Agami, N. Platzer, H. Sevestre, *Bull. Soc. Chim. Fr.* **1987**, *2*, 358.
108 C. Agami, H. Sevestre, *J. Chem. Soc., Chem. Commun.* **1984**, *21*, 1385.
109 C. Agami, J. Levisalles, H. Sevestre, *J. Chem. Soc., Chem. Commun.* **1984**, 418.
110 C. Agami, C. Puchot, H. Sevestre, *Tetrahedron Lett.* **1986**, *27*, 1501.
111 C. Agami, C. Puchot, *J. Mol. Cat.* **1986**, *38*, 341.
112 C. Agami, C. Puchot, *Tetrahedron* **1986**, *42*, 2037.
113 C. Agami, J. Levisalles, C. Puchot, *J. Chem. Soc., Chem. Commun.* **1985**, 441.
114 C. Agami, *Bull. Soc. Chim. FR.* **1988**, *3*, 499.
115 D. Rajagopal, M. S. Moni, S. Subramanian, S. Swaminathan, *Tetrahedron: Asymmetry* **1999**, *10*, 1631.
116 S. Bahmanyar, K. N. Houk, *J. Am. Chem. Soc.* **2001**, *123*, 12911.
117 K. L. Brown, L. Damm, J. D. Dunitz, A. Eschenmoser, R. Hobi, C. Kratky, *Helv. Chim. Acta* **1978**, *61*, 3108.
118 L. Hoang, S. Bahmanyar, K. N. Houk, B. List, *J. Am. Chem. Soc.* **2003**, *125*, 16.
119 C. Puchot, O. Samuel, E. Dunach, S. Zhao, C. Agami, H. B. Kagan, *J. Am. Chem. Soc.* **1980**, *108*, 2353.
120 C. Pidathala, L. Hoang, N. Vignola, B. List, *Angew. Chem. Int. Ed. Engl.* **2003**, *115*, 2891.
121 Also see: R. B. Woodward, E. Logusch, K. P. Nambiar, K. Sakan, D. E. Ward, et al., *J. Am. Chem. Soc.* **1981**, *103*, 3210.
122 C. Agami, N. Platzer, C. Puchot, H. Sevestre, *Tetrahedron* **1987**, *43*, 1091.
123 B. List, P. Pojarliev, C. Castello, *Org. Lett.* **2001**, *3*, 573.
124 I. Izquierdo, M. T. Plaza, R. Robles, A. J. Mota, F. Franco, *Tetrahedron: Asymmetry* **2001**, *12*, 2749.
125 S. Bahmanyar, K. N. Houk, H. J. Martin, B. List, *J. Am. Chem. Soc.* **2003**, *125*, 2475.
126 K. N. Rankin, J. W. Gauld, R. J. Boyd, *J. Phys. Chem. A* **2002**.
127 M. Arnó, L. R. Domingo, *Theor. Chem. Acc.* **2002**, 381.

128 L. Peng, H. Liu, T. Zhang, F. Zhang, T. Mei, Y. Li, *Tetrahedron Lett.* **2003**, *44*, 5107.
129 K. Sakthivel, W. Notz, T. Bui, C. F. Barbas, *J. Am. Chem. Soc.* **2001**, *123*, 5260.
130 S. Saito, M. Nakadai, H. Yamamoto, *Synlett* **2001**, *8*, 1245.
131 M. Nakadai, S. Saito, H. Yamamoto, *Tetrahedron* **2002**, *58*, 8167.
132 Z. Tang, F. Jiang, L.-T. Yu, Y. Cui, L.-Z. Gong, A.-Q. Mi, Y.-Z. Jiang, Y.-D. Wu, *J. Am. Chem. Soc.* **2003**, *125*, 5262.
133 Suggested by Professor K. N. Houk, UCLA.
134 J. Kofoed, J. Nielsen, J.-L. Reymond, *Biorg. Med. Chem. Lett.* **2003**, on-line article.
135 H. J. Martin, B. List, *Synlett*, **2003**, 1901.
136 T. P. Loh, L.-C. Feng, H. Y. Yang, J.-Y. Yang, *Tetrahedron Lett.* **2002**, *43*, 8741.
137 P. Kotrusz, I. Kmentová, B. Gotov, S. Toma, E. Solcániová, *Chem. Comm.* **2002**, 2510.
138 T. Kitazume, Z. Jiang, K. Kasai, Y. Mihara, M. Suzuki, *J. Fluorine Chem.* **2003**, *121*, 205.
139 A. Córdova, W. Notz, C. F. Barbas, *Chem. Commun.* **2002**, 3024.
140 Y.-Y. Peng, Q.-P. Ding, Z. Li, P. G. Wang, J. P. Cheng, *Tetrahedron Lett.* **2003**, *44*, 3871.
141 K. Kondo, T. Yamano, K. Takemoto, *Macromol. Chem.* **1985**, *186*, 1781.
142 M. Benaglia, G. Celentano, F. Cozzi, *Adv. Synth. Catal.* **2001**, *343*, 171.
143 M. Benaglia, M. Cinquini, F. Cozzi, A. Puglisi, *Adv. Synth. Catal.* **2002**, *344*, 533.
144 Proline-mediated aldehyde self-aldolizations have been mentioned by: F. Orsini, F. Pelizzoni, M. Forte, *J. Heterocycl. Chem.* **1989**, *26*, 837.
145 A. Córdova, W. Notz, C. F. Barbas, *J. Org. Chem.* **2002**, *67*, 301.
146 Also see: N. S. Chowdari, D. B. Ramachary, A. Córdova, C. F. Barbas, *Tetrahedron Lett.* **2002**, *43*, 9591.
147 A. Bogevig, N. Kumaragurubaran, K. A. Jorgensen, *Chem. Commun.* **2002**, 620.
148 A. B. Northrup, D. W. C. MacMillan, *J. Am. Chem. Soc.* **2002**, *124*, 6798.

5
Enzyme-catalyzed Aldol Additions

Wolf-Dieter Fessner

5.1
Introduction

As a supplement to classical chemical methodology, enzymes are finding increasing acceptance as chiral catalysts for the in-vitro synthesis of asymmetric compounds because they have been optimized by evolution for high selectivity and catalytic efficiency [1–11]. In particular, the high stereospecificity of aldolases in C–C bond-forming reactions gives them substantial utility as synthetic biocatalysts, making them an environmentally benign alternative to chiral transition metal catalysis of the asymmetric aldol reaction [12].

In contrast with most classical chemical operations, biocatalytic conversions can usually be performed under mild reaction conditions that are compatible with underivatized substrates, thus also obviating tedious and costly protecting group manipulation [13]. Several dozen aldolases have been identified in Nature [14, 15] and many of these are commercially available on a scale sufficient for preparative applications. Recent advances in molecular and structural biology have improved access to virtually any biocatalyst in large quantity and to its detailed functional topology which increasingly enables exploitation of its synthetic utility on rational grounds.

Evidently, enzyme catalysis is thus most attractive for the synthesis and modification of biologically relevant classes of organic compounds that are typically complex, multifunctional, and water soluble. Typical examples are those structurally related to amino acids [16, 17] or carbohydrates [18–24], which are difficult to prepare and handle by conventional methods of chemical synthesis. Because of the multitude of factors that might be critical to the success of an enzymatic conversion, and because of the empirical nature of their development, it is mandatory in the design of new biocatalytic processes to become familiar with the scope and limitations of synthetically useful enzymes, both as a source of inspiration and for reference. Thus, this overview attempts to outline the current status of development for the most important aldolase biocatalysts and their preparative potential for asymmet-

Modern Aldol Reactions. Vol. 1: Enolates, Organocatalysis, Biocatalysis and Natural Product Synthesis.
Edited by Rainer Mahrwald

ISBN: 3-527-30714-1

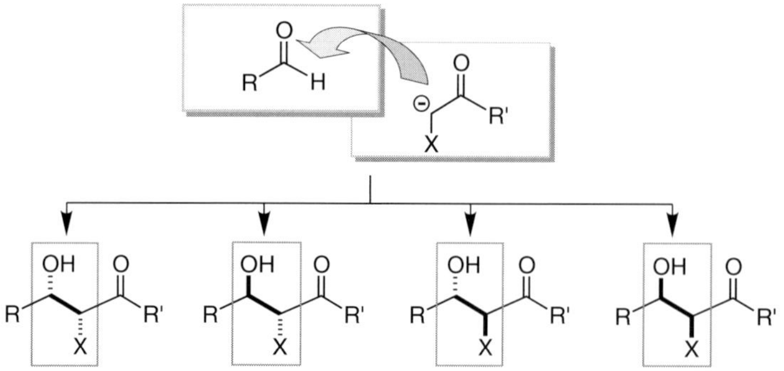

Fig. 5.1
Generation of stereo-diversity by aldol addition.

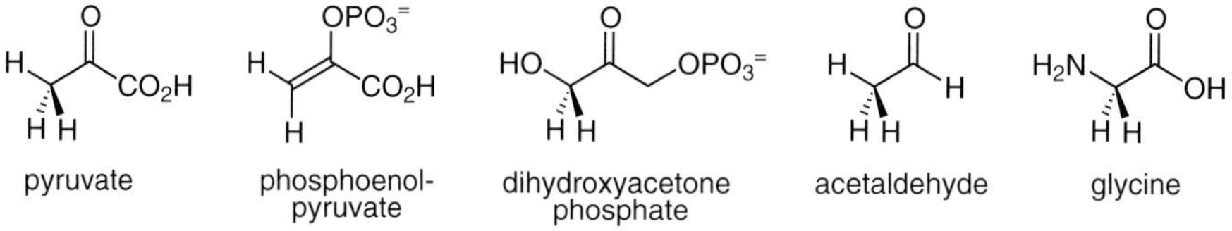

Fig. 5.2
Nucleophiles for preparatively useful aldolases.

ric synthesis. Catalytic C–C coupling is among the most useful synthetic methods in asymmetric synthesis because of its potential for stereoisomer generation by a convergent, "combinatorial" strategy [25]. Thus, attention is also paid to the feasibility of directed stereodivergent approaches by which multiple, diastereomeric products can be derived from common synthetic building blocks (Figure 5.1) [25, 26]. Obviously, such a synthetic strategy requires the prevalence of related, stereocomplementary enzymes which must have a similarly broad substrate tolerance.

5.2 General Aspects

5.2.1 Classification of Lyases

Most enzymes used by Nature for carbon–carbon bond formation and cleavage ("lyases") catalyze a crossed aldol reaction in the form of a reversible, stereocontrolled addition of a nucleophilic ketone donor to an electrophilic aldehyde acceptor. Synthetically the most useful and most extensively studied enzymes use aldol donors comprising 2-carbon or 3-carbon fragments and can be grouped into four categories depending on the structure of their nucleophilic component (Figure 5.2): (i) pyruvate-

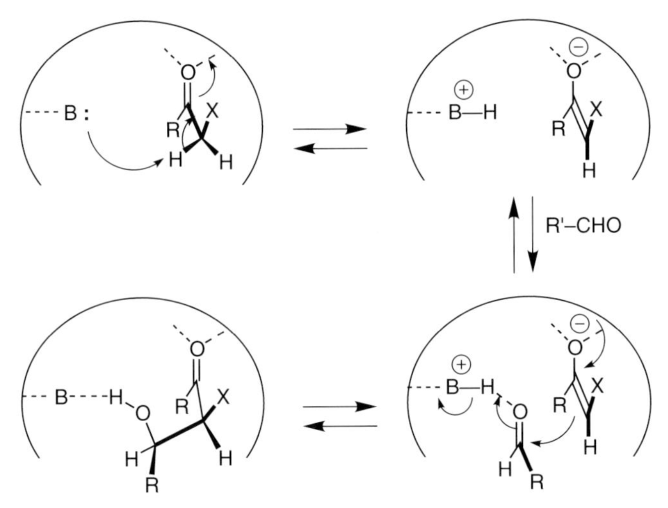

Fig. 5.3
Retention mechanism for deprotonation and C–C bond formation in enzymatic aldolization.

(and phosphoenolpyruvate-) dependent aldolases, (ii) dihydroxyacetone phosphate-dependent aldolases, (iii) an acetaldehyde-dependent aldolase, and (iv) glycine-dependent enzymes. Apart from possible mechanistic differences, members of the first and third types generate α-methylene carbonyl compounds and thereby generate a single stereocenter, whereas members of the other types form α-substituted carbonyl derivatives that contain two new vicinal chiral centers at the new C–C bond, which makes them particularly appealing for asymmetric synthesis.

Typically, lyases require highly specific nucleophilic donor components, because of mechanistic requirements. This includes the need for reasonably high substrate affinity and the general difficulty of binding and anchoring a rather small molecule in a fashion that restricts solvent access to the carbanionic site after deprotonation and shields one enantiotopical face of the nucleophile to secure correct diastereofacial discrimination (Figure 5.3) [27]. Usually, approach of the aldol acceptor to the enzyme-bound nucleophile occurs stereospecifically following an overall retention mechanism, whereas facial differentiation of the aldehyde carbonyl is responsible for the relative stereoselectivity. In this manner the stereochemistry of the C–C bond formation is completely controlled by the enzymes, in general irrespective of the constitution or configuration of the substrate, which renders the enzymes highly predictable. Few specific exceptions to this rule are currently known (vide infra).

Most lyases tolerate reasonably broad variation of the electrophilic acceptor component, usually an aldehyde, on the other hand. This feature, which nicely complements the emerging needs of combinatorial synthesis,

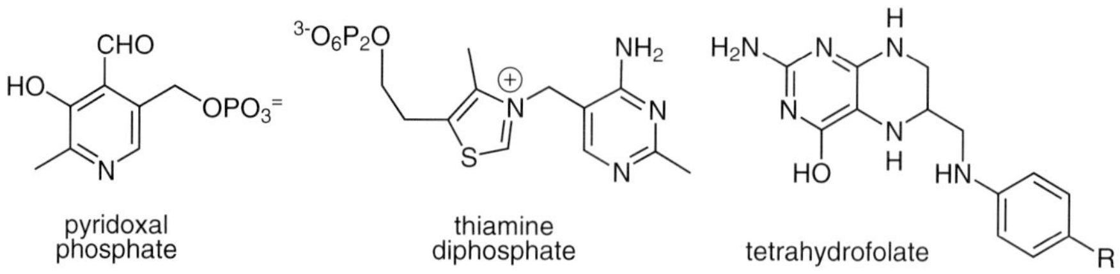

pyridoxal phosphate

thiamine diphosphate

tetrahydrofolate

Fig. 5.4
Types of cofactors for aldol and related reactions.

makes possible a stereodivergent strategy for synthesis of arrays of stereoisomeric compounds by employing stereo-complementary enzymatic catalysts to selectively produce individual diastereomers by design from a given starting material.

Biochemical nomenclature of enzymes [28] usually follows historical classifications that are different from the needs of synthetic organic chemists today. Whereas the common EC numbers are valuable for reference, in this overview an enzyme designation will be used that utilizes a three-letter code as reference to the natural substrate (from which usually stereochemistry is immediately obvious), followed by a single capital letter that indicates the conversion type (e.g., aldolases (A), synthases (S), isomerases (I), dehydrogenases (D), oxidases (O), kinases (K), (glycosyl)transferases (T), epimerases (E) etc.) [25].

In addition to the preparatively useful aldolases, several mechanistically distinct enzymes can be employed for synthesis of product structures identical with those accessible from aldolase catalysis. Such alternative enzymes (e.g. transketolase), which are actually categorized as transferases but also catalyze aldol-related additions with the aid of cofactors (Figure 5.4) such as pyridoxal 5-phosphate (PLP), thiamine pyrophosphate (TPP), or tetrahydrofolate (THF), are emerging as useful catalysts in organic synthesis. Because these operations often extend and/or complement the synthetic strategies open to aldolases, a selection of such enzymes and examples of their synthetic utility are included also in this overview.

5.2.2
Enzyme Structure and Mechanism

Mechanistically, enzymatic activation of the aldol donor substrates is achieved by stereospecific deprotonation along two different pathways (Figures 5.5 and 5.6) [29]. Class I aldolases bind their substrates covalently via imine-enamine formation with an active site lysine residue to initiate bond cleavage or formation (Figure 5.5); this can be demonstrated by reductive interception of the intermediate with borohydride which causes irreversible inactivation of the enzyme as a result of alkylation of the amine [30, 31].

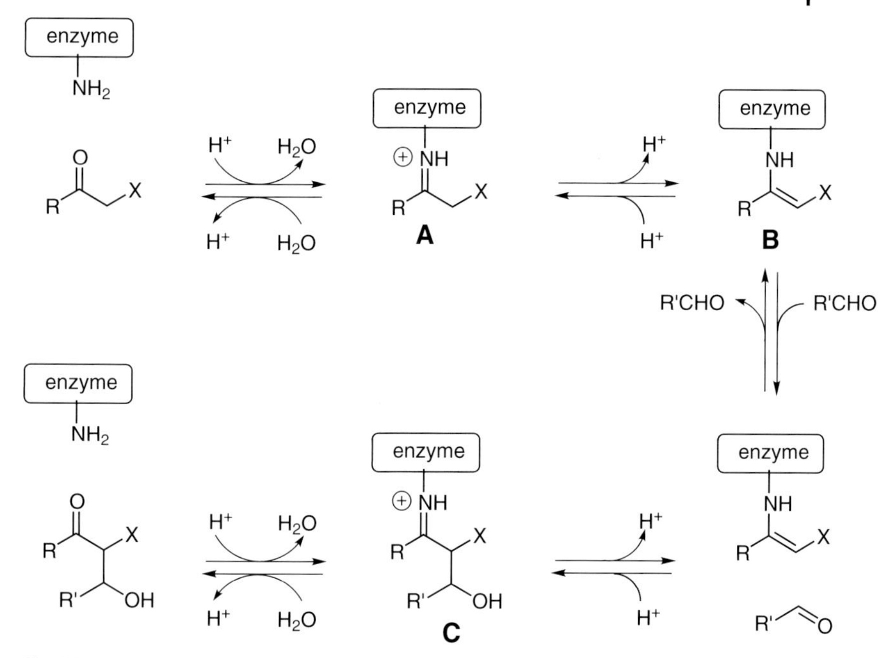

Fig. 5.5
Schematic mechanism for class I aldolases.

Fig. 5.6
Schematic mechanism for DHAP-dependent class II aldolases.

In contrast, class II aldolases utilize transition metal ions as a Lewis acid cofactor, which facilitates deprotonation by bidentate coordination of the donor to give the enediolate nucleophile (Figure 5.6) [32]. This effect is usually achieved by means of a tightly bound Zn^{2+} ion but a few other divalent cations can act instead. Evidently, aldolases of the latter class can be effectively inactivated by addition of strong complexing agents such as EDTA [29].

Mechanistic pathways for both classes of aldolases have been substantiated by several recent crystal structures of liganded enzymes that altogether provide a detailed insight into the catalytic cycles and the individual function of active site residues in the stereochemically determining events.

For class I type enzymes, the $(\beta\alpha)_8$-barrel structure of the class I fructose 1,6-bisphosphate aldolase (FruA, vide infra) from rabbit muscle was the first to be uncovered by X-ray crystal-structure analysis [33]; this was followed by those from several other species [34–37]. A complex of the aldolase with non-covalently bound substrate DHAP (dihydroxyacetone phosphate) in the active site indicates a trajectory for the substrate traveling towards the nucleophilic Lys229 N^{ε} [38, 39]. There, the proximity of side-chains Lys146 and Glu187 is consistent with their participation as proton donors and acceptors in Schiff base formation (**A**, **B**); this was further supported by site-directed mutagenesis studies [40].

Complementary information was gained from structural investigations of the *E. coli* transaldolase (EC 2.2.1.2) at 1.87 Å resolution [41], including those on its covalent complex with dihydroxyacetone, which had been trapped at the active site Lys N^{ε} by Schiff base reduction [42].

The three-dimensional structure of the neuraminic acid aldolase (NeuA, vide infra) from *Escherichia coli* has also been determined [43] after interception of the pyruvate Schiff base at Lys165 N^{ε} by borohydride reduction [44]. Further insight came from high-resolution structures (up to 1.6 Å) of the *hemophilus influenzae* enzyme in a complex with three substrate analogs [45]. Formation of a carbinolamine as the short-lived Schiff base precursor (**C**) could be trapped by flash freezing of 2-keto-3-deoxy-6-phosphogluconate aldolase in the presence of its natural donor substrate pyruvate [46]. Stereospecificity of the reaction seems to be ensured mostly by hydrophobic contact of the pyruvate methyl group in the active site.

For class II aldolases, X-ray structures of the zinc-dependent fuculose 1-phosphate aldolase (FucA, vide infra) [47, 48] and the rhamnulose 1-phosphate aldolase (RhuA, vide infra) [49] from *E. coli* have recently been solved and confirm close similarity in their overall fold. Both enzymes are homotetramers in which subunits are arranged in C_4-symmetry. The active site is assembled in deep clefts at the interface between adjacent subunits and the catalytic zinc ion is tightly coordinated by three His residues.

Phosphoglycolohydroxamate (PGH), a structural analog with the DHAP enediolate (**D**) that can be regarded as a mimic of an advanced intermediate or transition state, has been shown it to be a potent inhibitor of all currently accessible class II DHAP aldolases with K_i in the nM range [32].

Its metal chelating binding mode has been determined by liganded structures of native protein and active-site mutants. In concert, these studies have enabled derivation of a conclusive blueprint for the catalytic cycle of metal-dependent aldolases that successfully rationalizes all key stereochemical issues [50].

Structural details are also available for the class II $(\beta\alpha)_8$-barrel enzyme FruA from *E. coli* at excellent resolution (1.6 Å) [51, 52]. The homodimeric protein requires movement of the divalent zinc cofactor from a buried position to the catalytically effective surface position. Recent attempts to explore the origin of substrate discrimination of the structurally related *E. coli* aldolase with specificity for tagatose 1,6-bisphosphate by site-directed mutagenesis and structure determinations highlight the complexity of enzyme catalysis in this class of enzymes and the subtleties in substrate control [53–55].

5.2.3
Practical Considerations

Although most of the aldolases attractive for synthetic applications arise from catabolic pathways in which they function in the degradative cleavage of metabolites, the reverse C–C bond-forming processes are often favored by thermodynamic relationships [56]. Because of the bimolecular nature of the reaction, the fraction of product at equilibrium might be increased in less favorable cases by working at higher substrate concentrations or by driving the reaction with a higher concentration of one of the reactants. Individual choice will certainly depend primarily on the cost of starting materials and ease of separation from the product when used in large excess; critical factors such as enzyme inhibition by the substrate(s) or product must also be considered. The latter factor will be more obvious if one recognizes that for most of the lyases both the donor and acceptor components contain strong electrophilic sites such as aldehyde or ketone carbonyl groups, and that many lyases, including class I aldolases, involve covalent binding of substrate and product at – but not necessarily restricted to – the active site.

Most useful lyase families use substrates functionalized by anionically charged groups such as those present in pyruvate or dihydroxyacetone phosphate, which remain unaltered during catalysis. The charged group thereby introduced into products (phosphate, carboxylate) not only constitutes a handle for binding of the substrates by the enzymes but can also facilitate preparative isolation from an aqueous solution of the products and their purification by salt precipitation or ion-exchange techniques. One problem arising as a result of the affinity of the enzymes for anionic substrates is potential (competitive) inhibition by common buffer salts, e.g. inorganic phosphate.

For routine practical application most aldolases are sufficiently robust to enable their use in solution for an extended period of time, often several days. To enhance lifetime and to facilitate recovery of the biocatalysts after completion of the desired conversion, several options have been tested to immobilize the enzymes to or within insoluble matrices [57, 58], including cross-linking of enzyme crystals [59] or confining them in membrane reactors [60–62]. To facilitate conversion of less polar substrates, reaction media containing up to 30% organic cosolvent and even highly concentrated water-in-oil emulsions have been tested successfully [63]. Applications of other current techniques are indicated with the individual enzymes, if available.

Carbon–carbon bond construction is a pivotal process in asymmetric synthesis of complex molecular targets. Using enzyme catalysis, molecular complexity can be rapidly built up under mild conditions, without a need for protection of sensitive or reactive functional groups and with high chemical efficiency and usually uncompromised stereochemical fidelity [12, 14]. Within a synthetic strategy for an intricate target structure, subsequent chemoselective or regioselective differentiation of functional groups might pose a considerable difficulty that can easily outbalance the economical nature of the biocatalytic transformation. This strongly suggests consideration of schemes suitable for introduction and handling of protective groups enzymatically [13]. For this purpose, a broad array of enzymes is available, particularly such with ester and amide-forming or cleaving behavior, and the feasibility of selective operation has been well investigated [1–11]. Such technology has not yet been adopted in routine applications at its full potential, but certainly would foster a more practical interface between biocatalytic and chemical synthesis.

5.3 Pyruvate Aldolases

In vivo, pyruvate-dependent lyases mostly serve a catabolic function in the degradation of sialic acids and KDO (2-keto-3-deoxy-*manno*-octosonate) and in that of 2-keto-3-deoxy aldonic acid intermediates from hexose or pentose catabolism. Because these freely reversible aldol additions often have less favorable equilibrium constants [29], synthetic reactions are usually driven by excess pyruvate to achieve a satisfactory conversion.

A few related enzymes have been identified that use phosphoenolpyruvate in place of pyruvate which, by release of inorganic phosphate upon C–C bond formation, renders aldol additions essentially irreversible [14, 25]. Although attractive for synthetic applications, such enzymes have not yet been intensively studied for preparative applications [64].

5.3.1 *N*-Acetylneuraminic Acid Aldolase

N-Acetylneuraminic acid aldolase (or sialic acid aldolase, NeuA; EC 4.1.3.3) catalyzes the reversible addition of pyruvate (**2**) to *N*-acetyl-D-mannosamine

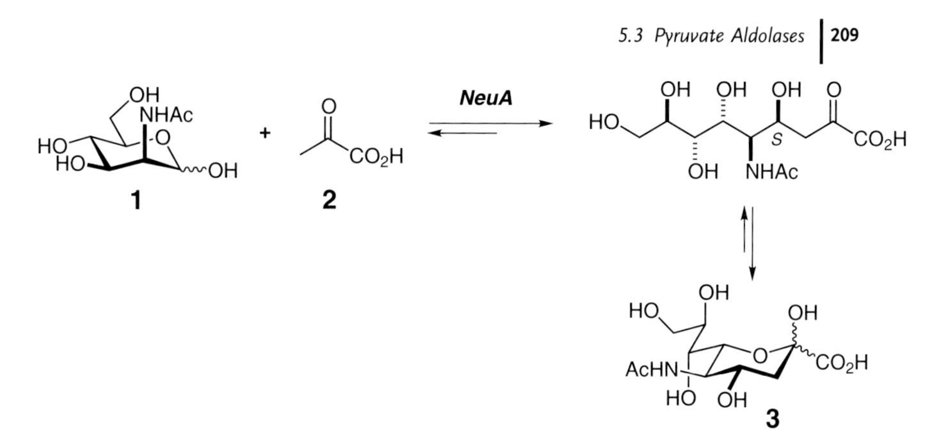

Fig. 5.7
Natural substrates of *N*-acetylneuraminic acid aldolase.

(ManNAc; **1**) in the degradation pathway of the parent sialic acid (**3**) (Figure 5.7). The NeuA lyases found in both bacteria and animals are type-I enzymes that form a Schiff base–enamine intermediate with pyruvate and promote *si*-face attack on the aldehyde carbonyl group with formation of a (4*S*) configured stereocenter. Enzyme preparations from *Clostridium perfringens* and *Escherichia coli* are commercially available, and the latter has been cloned for overexpression in *E. coli* [65, 66] or temperature-tolerant *Serratia liquefaciens* [67]. The enzyme has a broad pH optimum at approximately 7.5 and useful stability in solution at ambient temperature [68]. It has been used for synthetic applications in homogenous solution, in an immobilized form [68–72] or enclosed in a dialysis membrane [73, 74].

Because neuraminic acid is an important precursor to *Zanamivir*, an inhibitor of viral sialidases that is marketed for treatment of influenza infections, the large-scale synthesis of **3** has been developed as a prime example of an industrial aldolase bioconversion process at the multi-ton scale [75–77]. In this equilibrium-controlled bioconversion (equilibrium constant 12.7 M^{-1} in favor of the retro-aldolization), the expensive **1** can be produced by integrated enzymatic in situ isomerization of inexpensive *N*-acetylglucosamine **4** by a combination of NeuA with an *N*-acylglucosamine 2-epimerase (EC 5.1.3.8) catalyst in an enzyme membrane reactor [61, 78] (Figure 5.8). To facilitate product recovery, excessive **2** can be removed by formation of a separable bisulfite adduct, or by decomposition with yeast pyruvate decarboxylase into volatile compounds [75, 79]. The need for excess **2** might be circumvented altogether by coupling of the synthesis of **3** to a thermodynamically more favored process, e.g. combination with an irreversible sialyltransferase reaction to furnish sialyloligosaccharides [80, 81].

Extensive studies have indicated that only pyruvate is acceptable as the NeuA donor substrate, with the exception of fluoropyruvate [82], but that the enzyme has fairly broad tolerance of stereochemically related aldehyde

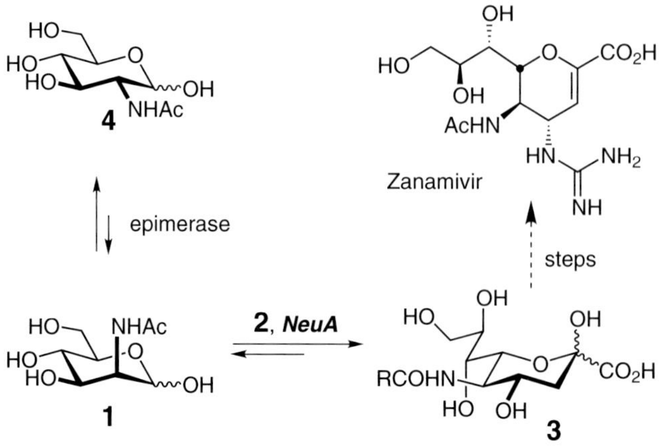

Fig. 5.8
Industrial process for the production of *N*-acetylneuraminic acid as a precursor to an influenza inhibitor.

substrates as acceptor alternatives, for example several sugars and their derivatives larger than or equal in size to pentoses (Table 5.1) [62, 68, 83]. Permissible variations include replacement of the natural D-*manno* configured substrate with derivatives containing modifications such as epimerization, substitution, or deletion at positions C-2, C-4, or C-6 [20, 25]. Epimerization at C-2, however, is restricted to small polar substituents at strongly reduced reaction rates [84, 85]. As an example of continuous process design, KDN (**6**) has been produced on a 100-g scale from D-mannose and pyruvate using a pilot-scale enzyme membrane reactor (EMR) with a space–time yield of 375 g L^{-1} d^{-1} and an overall crystallized yield of 75% (Figure 5.9) [86]. Similarly, L-KDO (**5**) can be synthesized from L-arabinose [62]. The broad substrate tolerance of the catalyst for sugar precursors has recently been exploited in the equilibrium generation of sialic acid and analogs for in-situ screening of a dynamic combinatorial library [87].

Because of the importance of sialic acids in a wide range of biological recognition events, the aldolase has become popular for chemoenzymatic synthesis of a multitude of other natural and non-natural derivatives or analogs of **3** (Figure 5.10). Many examples have been reported for sialic acid modifications at C-5/C-9 [20, 25], for example differently *N*-acylated derivatives **8** [88–90], including amino acid conjugates [91], or 9-modified analogs **9** [92–94], in search for new neuraminidase (influenza) inhibitors. Most notably, the *N*-acetyl group in **1** can be either omitted [84, 85] or replaced by sterically demanding substituents such as *N*-Cbz (**10**) [95, 96] or even a non-polar phenyl group [84] without destroying activity. Large acyl substituents are also tolerated at C-6, as shown by the conversion of a Boc-glycyl derivative **11** as a precursor to a fluorescent sialic acid conjugate [97].

Tab. 5.1
Substrate tolerance of neuraminic acid aldolase.

R₁	*R₂*	*R₃*	*R₄*	*R₅*	*Yield [%]*	*Rel. Rate [%]*	*Ref.*
NHAc	H	OH	H	CH_2OH	85	100	72, 68
NHAc	H	OH	H	CH_2OAc	84	20	72, 68
NHAc	H	OH	H	CH_2OMe	59	–	72
NHAc	H	OH	H	CH_2N_3	84	60	85
NHAc	H	OH	H	$CH_2OP(O)Me_2$	42	–	85
NHAc	H	OH	H	CH_2O(L-lactoyl)	53	–	72, 85
NHAc	H	OH	H	CH_2O(Gly-*N*-Boc)	47	–	97
NHAc	H	OH	H	CH_2F	22	60	85
NHAc	H	OMe	H	CH_2OH	70	–	72
NHAc	H	H	H	CH_2OH	70	–	100
$NHC(O)CH_2OH$	H	OH	H	CH_2OH	61	–	72
NHCbz	H	OH	H	CH_2OH	75	–	96
OH	H	OH	H	CH_2OH	84	91	79, 100
OH	H	H	H	CH_2OH	67	35	100
OH	H	H	F	CH_2F	40	–	85
OH	H	OH	H	H	66	10	100
H	F	OH	H	CH_2OH	30	–	85
H	H	OH	H	CH_2OH	36	130	85, 100
Ph	H	OH	H	CH_2OH	76	–	100

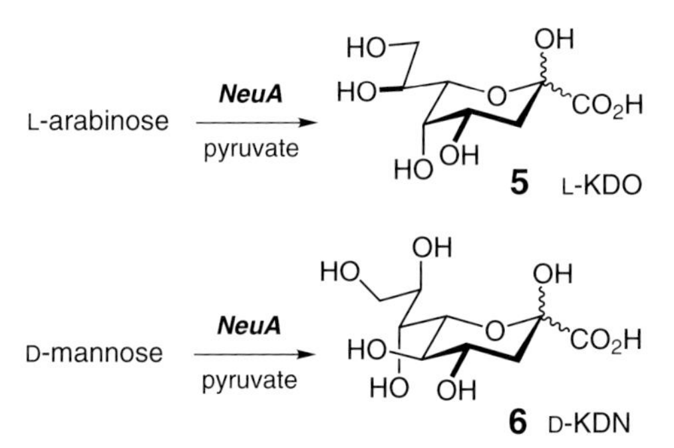

Fig. 5.9
Sialic acids prepared on large scale.

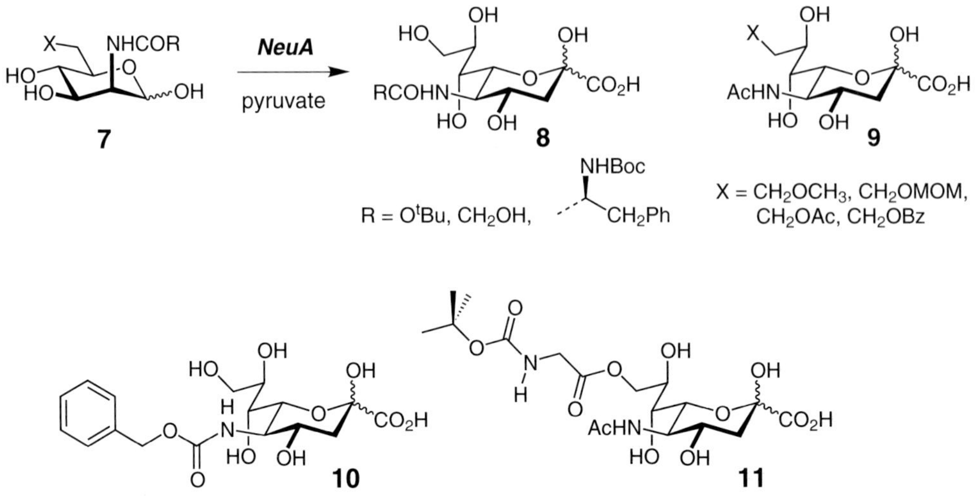

Fig. 5.10
Neuraminic acid derivatives accessible by NeuA catalysis.

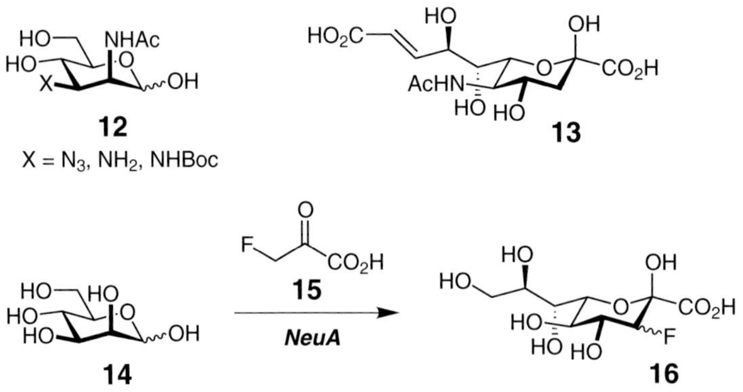

Fig. 5.11
Sugar derivatives not accepted by NeuA in direction of synthesis or cleavage, and use of fluoropyruvate for synthesis of fluorosialates.

No 3-azido, 3-amino, or Boc-protected mannosamine analogs **12** were accepted by the enzyme (Figure 5.11) [98] which suggests that the presence of a 3-hydroxyl group is a necessary precondition for substrates of the aldolase. Likewise, conformationally inflexible acrylate **13** was not accepted in cleavage direction. By use of fluoropyruvate **15** as the donor substrate a series of diastereomeric 3-deoxy-3-fluoro ulosonic acids such as **16** has been prepared in good yields ($>49\%$) from pentoses or hexoses [82]. Such products are attractive for non-invasive in-vivo pharmacokinetic studies by NMR tomography (^{19}F derivatives) or positron-emission spectroscopy (^{18}F derivatives).

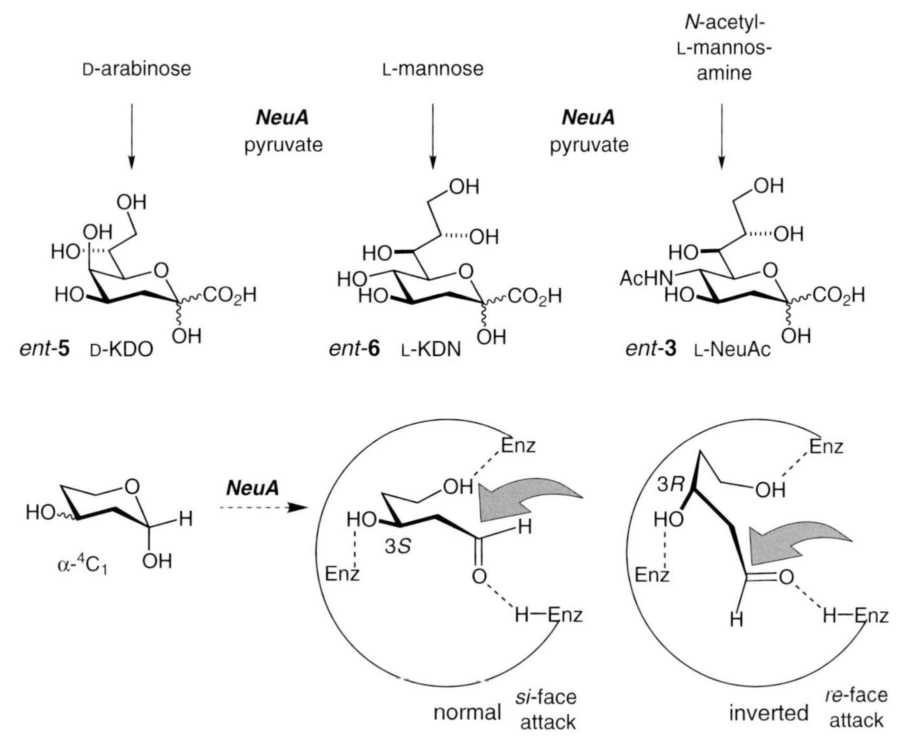

Fig. 5.12
NeuA-catalyzed synthesis of sialic acids bearing the unusual inverted (4*R*) configuration, and three-point binding model for prediction of NeuA stereoselectivity based on conformational analysis.

In most of the examples investigated so far a high level of asymmetric induction for the (4*S*) configuration is retained. Several carbohydrates were, however, also found to be converted with random or even inverse stereoselectivity for the C-4 configuration [79, 83, 99]. Products generated by the unusual *re*-face attack with (4*R*) stereochemical preference include a number of related higher ulosonic acids of biological importance such as D-KDO (*ent*-**5**) [62, 79, 100, 101] or the enantiomers of naturally occurring sugars such as L-KDN (*ent*-**6**) [68, 71, 75, 86], or L-NeuNAc (*ent*-**3**) (Figure 5.12) [79]. Ready access to compounds of this type can be particularly valuable for investigation of the biological activity of sialoconjugates containing non-natural sialic acid derivatives [102]. A critical and distinctive factor seems to be recognition by the enzyme catalyst of the configuration at C-3 in the aldehydic substrate [62, 83]. The unusual outcome has been interpreted as a result of thermodynamic control or, alternatively, of an inverse conforma-

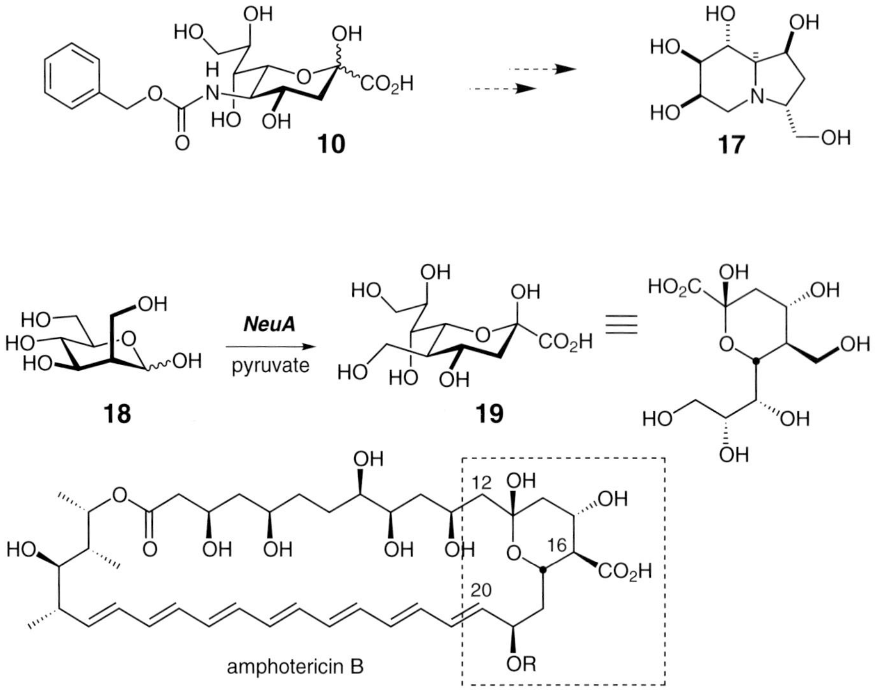

Fig. 5.13
NeuA-catalyzed preparation of an intermediate for alkaloid synthesis, and of a synthetic precursor to the macrolide antibiotic amphotericin B.

tional preference of the different substrate classes [62]. From an application-oriented perspective a three-point binding model has been proposed for the conversion of substrate analogs in the direction of synthesis (Figure 5.12) [25]. On the basis of the (3*S*)-α-$^4C_1$ structure of the natural substrate and conformational analysis of its analogs, this model can predict the occasionnally observed compromise in, or even total inversion of, the facial stereoselectivity of C–C bond formation.

Starting from the *N*-Cbz-protected aldolase product **10**, azasugar **17** has been obtained stereoselectively by internal reductive amination as an analog of the bicyclic, indolizidine-type glycosidase inhibitor castanospermine (Figure 5.13) [96]. It has also been recognized that the C-12 to C-20 sequence of the macrolide antibiotic amphotericin B resembles the *β*-pyranose tautomer of **19**. Thus, the branched-chain *manno*-configured substrate **18** was successfully chain-extended under the action of NeuA catalysis to yield the potential amphotericin B synthon **19** in good yield [103, 104].

A one-pot ^{13}C-labeling strategy has been developed for sialic acids by exploiting the reversible nature of the aldolase reaction (Figure 5.14) [105].

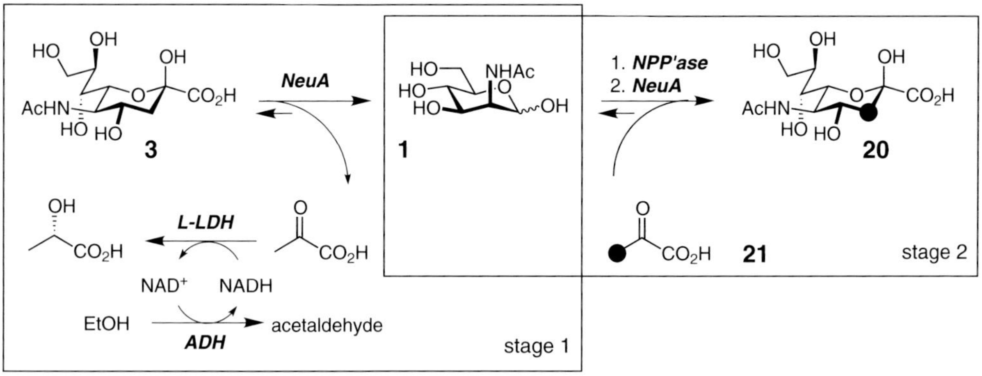

Fig. 5.14
Process for [^{13}C]-labeling of neuraminic acid (and analogs) by controlled reversible aldolizations (• denotes [^{13}C]-label).

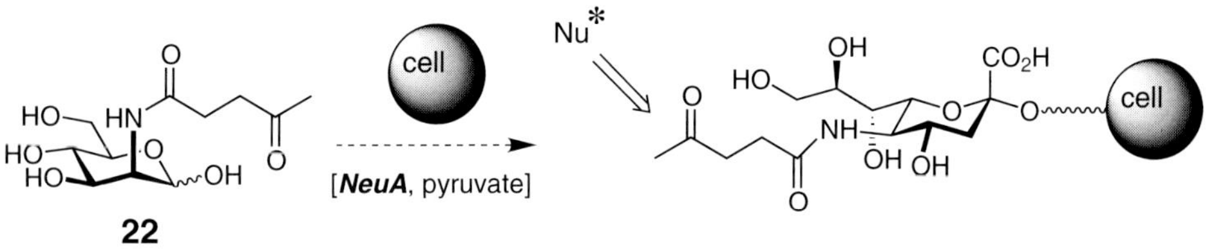

Fig. 5.15
Cellular synthesis of a modified sialic acid by exposure of human cells to *N*-levulinoyl D-mannosamine **22**. Resulting levulinoylated cell surface glycoproteins provide an opportunity for selective, covalent attachment of a variety of nucleophiles Nu*.

The procedure consists of stepwise enzymatic degradation of chemically prepared neuraminic acid derivatives (**3**) to give the corresponding modified ManNAc analogs (**1**), which is followed by subsequent neuraminic acid reconstruction in the presence of [3-^{13}C] labeled pyruvate (**21**). To ensure complete conversion in the desired direction the first step was promoted by cofactor-dependent reduction of pyruvate; the second, synthetic step, was uncoupled by destruction of the nucleotide cofactor before addition of labeled pyruvate. Sialic acids such as **20** with high label incorporation ($>87\%$) were obtained in good yields (46–76%).

Acceptance of the non-natural *N*-levulinoyl D-mannosamine **22**, containing a ketone moiety in the *N*-acyl group, by the cellular machinery in vivo has been used to produce cell-surface oligosaccharides modified in their neuraminic acid constituents (Figure 5.15) [106, 107]. The reactive ketone groups thus produced on the cell surface could then be used for versatile covalent cell redecoration under physiological conditions, for example by

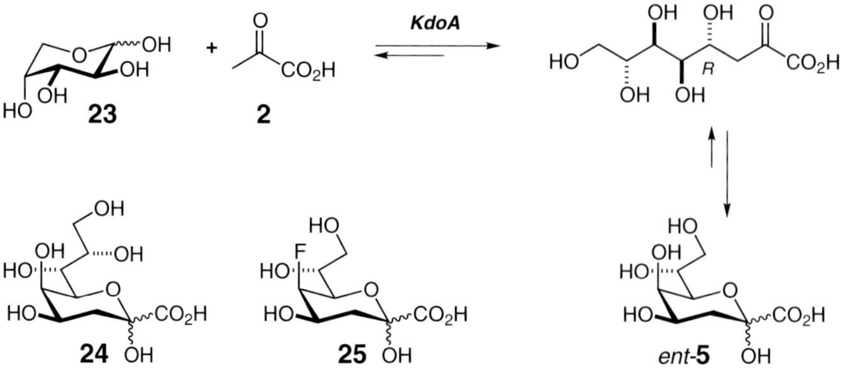

Fig. 5.16
Natural substrates of the 2-keto-3-deoxy-*manno*-octosonic acid aldolase, and non-natural sialic acids obtained by KdoA catalysis.

attaching functional nucleophiles (Nu*) such as fluorescent hydrazine markers or toxin conjugates.

5.3.2
KDO Aldolase

The functionally related 2-keto-3-deoxy-*manno*-octosonate (KDO) aldolase (KdoA; correctly termed 3-deoxy-D-*manno*-octulosonic acid aldolase, EC 4.1.2.23) is involved in the catabolism of the eight-carbon sugar D-KDO *ent*-**5**, a core constituent of the capsular polysaccharides (K-antigens) and outer membrane lipopolysaccharides (LPS, endotoxin) of Gram-negative bacteria [108], and of the cell wall of algae and a variety of plants [109]. The aldolase, which reversibly degrades *ent*-**5** to D-arabinose **23** and pyruvate (Figure 5.16), has been isolated from *Aerobacter cloacae*, *E. coli*, and an *Aureobacterium barkerei* strain in which the enzyme seems to be located in the cell wall or membrane fraction [110]. Partially purified KdoA preparations have been studied for synthetic applications [101, 111]; these studies have shown that, similar to NeuA, the enzyme has a broad substrate specificity for aldoses (Table 5.2) whereas pyruvate was found to be irreplaceable. As a notable distinction, KdoA was also active on smaller acceptors such as glyceraldehyde. Preparative applications, e.g. that for the synthesis of KDO (*ent*-**5**) and its homologs or analogs **24**/**25**, suffer from a less attractive equilibrium constant of 13 M^{-1} in direction of synthesis [56]. The stereochemical course of aldol additions generally seems to adhere to *re*-face attack on the aldehyde carbonyl which is complementary to that of NeuA. On the basis of the results published so far it can be concluded that a (3*R*) configuration is necessary (but not sufficient), and that stereochemical requirements at C-2 are less stringent

Tab. 5.2
Substrate tolerance of 2-keto-3-deoxy-*manno*-octosonic acid aldolase [101].

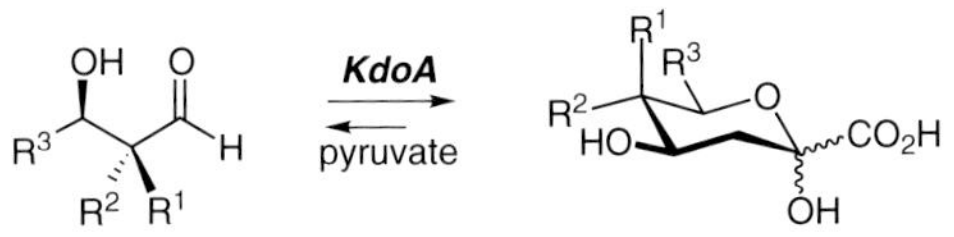

Substrate	R_1	R_2	R_3	*Yield [%]*	*Rel. Rate [%]*
D-Altrose	OH	H	(RR)-$(CHOH)_2$–CH_2OH	–	25
L-Mannose	H	OH	(SS)-$(CHOH)_2$–CH_2OH	61	15
D-Arabinose	OH	H	(R)-CHOH–CH_2OH	67	100
D-Ribose	H	OH	(R)-CHOH–CH_2OH	57	72
2-Deoxy-2-fluoro-D-ribose	F	H	(R)-CHOH–CH_2OH	19	46
2-Deoxy-D-ribose	H	H	(R)-CHOH–CH_2OH	47	71
5-Azido-2,5-dideoxy-D-ribose	H	H	(R)-CHOH–CH_2N_3	–	15
D-Threose	OH	H	CH_2OH	–	128
D-Erythrose	H	OH	CH_2OH	39	93
D-Glyceraldehyde	H	OH	H	11	23
L-Glyceraldehyde	OH	H	H	–	36

[101]. For simple product recovery excess pyruvate can be decomposed with pyruvate decarboxylase.

5.3.3 DAHP Synthase

In the biosynthesis of aromatic amino acids in microorganisms and plants, 3-deoxy-D-*arabino*-heptulosonic acid 7-phosphate synthase (DAHP synthase or AroS; EC 4.1.2.15) is a pivotal enzyme for carbon framework construction. The enzyme transfers a pyruvate moiety from phosphoenolpyruvate (PEP) to D-erythrose-4-phosphate (Ery4P) in an aldol-like reaction [112, 113]. Use of the activated donor renders formation of DAHP (**26**), the key metabolic intermediate, practically irreversible (Figure 5.17). The AroS enzyme has been purified from *E. coli* [114] and *Streptomyces rimosus* [115] and has been cloned from *E. coli*, *Salmonella typhimurium*, and potato [116, 117]. The structure of an *E. coli* isozyme has been determined in a complex with PEP at high resolution [118]. The cytosolic AroS isoenzyme prevalent in higher plants has been reported to be remarkably highly tolerant of acceptor substrates, including glyoxalate, glycolaldehyde, and most 3–5-carbon sugars or their phosphates [119].

For synthesis of **26** (Figure 5.17), an immobilized enzyme system was set up to generate the labile acceptor aldehyde in situ by ketol transfer from D-fructose 6-phosphate [120]. The twofold driving force from PEP used in the integrated scheme as both a phosphoryl and aldol donor secured complete

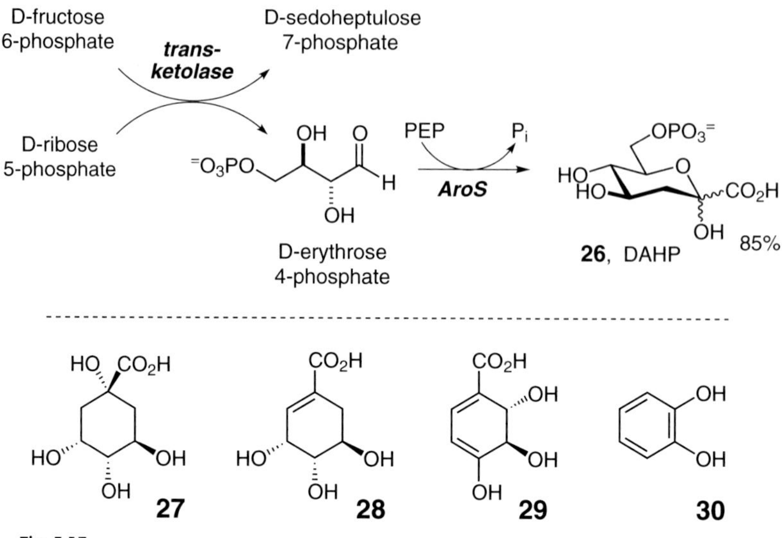

Fig. 5.17
Multi-enzymatic scheme for synthesis of 3-deoxy-*arabino*-heptulosonic acid 7-phosphate based on the catalysis of DAHP synthase, and products generated consecutively in vivo by advanced microbial pathway engineering.

conversion of the starting material. Generally, synthesis of **26** on a larger scale is more efficient and economical by using engineered whole cells that contain a plasmid coding for AroS [121]; thereby necessary substrates and enzymes are intrinsically provided by regular cell metabolism. Similar elaborate schemes have been designed to produce quinic acid (**27**) [122], shikimic acid (**28**) [123], cyclohexadiene *trans*-diols, for example **29** [208], or phenols such as **30** [124–126] by engineered microbial biosynthesis.

Purified enzyme from *Escherichia coli* has been studied for its reactivity with homologous five-carbon phosphosugars to yield KDO derivatives, and with modified nucleophiles. Thus, the stereochemically distinct fluoro-analogs of PEP (*Z*)- and (*E*)-**31** could be separately condensed with D-erythrose 4-phosphate to yield, stereospecifically, the corresponding (3*S*)- and (3*R*)-configured, fluoro-substituted DAHP derivatives **32** and **33**, respectively (Figure 5.18) [127]. This provides direct evidence that the enzyme catalyzes *si* face addition of the C-3 of PEP to the *re* face of the acceptor aldehyde.

5.3.4
KDPG Aldolase and Related Enzymes

An aldolase specific for cleavage of 2-keto-3-deoxy-6-phospho-D-gluconate (**35**) (KDPGlc aldolase or KdgA; EC 4.1.2.14) is produced by many species of bacteria for degradation of 6-phosphogluconate to give pyruvate and D-

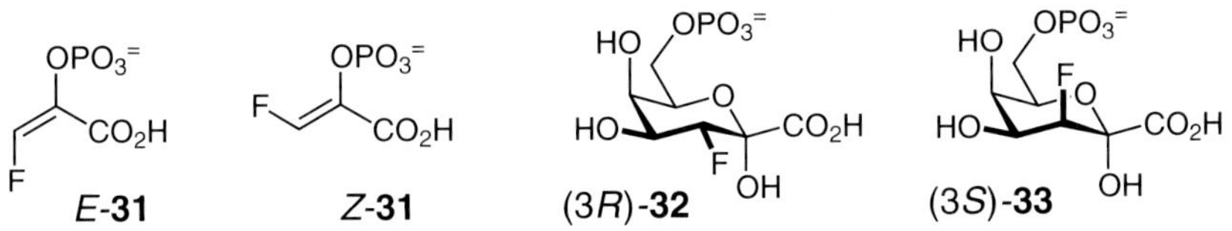

Fig. 5.18
Stereospecific conversion of 3-fluoro-labeled PEP analogs to substituted DAHP derivatives by use of DAHP synthase.

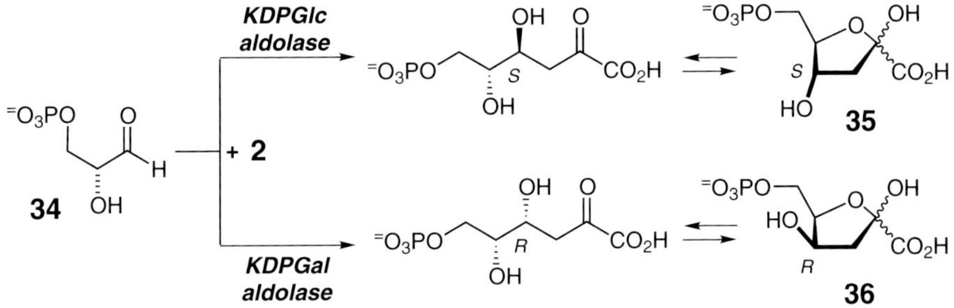

Fig. 5.19
Aldol reactions catalyzed in vivo by the 2-keto-3-deoxy-6-phospho-D-gluconate and 2-keto-3-deoxy-6-phospho-D-galactonate aldolases.

glyceraldehyde 3-phosphate **34** (Figure 5.19). The equilibrium constant favors synthesis (10^3 M^{-1}) [128]. Enzymes belonging to class I aldolases have been isolated [129–131] and cloned [132–135] from a variety of microbial sources.

Genome sequencing of *E. coli* revealed that the enzyme is identical with 2-keto-4-hydroxyglutarate aldolase (KHG aldolase or KhgA; EC 4.1.3.16), which is involved in the catabolism of hydroxyproline [136, 137]. This activity is responsible for the retroaldolization of 4-hydroxy-2-oxoglutaric acid to pyruvate **2** and glyoxalate. In both cases, the enzyme recognizes an identical (4*S*) configuration.

Enzyme preparations from liver or microbial sources were reported to have rather high substrate specificity [138] for the natural phosphorylated acceptor D-**34** and, but at much reduced reaction rates, to have rather broad substrate tolerance for polar, short-chain aldehydes (Table 5.3) [139–141]. Simple aliphatic or aromatic aldehydes are not converted. High stereoselectivity of the enzyme has been used in the preparation of compounds **39** and **40** and in a two-step enzymatic synthesis of **38**, the *N*-terminal amino acid portion of nikkomycin antibiotics (Figure 5.20) [142]. Thermophilic KdgA from *Thermotoga maritima* has been shown to have substrate tolerance with

Tab. 5.3
Substrate tolerance of 2-keto-3-deoxy-6-phospho-D-gluconate aldolase [140].

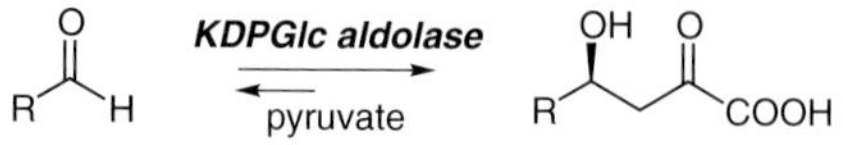

Substrate	*R*	*Rel. Rate [%]*
D-Glyceraldehyde 3-phosphate	D-$CH_2OH–CH_2OPO_3^{=}$	100.0
3-Nitropropanal	$CH_2–CH_2NO_2$	1.6
Chloroethanal	CH_2Cl	1.0
D-Glyceraldehyde	D-$CHOH–CH_2OH$	0.8
D-Lactaldehyde	D-$CHOH–CH_3$	0.2
D-Ribose 5-phosphate	D-*ribo*-$(CHOH)_3–CH_2OPO_3^{=}$	0.04
Erythrose	*erythro*-$(CHOH)_2–CH_2OH$	0.01
Glycolaldehyde	CH_2OH	0.01

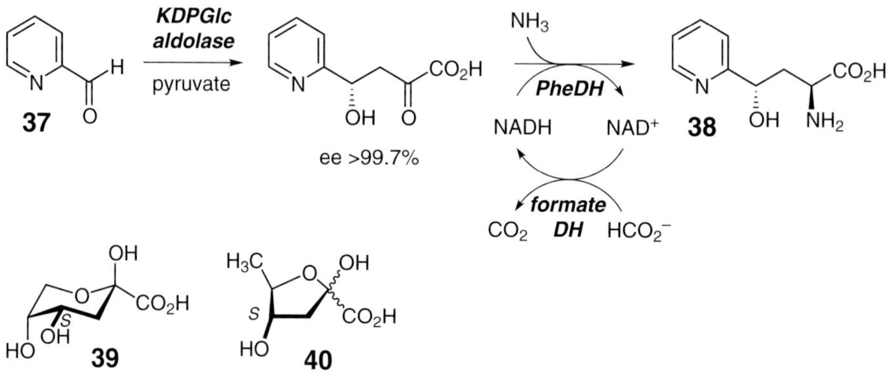

Fig. 5.20
Stereoselective synthesis of the amino acid portion of nikkomycin antibiotics and hexulosonic acids using KDPGlc aldolase.

a breadth that seems comparable to, but qualitatively distinct from, that of mesophilic enzymes, although with diminished stereoselectivity [135].

The aldolase from *E. coli* has been mutated for improved acceptance of non-phosphorylated and enantiomeric substrates towards facilitated enzymatic syntheses of both D and L sugars [143, 144].

Comparable with the situation for the sialic acid and KDO lyases (vide supra), a class I lyase complementary to the KDPGlc aldolase is known that has a stereopreference for the (4*S*) configuration (Figure 5.19). The aldolase, which acts on 2-keto-3-deoxy-6-phospho-D-galactonate (**36**) (KDPGal aldolase; EC 4.1.2.21) and is less abundant [145, 146], has recently been studied for synthetic applications [147].

FruA RhuA
3S,4R 3R,4S
42
D-fructose 1,6-bisphosphate
43
L-rhamnulose 1-phosphate
TagA FucA
3S,4S 3R,4R
44
D-tagatose 1,6-bisphosphate
45
L-fuculose 1-phosphate

Fig. 5.21
Aldol reactions catalyzed in vivo by the four stereo-complementary dihydroxyacetone phosphate-dependent aldolases.

5.4 Dihydroxyacetone Phosphate Aldolases

Whereas pyruvate aldolases form only a single stereogenic center, the aldolases specific for dihydroxyacetone phosphate (DHAP, **41**) as a nucleophile create two new asymmetric centers at the termini of the new C–C bond. Particularly useful for synthetic applications is that Nature has evolved a full set of four unique aldolases (Figure 5.21) to cleave all possible stereochemical permutations of the vicinal diol at C-3/C-4 of ketose 1-phosphates **42–45** during the retro-aldol cleavage [26]. These aldolases have proved to be exceptionally powerful tools for asymmetric synthesis, particularly stereocontrolled synthesis of polyoxygenated compounds, because of their relaxed substrate specificity, high level of stereocontrol, and commercial availability. In the direction of synthesis this situation formally enables generation of all four possible stereoisomers of a desired product in building-block fashion [22, 25, 26]. In this manner the deliberate preparation of a specific target molecule can be addressed simply by choosing the corresponding enzyme and suitable starting material, affording full control over constitution and absolute and relative configuration of the desired product. Alternatively, the stereocomplementary nature of enzymes also enables a combinatorial, stereodivergent approach (Figure 5.1) to the generation of all stereoisomers in a

Fig. 5.22
Natural glycolytic substrates of fructose 1,6-bisphosphate aldolases.

small parallel library of diastereoisomers, e.g. for bioactivity screening [25]. Properties of the individual enzymes are therefore discussed separately from more general treatment of their preparative applications.

The DHAP aldolases are quite specific for the phosphorylated nucleophile **41**, which must therefore be prepared independently or generated in situ (cf. Section 5.4.4). Initial aldol products will thus contain a phosphate ester moiety, which facilitates product isolation, for example by barium salt precipitation or by use of ion-exchange techniques. The corresponding phosphate free compounds can be easily obtained by mild enzymatic hydrolysis using an inexpensive alkaline phosphatase at pH 8–9 [148], whereas base-labile compounds might require working at pH 5–6 using a more expensive acid phosphatase [149].

5.4.1 FruA

The D-fructose 1,6-bisphosphate aldolase (FruA; EC 4.1.2.13) catalyzes, in vivo, the equilibrium addition of **41** to D-glyceraldehyde 3-phosphate (GA3P, **34**) to give D-fructose 1,6-bisphosphate (**42**). The equilibrium constant for this reaction of $10^4\ \text{M}^{-1}$ strongly favors synthesis [56]. The enzyme occurs ubiquitously and has been isolated from a variety of prokaryotic and eukaryotic sources, both as class I and class II forms [29]. Corresponding FruA genes have been cloned from a variety of sources and overexpressed [25]. Typically, class I FruA enzymes are tetrameric whereas the class II FruA are dimers, with subunits of ~40 kDa. The microbial class II aldolases are usually much more stable in solution (half-lives of several weeks to months) than their mammalian counterparts of class I (a few days) [149–151].

The class I FruA isolated from rabbit muscle ("RAMA") is the aldolase used for preparative synthesis in the widest sense, because of its commercial availability and useful specific activity of ~20 U mg^{-1}. Its operating stability in solution is limiting, but recently more robust homologous enzymes have been cloned, e.g. from *Staphylococcus carnosus* [152] or from the extremophilic *Thermus aquaticus* [153], which promise to be unusually stable in synthetic applications [154]. Attempts at catalyst immobilization have been performed with rabbit muscle FruA, which has been covalently attached to microcarrier beads [58], cross-linked in enzyme crystals (CLEC) [59], or enclosed in a dialysis membrane [73]. It was recently shown that

Tab. 5.4
Substrate tolerance of fructose 1,6-bisphosphate aldolase.

FruA
DHAP
$OPO_3^=$

R	*Rel. Rate [%]*	*Yield [%]*	*Ref.*
D-$CHOH–CH_2OPO_3^=$	100	95	149, 177
H	105	–	149
CH_3	120	–	149
CH_2Cl	340	50	149, 212
$CH_2–CH_3$	105	73	149
$CH_2–CH_2–COOH$	–	81	177
$CH_2OCH_2C_6H_5$	25	75	149
D-$CH(OCH_3)–CH_2OH$	22	56	149
CH_2OH	33	84	149, 177
D-$CHOH–CH_3$	10	87	149, 207
L-$CHOH–CH_3$	10	80	149
DL-$CHOH–C_2H_5$	10	82	149
$CH_2–CH_2OH$	–	83	206, 207
$CH_2–C(CH_3)_2OH$	–	50	205
DL-$CHOH–CH_2F$	–	95	207
DL-$CHOH–CH_2Cl$	–	90	206
DL-$CHOH–CH_2–CH{=}CH_2$	–	85	206

less polar substrates can be converted as highly concentrated water-in-oil emulsions [63].

Literally hundreds of aldehydes have so far been tested successfully by enzymatic assay and preparative experiments as a replacement for **34** in rabbit muscle FruA catalyzed aldol additions [18, 25], and most of the corresponding aldol products have been isolated and characterized. A compilation of selected typical substrates and their reaction products is provided in Table 5.4, and further examples are indicated in Section 5.4.5. In comparison, metal dependant FruA enzymes are more specific for phosphorylated substrates and accept non-phosphorylated substrate analogs only with much reduced activity ($< 1\%$).

For chiral substituted aldehydes, racemic precursors are most often more readily available than enantiomerically pure material, but must be resolved either before or during the aldol step. Rabbit FruA discriminates between the enantiomers of its natural substrate with a 20:1 preference for D-GA3P (**34**) over its L antipode [149]. Assistance from anionic binding was revealed by a study on a homologous series of carboxylated 2-hydroxyaldehydes for which enantioselectivity was optimum when the distance of the charged group equaled that of **34** (Figure 5.23) [155]. The resolution of racemic substrates is not, however, generally useful, because the kinetic enantiose-

46 E > 10 **34** E > 10

Fig. 5.23
Kinetic enantiopreference of rabbit muscle FruA.

lectivity for non-ionic aldehydes is rather low [149]. 3-Azido substituents (**95**) can lead to up to ninefold preference for enantiomers in kinetically controlled experiments [156] whereas hydroxyl (usually some preference for the D antipodes) and derived functionality, or chiral centers at a larger distance, rarely enable more than statistical diastereomer formation.

5.4.2
TagA

The D-tagatose 1,6-bisphosphate aldolase (TagA; EC 4.1.2.n) is involved in the catabolism of D-*galacto*-configured carbohydrates and catalyzes the reversible cleavage of D-tagatose 1,6-bisphosphate (**44**) to D-glyceraldehyde 3-phosphate (**34**) and dihydroxyacetone phosphate (**41**). Enzymes of class I that occur in a variety of *Coccus* species seem to have apparently no stereochemical selectivity with regard to distinction between **42** and **44** [157], whereas class II aldolases from Gram-positive microorganisms are highly stereoselective for the natural substrate in both cleavage and synthesis directions [158, 159]. The genes coding for TagA have been cloned from *Coccus* strains [160] and from *E. coli* [53, 161].

The reverse, synthetic reaction can be used to prepare ketose bisphosphates, as has been demonstrated by an expeditious multienzymatic synthesis of the all-*cis* (3*S*,4*S*)-configured D-tagatose 1,6-bisphosphate **44** from dihydroxyacetone including its cofactor-dependent phosphorylation, using the purified TagA from *E. coli* (Figure 5.24) [158, 159]. The aldolase also accepts a range of unphosphorylated aldehydes as substrates but produces diastereomeric mixtures only. This lack of stereoselectivity with generic substrate analogs, which makes native TagA enzymes synthetically less useful, has stimulated recent X-ray structure determination and protein engineering to improve its properties [53, 54, 162].

5.4.3
RhuA and FucA

The L-rhamnulose 1-phosphate aldolase (RhuA; EC 4.1.2.19) and the L-fuculose 1-phosphate aldolase (FucA; EC 4.1.2.17) are found in many microorganisms where they are responsible for the degradation of deoxysugars L-rhamnose and L-fucose to give **41** and L-lactaldehyde (Figure 5.25). RhuA is specific for cleavage and synthesis of a L-*threo* diol unit whereas FucA

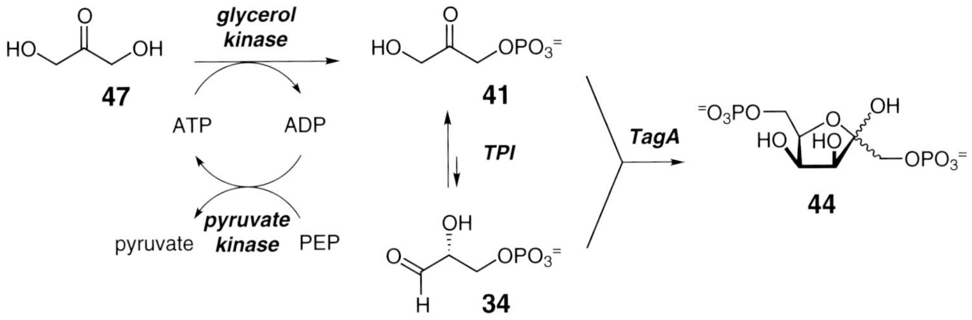

Fig. 5.24
Enzymatic one-pot synthesis of tagatose 1,6-bisphosphate based on the stereoselective TagA from *E. coli*.

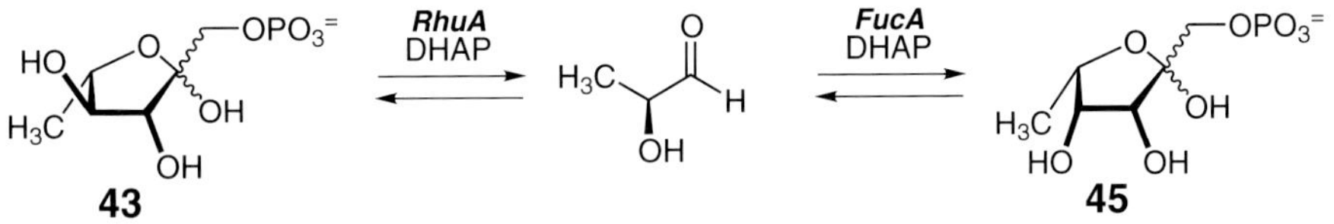

Fig. 5.25
Natural substrates of microbial deoxysugar phosphate aldolases.

recognizes the corresponding D-*erythro* configuration. Both enzymes have been isolated from several sources [163–165], and the proteins from *E. coli* have been shown to be homotetrameric Zn^{2+}-dependent aldolases with subunit molecular weights of ~25 kDa and ~30 kDa, respectively [150]. Cloning of RhuA [166, 167] and FucA [168] enzymes from several microorganisms has been reported. Efficient overexpression [150, 160, 169] has set the stage for X-ray structure determinations of both *E. coli* proteins [48, 49]. Like a number of other aldolases, both the RhuA and FucA enzymes are commercially available.

Overall practical features make the RhuA and FucA enzymes quite similar for synthetic applications. Both metalloproteins have very high stability in the presence of low Zn^{2+} concentrations with half-lives in the range of months at room temperature, and the enzymes even tolerate the presence of large proportions of organic cosolvents ($\geq 30\%$) [150]. Both have very broad substrate tolerance for variously substituted aldehydes, which is very similar to that of the FruA enzymes, with conversion rates generally being usefully high (Table 5.5). Characteristically, of all the DHAP aldolases yet investigated the RhuA has the greatest tolerance of sterically congested acceptor substrates, as exemplified by the conversion of the tertiary aldehyde 2,2-dimethyl-3-hydroxypropanal **48** (Figure 5.26) [25]. Aldehydes carrying an anionically charged group close to the carbonyl group, for example gly-

Tab. 5.5
Substrate tolerance of L-rhamnulose 1-phosphate and L-fuculose 1-phosphate aldolases [150, 170].

R	Rel. Rate [%]	RhuA Selectivity threo:erythro	Yield [%]	Rel. Rate [%]	FucA Selectivity threo:erythro	Yield [%]
L-CH_2OH–CH_3	100	>97:3	95	100	<3:97	83
CH_2OH	43	>97:3	82	38	<3:97	85
D-CHOH–CH_2OH	42	>97:3	84	28	<3:97	82
L-CHOH–CH_2OH	41	>97:3	91	17	<3:97	86
CH_2–CH_2OH	29	>97:3	73	11	<3:97	78
CHOH–CH_2OCH_3	–	>97:3	77	–	<3:97	83
CHOH–CH_2N_3	–	>97:3	97	–	<3:97	80
CHOH–CH_2F	–	>97:3	95	–	<3:97	86
H	22	–	81	44	–	73
CH_3	32	69:31	84	14	5:95	54
$CH(CH_3)_2$	22	97:3	88	20	30:70	58

Fig. 5.26
Substrate tolerance of RhuA for tertiary hydroxyaldehyde.

ceraldehyde phosphates, other sugar phosphates, or glyoxylic acid, are not converted by the deoxysugar aldolases [25, 148].

The stereospecificity of both enzymes for the absolute (3*R*) configuration is mechanism-based (vide supra), and RhuA generally directs attack of the DHAP enolate to the *re* face of an approaching aldehyde carbonyl and is thereby specific for synthesis of a (3*R*,4*S*)-*trans* diol unit [150] whereas FucA controls *si* face attack to create the corresponding (3*R*,4*R*)-*cis* configuration [150, 169]. This specificity for a vicinal configuration is, however, somewhat substrate-dependent, in that simple aliphatic aldehydes can give rise to a certain fraction of the opposite diastereomer [25, 150]. In general, stereochemical fidelity is usually higher, and diastereospecific results are observed more often with the FucA than with the RhuA enzyme. Stereocontrol is, in general, usually highly effective with aldehydes carrying a 2- or 3-hydroxyl group (Table 5.5). In addition, both aldolases have strong kinetic

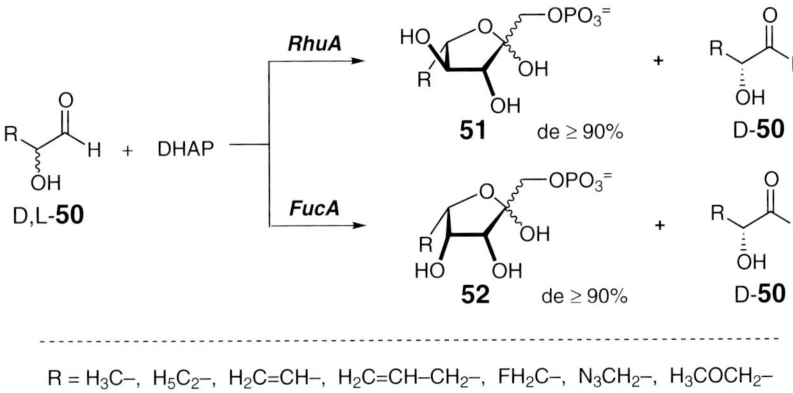

R = H_3C–, H_5C_2–, H_2C=CH–, H_2C=CH–CH_2–, FH_2C–, N_3CH_2–, H_3COCH_2–

Fig. 5.27
Kinetic enantiopreference of class II DHAP aldolases useful for racemic resolution of α-hydroxyaldehydes.

preference for L-configured enantiomers of 2-hydroxyaldehydes **50** (Figure 5.27), which facilitates racemate resolution [170, 171]. Essentially, this feature enables concurrent determination of three contiguous chiral centers in final products **51** or **52** having an L configuration (d.e. ≥95) even when starting from more readily accessible racemic material.

5.4.4
DHAP Synthesis

Apparently, all DHAP aldolases are highly specific for **41** as the donor component for mechanistic reasons [32, 50], which necessitates economical access to this compound for synthetic applications. Owing to the limited stability of **41** in solution, particularly at alkaline pH [172, 173], it is preferentially generated in situ to avoid high stationary concentrations.

The most convenient method is formation of two equivalents of **41** by retro-aldol cleavage from commercially available fructose 1,6-bisphosphate (FBP, **42**) by the combined action of FruA and triose phosphate isomerase (Figures 5.22 and 5.28 inset) [149]. This scheme has been extended into a highly integrated, multienzymatic scheme for efficacious in-situ preparation of **41** from inexpensive glucose, fructose, or sucrose by using an "artificial metabolism" in vitro made up from up to seven inexpensive enzymes (Figure 5.28) [174]. Complications can arise from incomplete conversion, depending on the thermodynamic stability of the final product relative to **42**, because of the overall equilibrium nature of the system, and from difficulties separating the products from other phosphorylated components. Dihydroxyacetone **47** can be enzymatically phosphorylated using a glycerol kinase with ATP regeneration (see, for example, Figure 5.24) [175], or by transphosphorylation from phosphatidyl choline using phospholipases

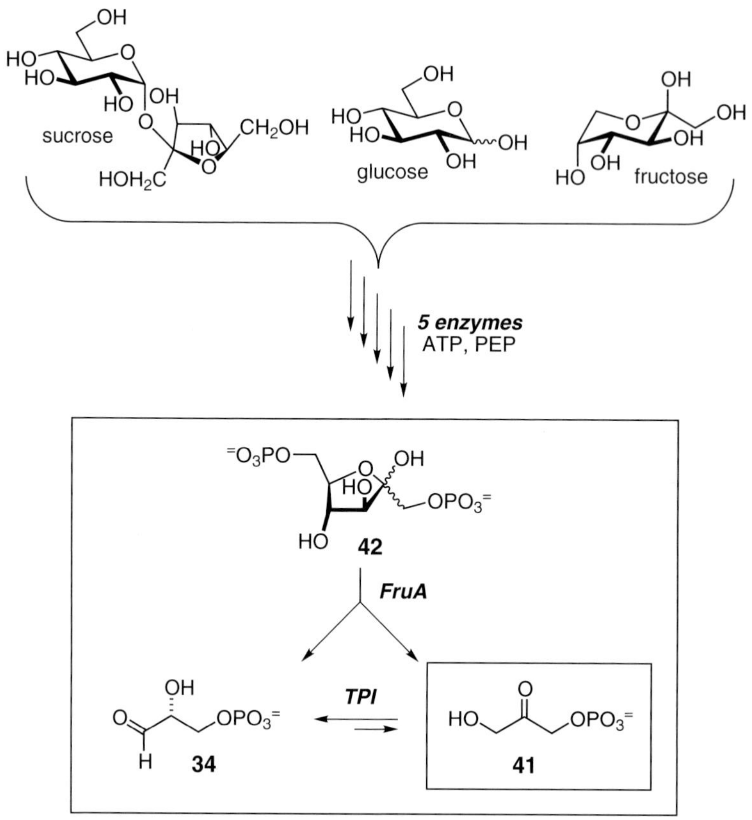

Fig. 5.28
Multi-enzymatic "artificial metabolism" for in-situ generation of dihydroxyacetone phosphate from inexpensive sugars.

[176]. Alternatively, **41** can be formed from glycerol by successive phosphorylation and oxidation effected by a combination of glycerol kinase and glycerol phosphate dehydrogenase, with an integrated double ATP/NAD^+ cofactor recycling system [148].

A more advanced technique for clean generation of **41** in situ is based on oxidation of L-glycerol 3-phosphate (**74**) catalyzed by microbial flavine-dependent glycerol phosphate oxidases (GPO; Figure 5.29, box) [177]. This method generates **41** practically quantitatively and with high chemical purity without a need for separate cofactor regeneration. Both oxygen from air or from a H_2O_2–catalase system can be used to sustain oxygenation. Because DHAP aldolases were found to be insensitive to oxygenated solutions, the oxidative generation of **41** can be smoothly coupled to synthetic aldol reactions [177]. This method has recently been extended to include reversible glycerol phosphorylation from inexpensive pyrophosphate (Figure 5.29).

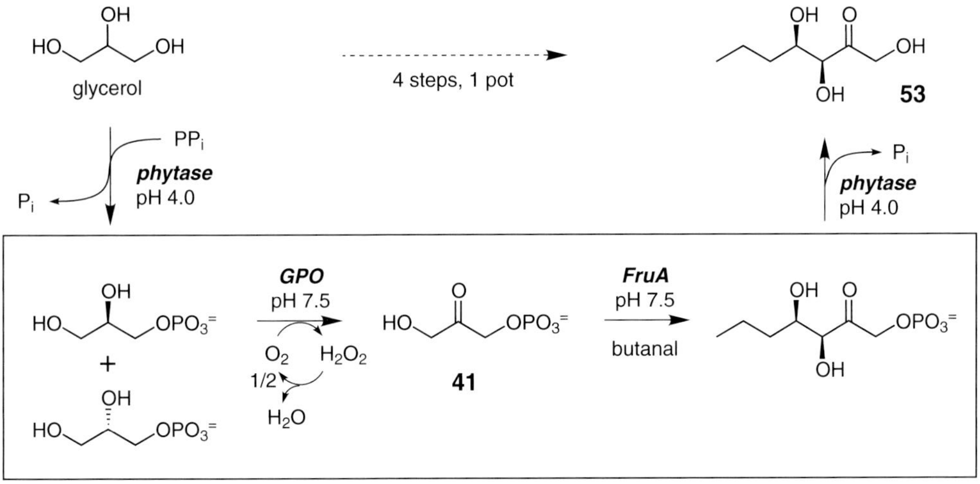

Fig. 5.29
Enzymatic in-situ generation of dihydroxyacetone phosphate for stereoselective aldol reactions using DHAP aldolases (box), and extension by pH-controlled, integrated precursor preparation and product liberation.

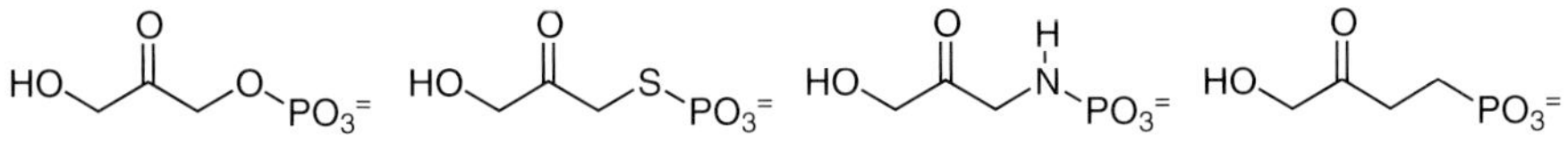

Fig. 5.30
Substrate analogs of dihydroxyacetone phosphate accessible by the GPO oxidation method.

Because phytase, an inexpensive acid phosphatase, is only active at low pH but virtually inactive at pH 7.5 in which aldolases have their catalytic optimum, this enables the independent staging of a one-pot synthetic cascade between (1) transphosphorylation, (2) aldolization, and (3) product dephosphorylation simply by switching the pH [178].

The GPO procedure can also be used for preparative synthesis of the corresponding phosphorothioate (**54**), phosphoramidate (**55**), and methylene phosphonate (**56**) analogs of **41** (Figure 5.30) from suitable diol precursors [179] which are used as aldolase substrates [177]. Such isosteric replacements of the phosphate ester oxygen were found to be tolerated by several class I and class II aldolases, and only some specific enzymes failed to accept the less polar phosphonate **56** [180]. Thus, sugar phosphonates (for example, **88** and **89**) that mimic metabolic intermediates but are hydrolytically stable to phosphatase degradation can be rapidly synthesized (see Figure 5.42).

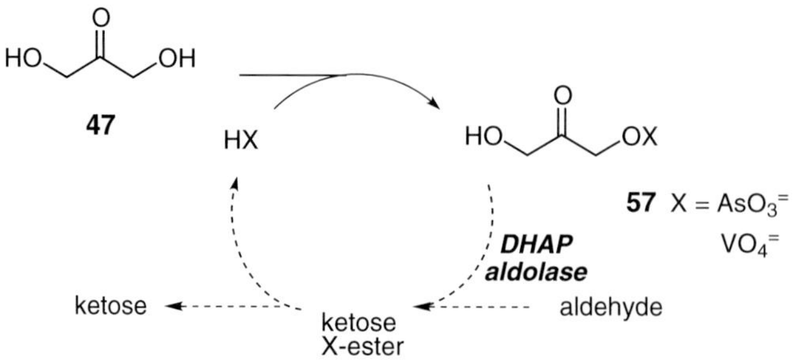

Fig. 5.31
Spontaneous, reversible formation of arsenate and vanadate analogs of dihydroxyacetone phosphate in situ for enzymatic aldol additions.

Interestingly enough, dihydroxyacetone **47** in the presence of higher concentrations of inorganic arsenate reacts reversibly to form the corresponding arsenate ester **57** in situ which can replace **41** as a donor in enzyme-catalyzed aldol reactions (Figure 5.31) [181, 182]. This procedure suffers from rather low reaction rates and the high toxicity of arsenates, however. Inorganic vanadate also spontaneously forms the corresponding vanadate ester analog under conditions that reduce its oxidation potential; so far only RhuA has been shown to accept the vanadate mimic **57** for preparative conversions [25]. Conversely, the dihydroxyacetone sulfate analog cannot be used in C–C bond-formation [183], and no replacement of the free hydroxy function in **41** for other electron-withdrawing substituents is tolerated [149].

Several alternative procedures have been developed for chemical synthesis of **41**. Acetals of the dimer of **47** can be phosphoryated using phosphoroxy chloride [184] or diphenyl chlorophosphate [185], or by phosphitylation then oxidation [186]. After phosphate deprotection, free **41** is obtained, by acid hydrolysis, in overall yields of up to 60%. New procedures have recently been developed by controlled successive substitutions of dibromoacetone [187, 188] or by controlled hydrolysis of a cyclic phosphate diester of **47** [189].

5.4.5
Applications

Four types of DHAP aldolase with distinct stereospecificity are available that have rather similar broad substrate tolerance (Figure 5.21). Despite having a rather unique physiological function in vivo, each enzyme will accept most differently substituted or unsubstituted aliphatic aldehydes in vitro as the acceptor component at a synthetically useful rate. Thus, from one common aldehyde substrate a full set of four possible diastereomeric

Fig. 5.32
General scheme for formation of ketofuranoses and ketopyranoses from 2- and 3-hydroxyaldehydes.

products becomes synthetically accessible in a stereodivergent manner (Figure 5.1). This building block-type feature renders the DHAP aldolases particularly powerful catalysts for asymmetric synthesis [25]. The FruA from rabbit muscle has been most extensively applied in synthesis, because of its commercial availability but results discussed below should be considered as exemplary demonstrations only, and complementary conversions can probably also be achieved by use of one of the other enzymes with different aldol stereospecificity.

A vast number of aldehydes have so far been used as substrates of this set of aldolases in preparative experiments [20, 24, 25, 190]. With the exception of generic aldehydes, acceptor components must be prepared by chemical synthesis. In general, ozonolysis of suitable olefins (with appropriate removal of the second fragment if this also is a substrate) or acid-catalyzed acetal deprotection are convenient routes for generation of aldehyde substrates under mild conditions. Chiral aldehydes require either asymmetric synthesis of the respective enantiomer or separation of diastereomeric products produced from racemic material. In specific cases racemate resolution can be effected by the enantiomer selectivity of an aldolase (kinetic resolution; Figure 5.27) or when isomeric products have significantly different stability (thermodynamic resolution; vide infra).

Typical applications of the DHAP aldolases include the synthesis of monosaccharides and derivatives of sugars from suitable functionalized aldehyde precursors. High conversion rates and yields are usually achieved with 2- or 3-hydroxyaldehydes, because for these compounds reaction equilibria benefit from the cyclization of the products in aqueous solution to give more stable furanose or pyranose isomers (Figure 5.32). For example, enantiomers of glyceraldehyde are good substrates, and stereoselective addition of dihydroxyacetone phosphate produces enantiomcrically pure ketohexose

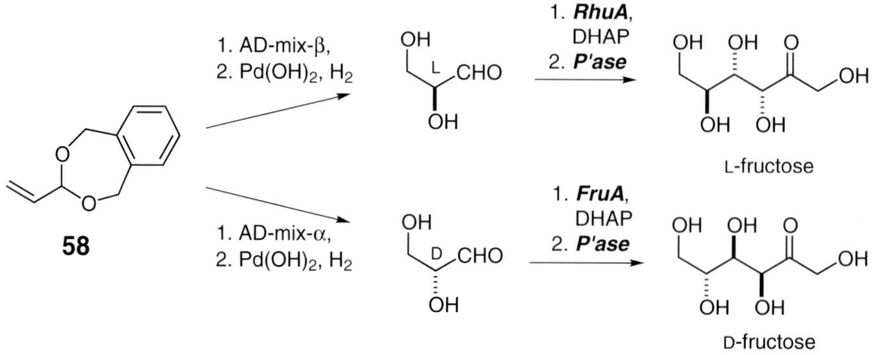

Fig. 5.33
Combination of metal-assisted asymmetric synthesis and enzymatic aldolization.

1-phosphates in high yield [149, 150, 170]; from these the free ketosugars are obtained by enzymatic dephosphorylation.

Only TagA catalysis is an exception, because this enzyme is not *erythro*-selective with unphosphorylated substrates [158, 159]. Using RhuA and FucA, the less common L-configured ketohexoses L-fructose and L-tagatose can be obtained as pure diastereomers from racemic glyceraldehyde also, because of the high kinetic enantiomer selectivity of these catalysts (Figure 5.27) [170]. Alternatively, non-racemic hydroxyaldehydes can be prepared for use in aldolase reactions by Sharpless asymmetric dihydroxylation from suitably protected precursors **58** (Figure 5.33) [191].

This general approach has been followed for the de novo synthesis of a multitude of differently substituted, unsaturated [192, 193], or regiospecifically labeled sugars [194]. Unusual branched-chain (**60**, **61**) and spiro-anulated sugars (**63**, **64**) have been synthesized from the corresponding aldehyde precursors (Figure 5.34) [174]. 6-Substituted D-fructofuranoside derivatives such as aromatic sulfonamide **62** (a low nanomolar *Trypanosoma brucei* inhibitor) [197] are accessible via 6-azido-6-deoxyfructose from 3-azido-2(*R*)-hydroxypropanal **95** by FruA catalysis [151, 196]. In an approach resembling the "inversion strategy" (vide infra) an α-*C*-mannoside **65** has been prepared from D-ribose 5-phosphate [216]. The synthesis of 6-*C*-perfluoroalkyl-D-fructose **66** met challenges from the strong hydrophobicity and electron-withdrawing capacity of a fluorous chain, and the product's potential surfactant properties [198]. The L-*sorbo*-configured homo-*C*-nucleoside analog **67** has been synthesized as a structural analog of adenosine from an enantiomerically pure (*S*)-aldehyde precursor [195].

FruA-based, diastereoselective chain extension of chiral pool carbohydrates or their corresponding phosphates has been achieved in the synthesis of high-carbon ketose 1-mono- or 1,*n*-bisphosphates [199]. Such heptulose to nonulose derivatives are of biological importance as phosphorylated intermediates of the pentose phosphate pathway or as sialic acid analogs (for

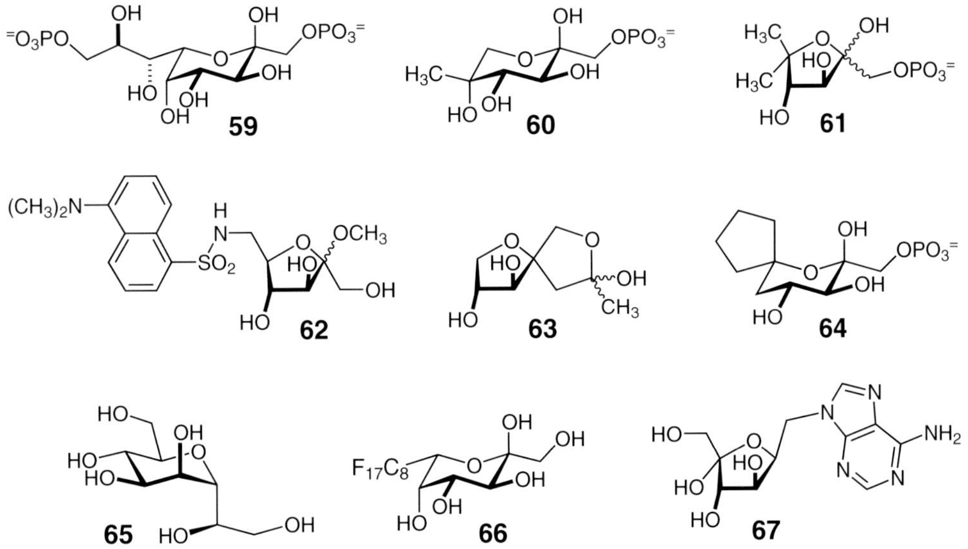

Fig. 5.34
Examples of product structures accessible by enzymatic aldolization.

example, **59**) that are difficult to isolate from natural sources. Synthesis of sugar bisphosphates is rather unique for FruA catalysis, because the deoxy sugar aldolases (FucA, RhuA) will not usually accept anionically charged substrates.

On the basis of FruA-catalyzed aldol reactions 3-deoxy-D-*arabino*-heptulosonic acid 7-phosphate (DAHP, **26**), an intermediate of the shikimic acid pathway, has been synthesized from *N*-acetylaspartic semialdehyde **68** (Figure 5.35) [200]. Precursors to KDO (Section 5.3.2) and its 4-deoxy analog (**71**) have been prepared by FruA catalysis from aldehydes **70** that incorporate an acrylic moiety for further functionalization [201].

Pendant anionically charged chains have been extended from O- or C-glycosidic aldehydes (**72**) to furnish low-molecular-weight mimics of the sialyl Lewis X tetrasaccharide, for example **73** (Figure 5.36) [202]. Other higher-carbon sugar derivatives such as the bicyclic sugar **76** have been prepared by diastereoselective chain extension from simple alkyl galactosides (**75**; Figure 5.36) after their terminal oxidation in situ by use of a galactose oxidase (GalO; EC 1.1.3.9). The whole scheme can be conveniently effected as a one-pot operation including the parallel generation of **41** by the GPO method [203]. Further bicyclic carbohydrate structures similar to **76** have also been prepared by uni- [152] and bidirectional extension of dialdehyde substrates (see Figure 5.48) [204].

The Zn^{2+}-dependent aldolases facilitate effective kinetic resolution of racemic 2-hydroxyaldehydes (*rac*-**50**) as substrates by an overwhelming preference (d.e. $\geq$95) for the L-configured enantiomers L-**50**; this enables control

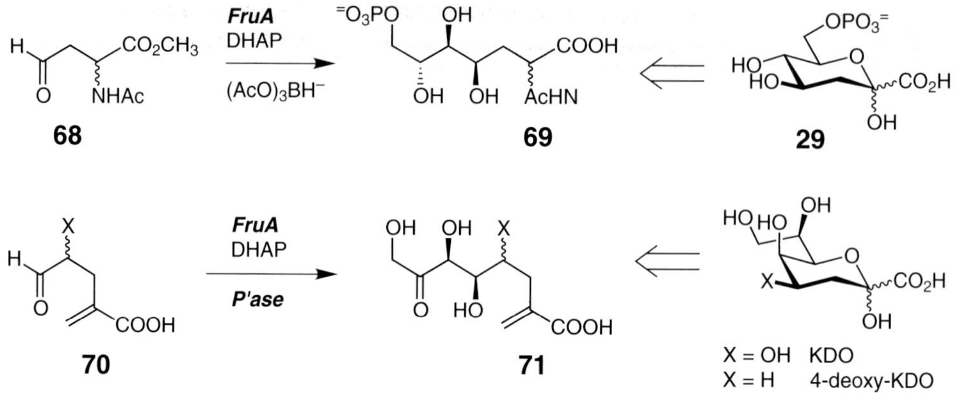

Fig. 5.35
Synthetic approaches to DAHP and KDO by a "backbone inversion" strategy using FruA catalysis.

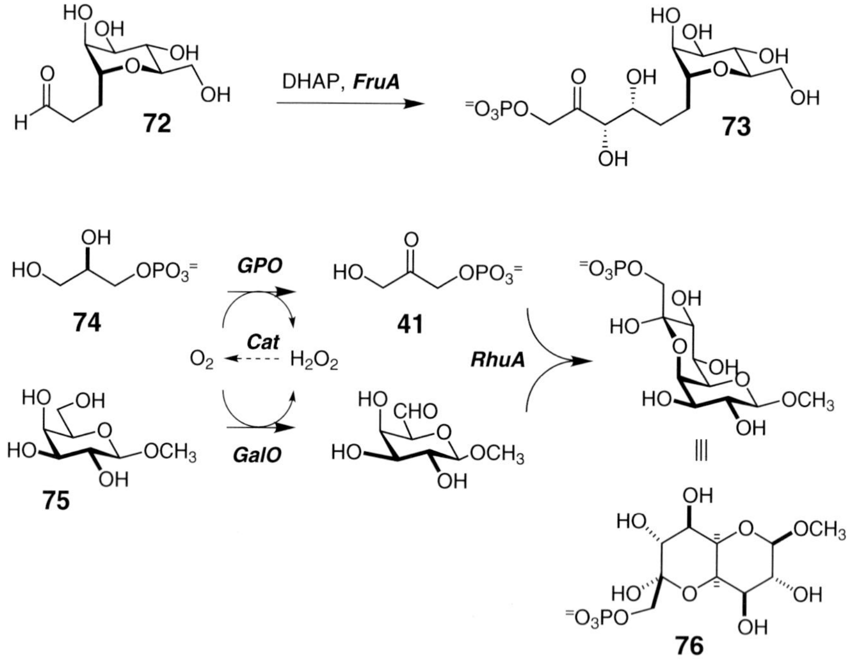

Fig. 5.36
Sialyl Lewis^{x}-related selectin inhibitor by chain extension of *C*-glycosyl aldehydes using enzymatic aldolization, and multienzymatic oxidation–aldolization strategy for synthesis of bicyclic higher-carbon sugars.

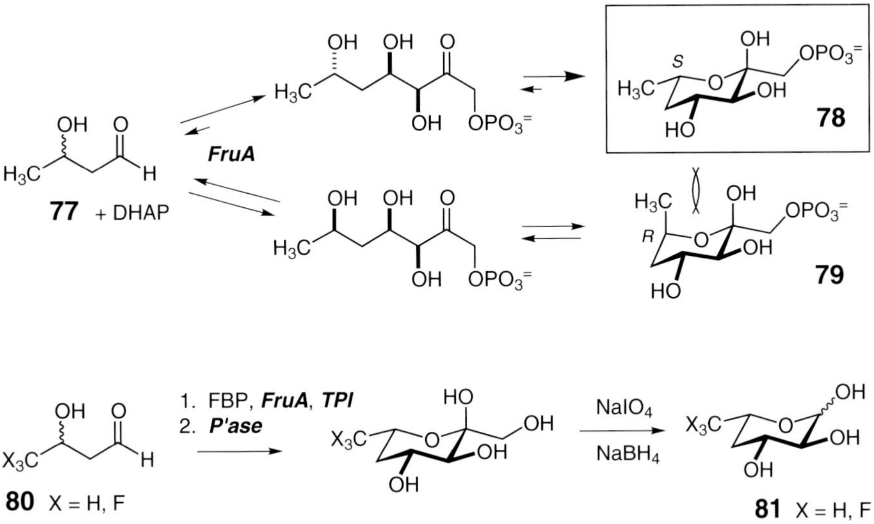

Fig. 5.37
Diastereoselectivity in FruA catalyzed aldol additions to 3-hydroxyaldehydes under thermodynamic control, and synthesis of L-fucose derivatives based on thermodynamic preference.

of three contiguous centers of chirality in the products (**51** and **52**; Figure 5.27) [170, 171]. This feature has been exploited in highly stereoselective syntheses of several rare ketose phosphates and of naturally occurring or related non-natural L sugars. For example, the kinetic enantioselectivity of FucA was not impaired by steric bulk or degree of unsaturation in the aldehyde substrates (*de* $\geq$90% for **50**) in syntheses of higher homologs and unsaturated analogs of L-fucose **84** (see Figure 5.39) [193]. For comparison, the kinetic enantioselectivity of class I aldolases is limited to aldehydes carrying an anionic charge, and distinction of uncharged substrates is insignificant (Figure 5.23) [149, 155].

Under fully equilibrating conditions, however, diastereoselectivity of aldolase reactions can be steered by thermodynamic control to favor the energetically most stable product [149, 174, 205, 206]. Particularly strong discrimination results from utilization of 3-hydroxylated aldehydes such as **77**, because of the cyclization of products to form pyranoid rings in water (Figure 5.37). The pronounced conformational destabilization by diaxial repulsion (**79**) strongly supports those diastereoisomers with a maximum of equatorial substituents [204, 206]. Thus, in FruA-catalyzed reactions (3*S*)-configured hydroxyaldehydes are the preferred substrates giving the most stable all-equatorial substitution in the product (for example **78**) with a de of up to 95%. Similarly, 2-alkylated aldehydes can be resolved, because of the high steric preference of an alkyl group for an equatorial position [205].

Fig. 5.38
"Inverted" approach for aldose synthesis using FruA catalysis, and application of the strategy for deoxysugar synthesis based on a phosphorothioate analog.

Under conditions of thermodynamic control the enantio-complementary nature of the FruA–RhuA biocatalyst pair enables construction of mirror image products **78** and *ent*-**78** from racemic 3-hydroxybutanal **77** with similar selectivity, but preference for opposite enantiomers [25]. The all-equatorial substitution in the predominant product can facilitate its separation by crystallization so that the remaining mixture can be re-subjected to further equilibration to maximize the yield of the preferred isomer **78** [177]. This general technique has recently found an application in a novel approach for the de novo synthesis of 4,6-dideoxy sugars such as 4-deoxy-L-fucose **81** or its trifluoromethylated analog (Figure 5.37) [25].

Conversely, FucA will produce diastereomers at low de, because of the more balanced stability relationships for pyranoid products sharing a *cis*-diol substitution pattern. With racemic 2-hydroxylated aldehydes, thermodynamic control in FruA- or RhuA-catalyzed reactions favors (2*R*) configured enantiomers but also with lower discrimination (de up to 55%), because of the higher flexibility of the corresponding ketofuranose rings [25, 206].

Because of the structure of the DHAP nucleophile, the enzymatic aldolization technique is ideal for direct synthesis of ketose monosaccharides and related derivatives or analogs (Figure 5.32). For an entry to aldoses an "inversion strategy" has been developed (Figure 5.38) which utilizes monoprotected dialdehydes (for example, **82** or **83**) for aldolization and, after stereoselective ketone reduction, provides free aldoses on deprotection of the masked aldehyde function [209]. In this respect, the phosphorothioate analog of DHAP **54** makes terminally deoxygenated sugars accessible via a sequence of FruA-catalyzed aldolization followed by reductive desulfurization, as illustrated by the preparation of D-olivose along this "inversion strategy"

1. **FucA**, DHAP
2. **P'ase**

Fucl

50

84

R = CH_3, C_2H_5, $CH{=}CH_2$, $C{\equiv}CH$

Fig. 5.39
Short enzymatic synthesis of L-fucose and hydrophobic analogs by aldolization–ketol isomerization, including kinetic resolution of racemic hydroxyaldehyde precursors.

[219]. Otherwise, deoxy sugars are usually only obtained when the deoxy functionality is introduced with the aldehyde.

More general access to biologically important and structurally more diverse aldose isomers makes use of the enzymatic interconversion of ketoses and aldoses that in Nature is catalyzed by ketol isomerases. For full realization of the concept of enzymatic stereodivergent carbohydrate synthesis the stereochemically complementary L-rhamnose (RhaI; EC 5.3.1.14) and L-fucose (FucI; EC 5.3.1.3) isomerases from *E. coli* have recently been shown to have broad substrate tolerance [25, 170, 193, 210]. Both enzymes convert sugars and their derivatives that have a common (3*R*)-OH configuration but might deviate in stereochemistry or substitution pattern at subsequent positions of the chain [25, 26]. Because ketose products from RhuA- and FucA-catalyzed aldol reactions share the (3*R*) specificity, they can both be converted by the isomerases to corresponding aldose isomers; this enables access to a broad segment of aldose configurational space in a stereospecific, building block manner [26, 211]. This strategy has been illustrated by the synthesis of several L-configured aldohexoses using different enzyme combinations, and by tandem FucA–FucI catalysis in the synthesis of new L-fucose analogs **84** having tails with increased hydrophobicity and reactivity (Figure 5.39), starting from simple higher homologs of lactaldehyde and unsaturated analogs **50** [193, 210]. Similar results have been achieved by using a glucose isomerase (GlcI; EC 5.3.1.5) which is an industrially important enzyme for isomerization of D-glucose to D-fructose (Figure 5.40) but has a more narrow specificity. This enzyme also accepts derivatives and analogs of D-fructose and has been used in combined enzymatic syntheses, particularly of 6-modified D-glucose derivatives [207].

Further structural diversification of FruA products has been investigated by enzymatic reduction to corresponding alditols using stereochemically complementary alditol dehydrogenases (Figure 5.40). Indeed, stereospecific (2*S*)- and (2*R*)-specific reduction of simple derivatives of D-fructose could be achieved by NADH-dependent catalysis of sorbitol dehydrogenase (EC 1.1.1.14) [209] or mannitol dehydrogenase (EC 1.1.1.67), respectively [26].

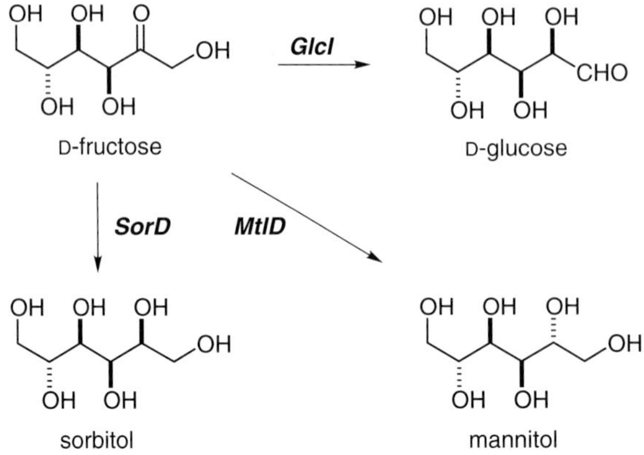

Fig. 5.40
Stereospecific diversification of aldol products by ketol isomerization or carbonyl reduction.

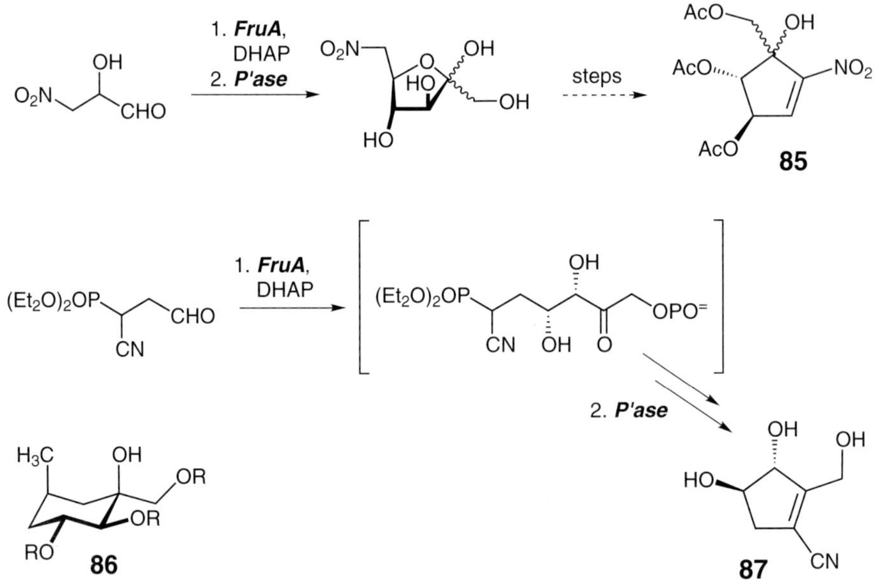

Fig. 5.41
Preparation of cyclitols by chemoenzymatic tandem reactions.

Several cyclitols (e.g. **85–87**) have been prepared from aldol products carrying suitably positioned halogen, nitro, or phosphonate functionality by subsequent radical or nucleophilic cyclization reactions (Figure 5.41) [212–214]. A cyclitol product was also found to be correctly configured to serve as a precursor to the spirocyclic *Streptomyces* metabolite sphydrofuran **63** [215].

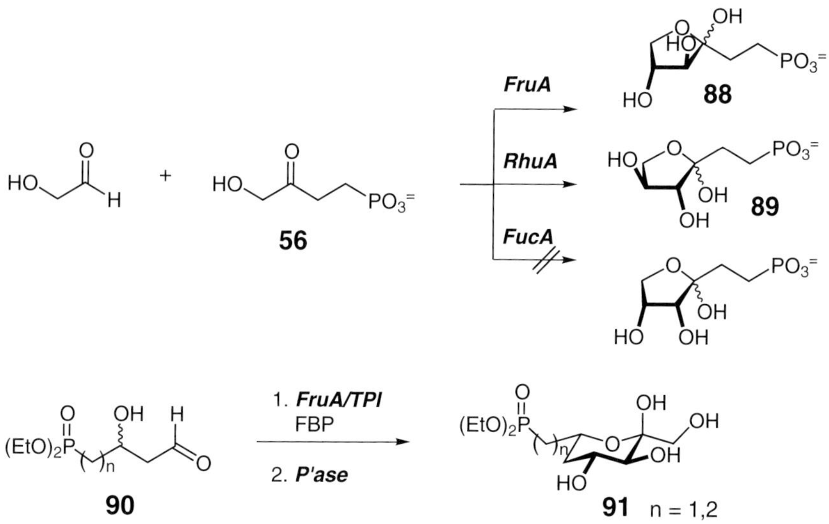

Fig. 5.42
Stereoselective synthesis of hydrolytically stable sugar phosphonates either from the bio-isosteric phosphonate analog of DHAP or from phosphonylated aldehydes.

Phosphonate analogs to phosphate esters, in which the P–O bond is formally replaced by a P–C bond, have attracted attention because of their stability toward the hydrolytic action of phosphatases, which renders them potential inhibitors or regulators of metabolic processes. Introduction of the phosphonate moiety by enzyme catalysis might, in fact, be achieved by two alternative pathways. The first employs the bioisosteric methylene phosphonate analog **56**, which gives products related to sugar 1-phosphates such as **88** and **89** (Figure 5.42) [177, 180]. This strategy is rather effective because of the inherent stability of **56** as a replacement for **41** but depends on the individual tolerance of the aldolase for structural modification close to the reactive center. The second option is a suitable choice for a phosphonylated aldehyde such as **90**, which gives rise to analogs of sugar ω-phosphates **91** [217, 218].

When placing thiol substitution in the aldehyde component for aldolase catalysis, the reaction products can cyclize to form rather stable cyclic hemiacetal structures. Such thiosugars are a structural variation of carbohydrates that have interesting biological properties, for example glycosidase inhibition. From 2- and 3-thiolated aldehydes stereochemical sets of furanoid or pyranoid thiosugars such as **92–94** have been prepared using different DHAP aldolases (Figure 5.43) [148, 220, 221]. It is worth noting that the observed unbiased stereoselectivity indicates the full equivalence of OH and SH substituents for correct substrate recognition.

The structural resemblance of "azasugars" (1-deoxy sugars in which an

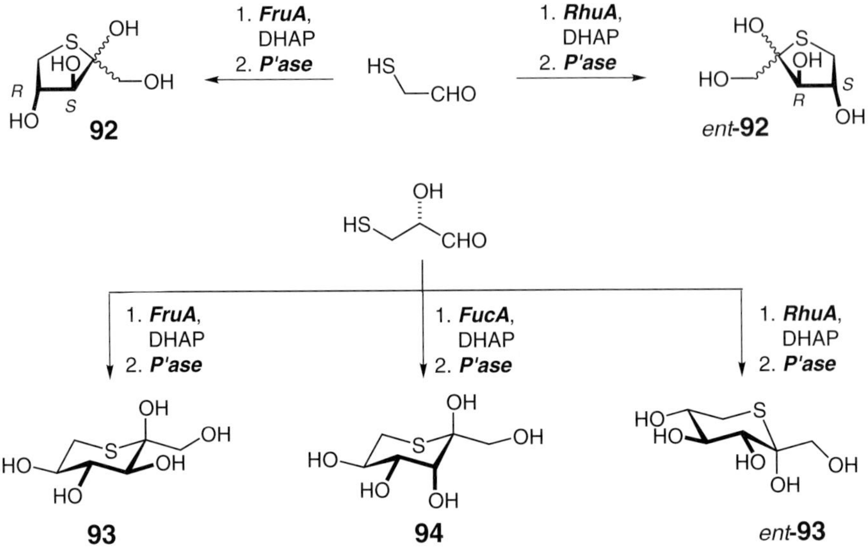

Fig. 5.43
Stereodivergent enzymatic synthesis of thiosugars.

imino group replaces the ring oxygen) to transition states or intermediates of glyco-processing enzymes has made these compounds an attractive subject of research, because of their potential value as enzyme inhibitors for therapeutic applications. An important and flexible synthetic strategy has been developed which consists in stereoselective enzymatic aldol addition to an azido aldehyde followed by azide hydrogenation with intramolecular reductive amination [222, 223]. Particularly noteworthy are the stereodivergent chemoenzymatic syntheses of diastereomers of the nojirimycin type from 3-azidoglyceraldehyde **95** that have been developed independently by several groups (Figure 5.44) [151, 196, 224–227]. Because of the low kinetic selectivity of FruA for 2-hydroxyaldehydes, use of enantiomerically pure aldehyde proved superior to the racemate for preparation of the parent 1-deoxy-D-nojirimycin.

An extensive array of further 5-, 6- and 7-membered ring alkaloid analogs have since been made by following the same general strategy. For structural variation, as exemplified by **96–98**, differently substituted azido aldehydes of suitable chain length were converted by the distinct DHAP aldolases (Figure 5.45) [151, 156, 225, 228–230]. Stereocontrol during the reductive cyclization seems to be best effected by Pd-catalyzed hydrogenation.

The technique has been extended to the bifunctional class of azasugar phosphonic acids, for example **99**, by exploiting the tolerance of rabbit FruA for the bioisosteric phosphonate nucleophile **56** [231]. The resulting heterocycles are a minimum structural motif of glycosyltransferase transition-state analogs. In a strategy inverse of that employed for compound **99**, FucA

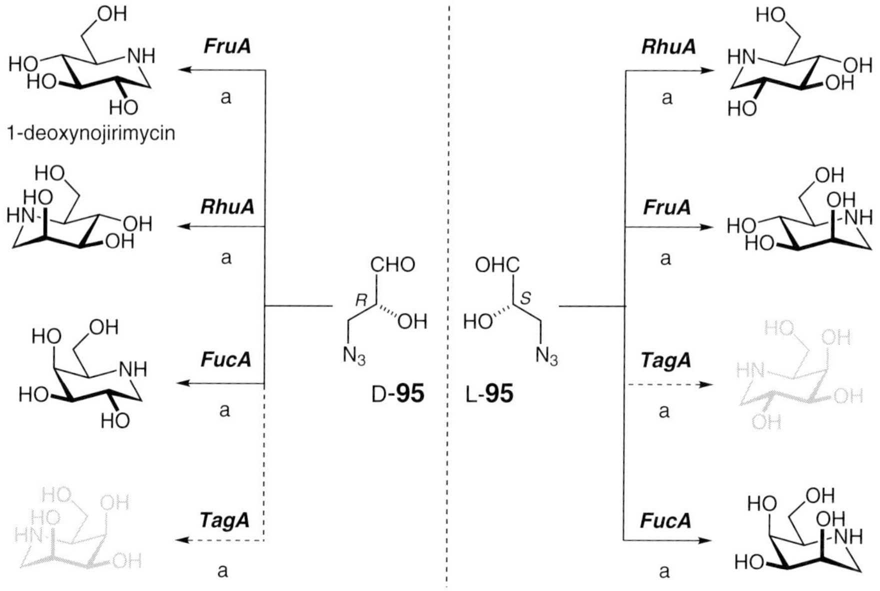

Fig. 5.44
Stereodivergent synthesis of 1-deoxy azasugars of the nojirimycin type by two-step enzymatic aldolization/catalytic reductive amination (a: DHAP; P'ase; H_2/Pd–C).

and FruA were employed in the chemoenzymic synthesis of six-membered iminocyclitol phosphonic acids [232]. Another illustrative example of the azasugar synthetic strategy is the chemoenzymatic synthesis of the naturally occurring australine, 3-epiaustraline, and 7-epialexin (Figure 5.46) [233]. The bicyclic pyrrolizidine core structure resulted from twofold reductive amination of a linear precursor **101** in which the asymmetric hydroxylation sites had been installed during an aldolase-catalyzed chain extension from aminoaldehyde **100**. A bidirectional aldolization approach furnished the *C*-glycosidically linked azadisaccharide **102** as an example of a disaccharide mimic. Ozonolysis of a racemic azido-substituted cyclohexenediol precursor was followed by tandem DHAP addition to both aldehydic termini to yield an intermediate azido-substituted dipyranoid 2,11-diulose which, when hydrogenated over Pd catalyst, gave the aza-*C*-disaccharide **102** highly selectively as a single diastereomer [25].

Such aldolase-catalyzed bi-directional chain elongation ("tandem" aldolization) of simple, readily available dialdehydes has been developed into an efficient method for generation of higher-carbon sugars (for example **104**, **106**, **108**) by simple one-pot operations (Figure 5.47) [204, 234]. The furanoid (**104**) or pyranoid (**106**) nature of the products can be determined by use of a suitable hydroxyl substitution pattern in a corresponding cyclo-

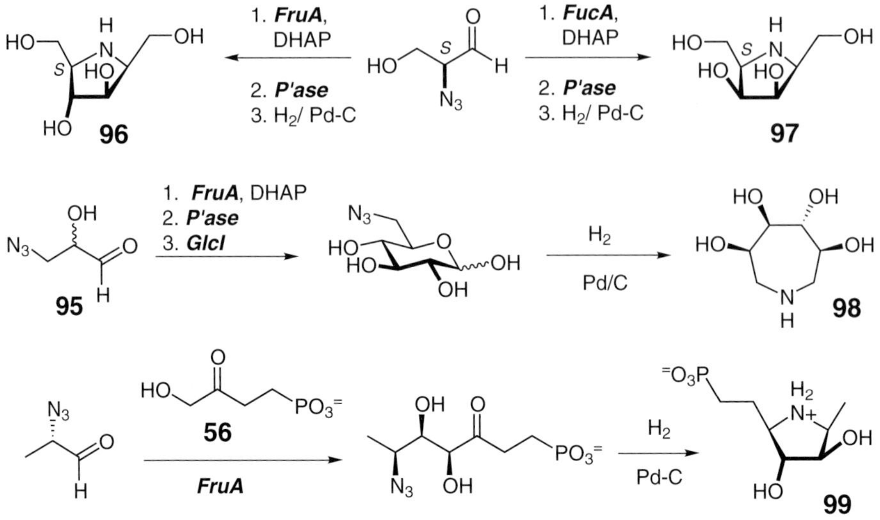

Fig. 5.45
Stereoselective synthesis of 5- and 7-membered ring azasugars and of novel azasugar phosphonates.

olefinic precursor (e.g. **103** compared with **105**). The overall specific substitution pattern in the carbon-linked disaccharide mimetics is deliberately addressable by the relative hydroxyl configuration and choice of the aldolase. Single diastereomers can be obtained in good overall yield from racemic precursors if the tandem aldolizations are conducted under thermodynamic control (Figure 5.37). The thermodynamic advantage much favors a twofold *trans* (**104**) or equatorial connectivity (**106**), so that a C_2-symmetrical diastereomer **108** is obtained selectively, even from linear *rac*/*meso* diol mixtures **107**.

Similarly, highly complex structures like anulated (**109**) and spirocyclic (**110**) carbohydrate mimics can be obtained from appropriately customized precursors (Figure 5.48). In suitable circumstances use of two aldolases that afford distinct selectivity can effect kinetic regioselective discrimination of the independent addition steps, and thus even enable a stereochemical terminus differentiation [204].

DHAP aldolases typically yield carbohydrates or carbohydrate-derived materials, because of the nature of the reactive components, but can also be advantageously used in the construction of stereochemically homogenous fragments of non-carbohydrate natural products. An impressive illustration is the FruA-based chemoenzymatic syntheses of (+)-*exo*-brevicomin (**111**), the aggregation pheromone of the Western pine bark beetle *Dendroctonus*

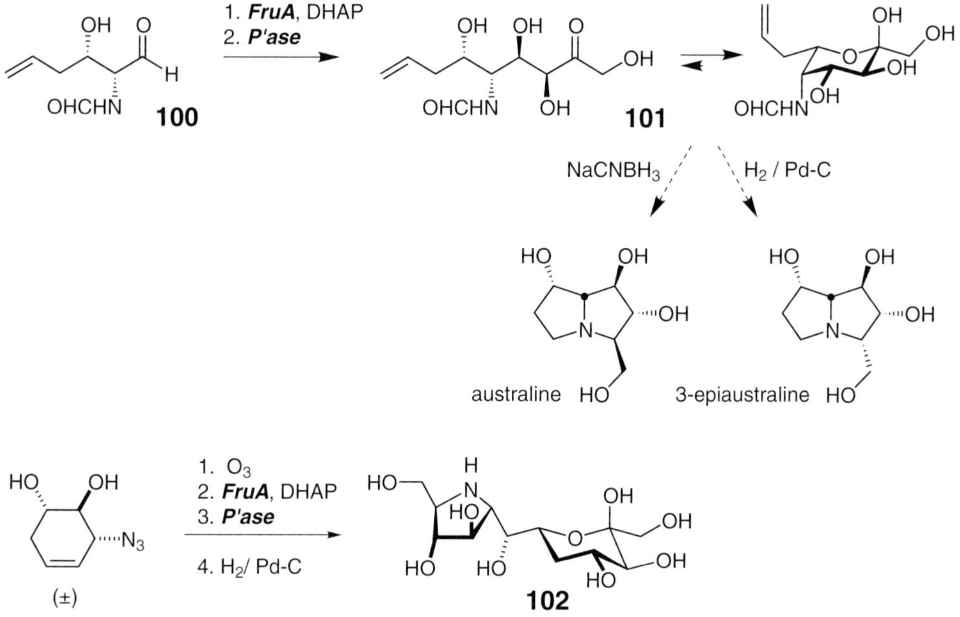

Fig. 5.46
Synthetic route to oxygenated pyrrolizidine alkaloids, and an aza-C-disaccharide as glycosidase inhibitors.

brevicomis [235]. Addition of DHAP to 5-oxohexanal generated an enantiomerically pure vicinal *syn*-diol structure containing the only independent stereogenic centers of brevicomin. A complementary inverse approach to **111** (convergent at an intermediate from FruA/propanal) has also been followed using transketolase [236].

Application of an aldolase to the synthesis of the tricyclic microbial elicitor (−)-syringolide (Figure 5.50) is another excellent illustration that enzyme-catalyzed aldolizations can be used to generate sufficient quantities of enantiomerically pure material in multi-step syntheses of complex natural and non-natural products [237]. Remarkably, the aldolase reaction established the absolute and relative configuration of the only chiral centers that needed to be externally induced in **113**; during the subsequent cyclization all others seemed to follow by kinetic preference.

A FruA-mediated stereoselective DHAP addition to a suitable aldehyde precursor **114** (Figure 5.51) served as the key step in the synthesis of the "non-carbohydrate", skipped polyol C9–C16 chain fragment **115** of the macrolide antibiotic pentamycin [238, 239]. Using the same enzyme, compound **117** has been stereoselectively prepared as a synthetic equivalent to the C3–C9 fragment of (+)-aspicillin, a lichen macrolactone (Figure 5.51) [240].

(±) 103
1. O_3
2. ***FruA***, DHAP
3. ***P'ase***
104

(±) 105
1. O_3
2. ***FruA***, DHAP
3. ***P'ase***
106

(±) 107
1. O_3
2. ***FruA***, DHAP
3. ***P'ase***
108

Fig. 5.47
Applications of bidirectional chain synthesis to the generation of disaccharide mimetics using tandem enzymatic aldol additions, including racemate resolution under thermodynamic control.

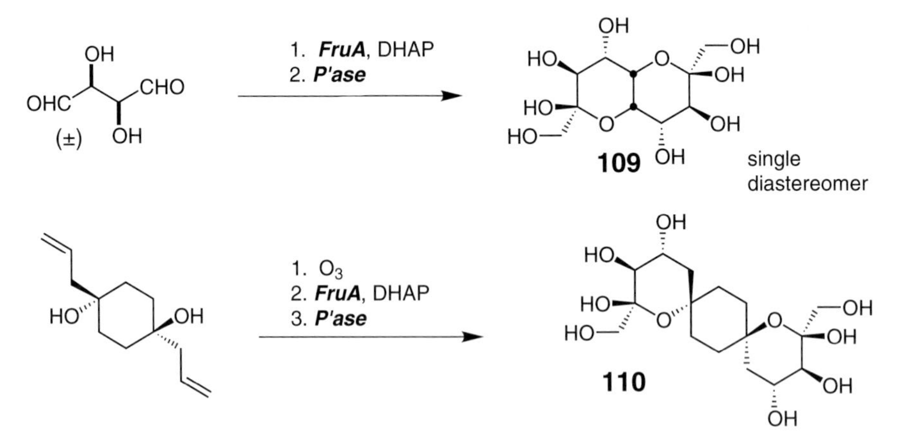

Fig. 5.48
Bidirectional chain extension for synthesis of anulated and spirocyclic oligosaccharide mimetics by DHAP aldolase catalysis.

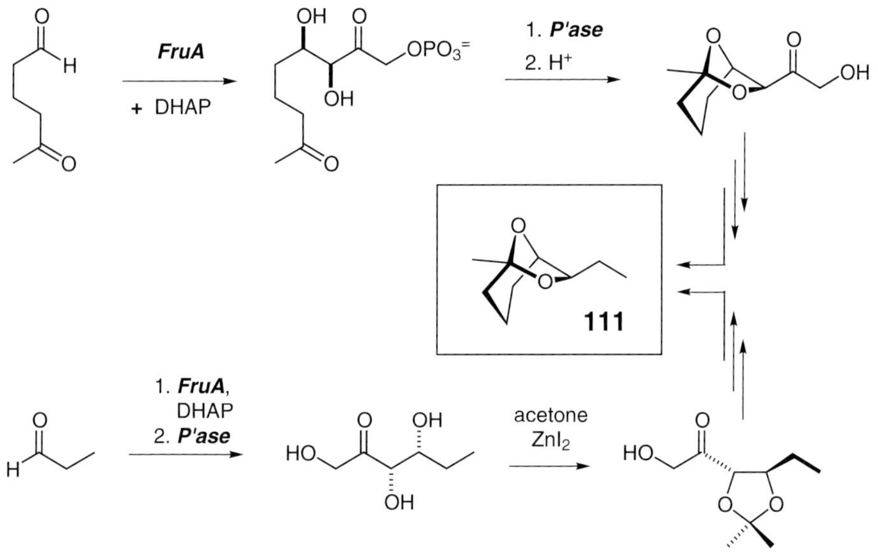

Fig. 5.49
Complementary, inverse approaches for the FruA-based chemoenzymatic synthesis of the insect pheromone (+)-*exo*-brevicomin.

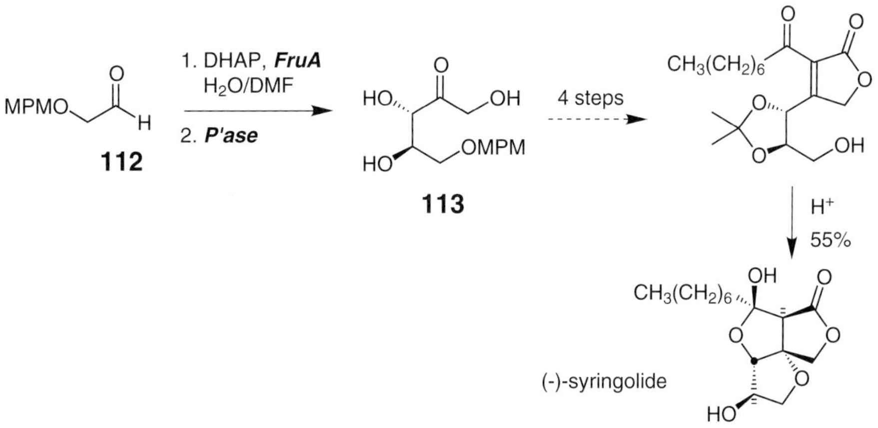

Fig. 5.50
Aldolase-based creation of independent centers of chirality in a short total synthesis of (−)-syringolide, a structurally complex microbial elicitor.

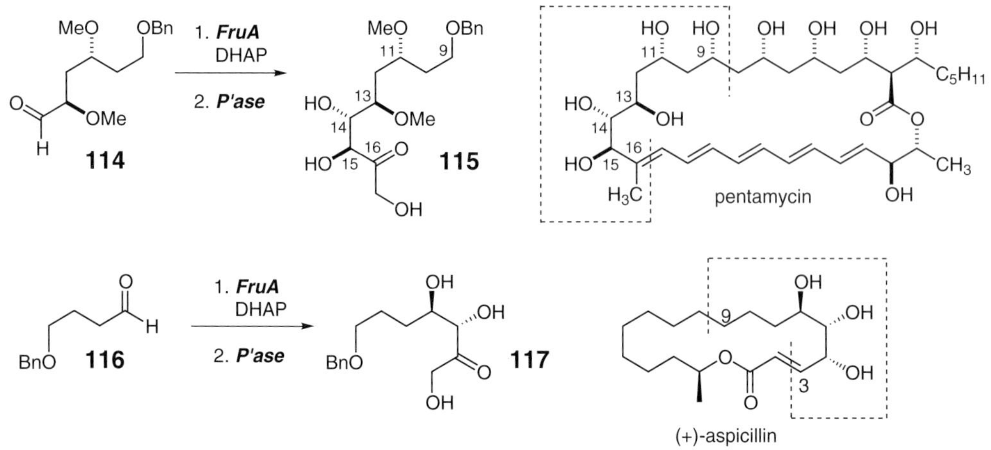

Fig. 5.51
Stereoselective generation of synthetic precursors to the macrolide antibiotic pentamycin and the lichen macrolactone (+)-aspicillin using FruA catalysis.

5.4.6
Aldol Transfer Enzymes

Transaldolase (EC 2.2.1.2) is an enzyme involved in the pentose phosphate pathway where it transfers a dihydroxyacetone unit between several phosphorylated metabolites [29]. The lyase belongs to class I aldolases and has been purified from several sources, cloned, and structurally characterized [41, 42]. In preparative studies yeast transaldolase, which is commercially available, has been shown to accept unphosphorylated aldehydes as the acceptor component [241–243].

The enzyme has also been employed in a multi-enzymatic scheme for conversion of starch into D-fructose; in this scheme transaldolase was used to formally dephosphorylate fructose 6-phosphate by transferring a dihydroxyacetone moiety to D-glyceraldehyde with formation of D-fructose and **34** [244].

The fructose 6-phosphate aldolase (FSA) from *E. coli* is a novel class I aldolase that catalyzes the reversible formation of fructose 6-phosphate from dihydroxyacetone and D-glyceraldehyde 3-phosphate; it is, therefore, functionally related to transaldolases [245]. Recent determination of the crystal structure of the enzyme showed that it also shares the mechanistic machinery [246]. The enzyme has been shown to accept several aldehydes as acceptor components for preparative synthesis. In addition to dihydroxyacetone it also utilizes hydroxyacetone as an alternative donor to generate 1-deoxysugars, for example **118**, regioselectively (Figure 5.52) [247].

D-fructose 6-phosphate

transaldolase

34

D-glyceraldehyde

D-fructose

FSA

1-deoxy-D-xylulose **118**

Fig. 5.52
Transaldolase catalysis used for an apparent "transphosphorylation" strategy, as incorporated in a synthesis of fructose from starch, and deoxysugar synthesis by related FSA using a non-natural donor.

5.5
Transketolase and Related Enzymes

Transketolase (EC 2.2.1.1) is involved in the oxidative pentose phosphate pathway in which it catalyzes the reversible transfer of a hydroxyacetyl nucleophile between a variety of sugar phosphates. The enzyme, which requires thiamine diphosphate and divalent Mg as cofactors [248], is commercially available from baker's yeast and can be readily isolated from many natural or recombinant sources [249, 250]. The yeast enzyme has been structurally well characterized [251], including protein with a carbanion intermediate covalently bound to the cofactor [252]. Large-scale enzyme production has been investigated for the transketolase from *Escherichia coli* [253–255]. Immobilization was shown to significantly increase stability against inactivation by aldehyde substrates [256]. The enzyme is quite tolerant to organic cosolvent, and preparative reactions have been performed continuously in a membrane reactor [255], with potential in-situ product removal via borate complexation [257].

Enzymes from yeast, spinach, and *Escherichia coli* have been shown to tolerate a broad substrate spectrum, with the newly formed chiral center always having an absolute (*S*) configuration as a result of *re*-face attack [258].

Tab. 5.6
Substrate tolerance of transketolase.

R	*Rel. Rate [%]*	*Yield [%]*	*Ref.*
H	100	70	236, 262, 302
CH_2OH	37	57	302
$CH_2OPO_3^{=}$	–	82	264
CH_2OCH_3	27	76	260, 261
$CH_2OCH_2C_6H_5$	–	79	260
CH_2F	47	79	260
CH_2N_3	–	71	196
CH_2CN	–	82	260
CH_2SH	–	78	220, 260
CH_2SCH_3	33	–	261
$CH_2SCH_2CH_3$	–	74	260
CH_2CH_2OH	<10	14	261
$CHOH–CH_3$	35	50	303
CH_3	44	88	260–262
CH_2CH_3	33	90	260, 261
$CH_2CH_2CH_3$	22	78	261
$C(CH_3)_3$	11	–	261
$CH{=}CH_2$	56	60	261
$CH_2CH{=}CH_2$	28	90	261
(*S*)-$CHOH–CH{=}CH_2$	36	60	261
(*R*)-$CHOH–CH{=}CH_2$	32	63	261

Fig. 5.53
Kinetic resolution by transketolase, and non-equilibrium C–C bond formation by decomposition of hydroxypyruvate.

Although generic aldehydes are converted with full stereocontrol, and even α,β-unsaturated aldehydes are acceptable to some extent, hydroxylated acceptors are usually converted at higher rates (Table 5.6) [259].

In addition, antipodes of racemic 2-hydroxyaldehydes **50** with a (2*R*)-hydroxyl group are discriminated with complete enantioselectivity; this enables efficient kinetic resolution (Figure 5.53) [260, 261]. Vicinal diols of (3*S*,4*R*) configuration are thereby generated with high stereocontrol. This

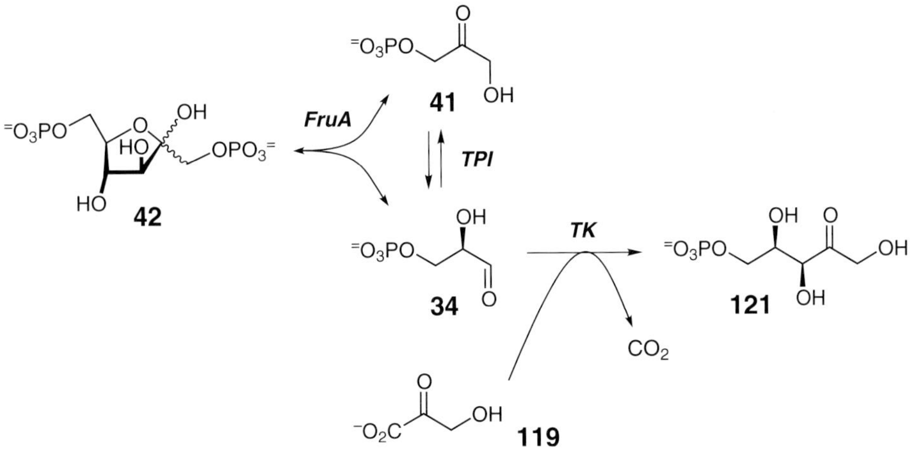

Fig. 5.54
Multi-enzymatic scheme for stereoselective synthesis of two equivalents of xylulose 5-phosphate from fructose 1,6-bisphosphate.

two-carbon chain-elongation method thus provides D-*threo* products **120** equivalent to those created by the respective three-carbon elongation reaction using FruA (or transaldolase) catalysis.

For synthetic purposes hydroxypyruvate **119** can effectively replace the natural donor components [258]. Its covalent activation occurs at a reduced rate of about 4% relative to xylulose 5-phosphate (**121**) but is accompanied by spontaneous decarboxylation [262]. Thus, loss of carbon dioxide renders synthetic reactions irreversible whereas alternative donors, for example L-erythrulose, require coupling to cofactor recycling to shift the overall equilibrium [263]. The thermodynamic driving force from decarboxylation of **119** is particularly useful with equilibrating multi-enzyme systems such as that used in the gram-scale synthesis of two equivalents of **121** from **42** (Figure 5.54) [264].

Transketolase has been used for the key steps in chemoenzymatic syntheses of (+)-*exo*-brevicomin **111** from racemic 2-hydroxybutyraldehyde [236], and of the azasugars 1,4-dideoxy-1,4-imino-D-arabinitol [196] or *N*-hydroxypyrrolidine **124** [265] from 3-azido (**95**) and 3-*O*-benzyl (**122**) derivatives, respectively, of glyceraldehyde (Figure 5.55). Such syntheses were all conducted with intrinsic racemate resolution of 2-hydroxyaldehydes and profited from utilization of **119**. Further preparative applications include the synthesis of valuable ketose sugars, particularly fructose analogs [258]. Transketolase has also been used for in-situ generation of erythrose 4-phosphate from fructose 6-phosphate in a multi-enzymatic synthesis of DAHP (**26**; Figure 5.17) [131].

1-Deoxy-D-xylulose 5-phosphate (**125**) is an intermediate in the recently discovered non-mevalonate pathway of terpene biosynthesis. This carbohy-

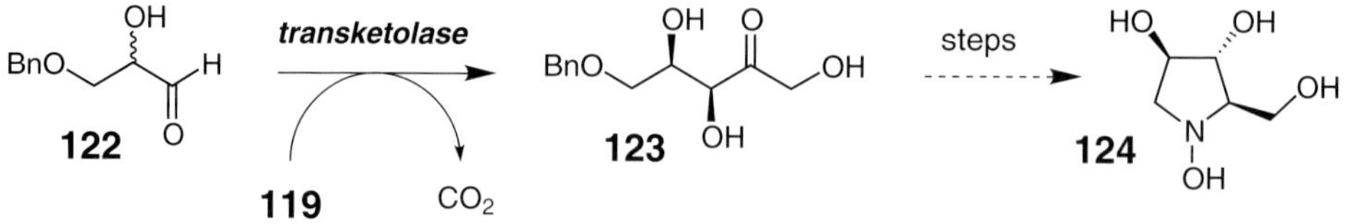

Fig. 5.55
Synthesis of a novel *N*-hydroxypyrrolidine on the basis of transketolase catalysis.

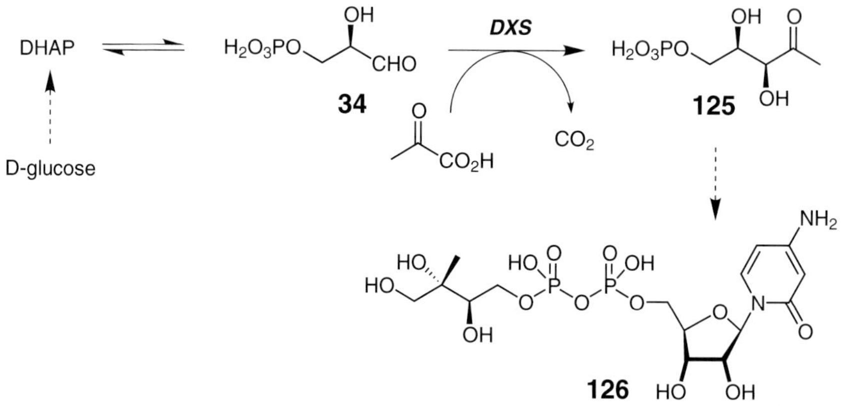

Fig. 5.56
Preparation of intermediates in the non-mevalonate pathway of terpene biosynthesis.

drate derivative can be efficiently produced from pyruvate and **34** by the catalytic action of the thiamine diphosphate-dependent 1-deoxy-D-xylulose 5-phosphate synthase (DXS) that has been cloned from *E. coli* [266]. The recombinant enzyme from *Bacillus subtilis* also has recently been applied in multi-enzymatic syntheses of different ^{13}C- and ^{14}C-labeled isotopomers of **125** (Figure 5.56) and of labeled 4-diphosphocytidyl-2*C*-methyl-D-erythritol **126**, a metabolic intermediate further downstream of this pathway [267, 268].

5.6 2-Deoxy-D-ribose 5-Phosphate Aldolase

The 2-deoxy-D-ribose 5-phosphate aldolase (RibA or "DERA"; EC 4.1.2.4) is a class I enzyme that, in vivo, catalyzes the reversible addition of acetaldehyde to D-glyceraldehyde 3-phosphate (**34**; Figure 5.57) in the metabolic degradation of **127** from deoxyribonucleosides [269], with an equilibrium constant for synthesis of 2×10^{-4} M [56]. It is, therefore, unique among the aldolases in that it uses an aldehyde rather than a ketone as the aldol donor. RibA has been isolated from eukaryotic and prokaryotic sources [270, 271],

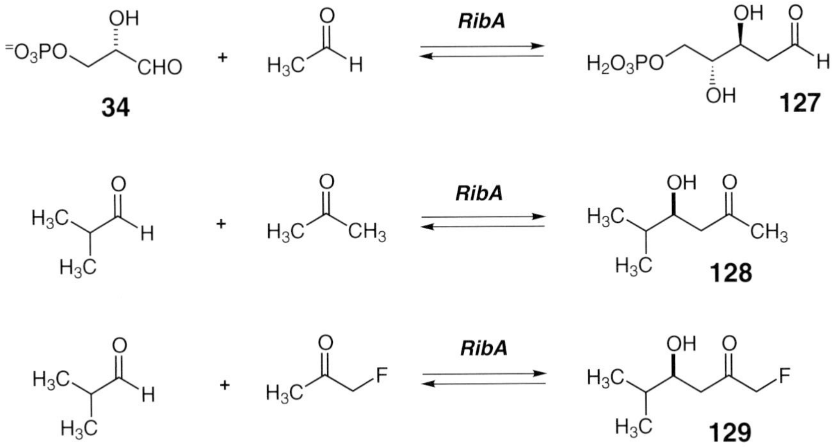

Fig. 5.57
Natural aldol reaction catalyzed by RibA, and acceptance of non-natural aldol donors.

and the enzyme from *E. coli* has been cloned [272] and overexpressed [273]. Spatial enzyme structures at high resolution have been determined for RibA from *E. coli*, including covalently bound intermediates [274], and from the hyperthermophilic archaea *Aeropyrum pernix* [275].

Interestingly, the enzyme's relaxed acceptor specificity enables substitution of both cosubstrates, albeit at strongly reduced rates ($< 1\%$ of v_{max}). Propionaldehyde, acetone, or fluoroacetone can replace acetaldehyde as the donor in the synthesis of variously substituted 3-hydroxyketones for example **128** or **129** (Figure 5.28) [273, 276]. It is worthy of note that from reactions with propionaldehyde as the donor (which leads to formation of a second stereocenter) only a single diastereomer of absolute (2*R*,3*S*) configuration results (for example **131**; Figure 5.59) [276]. This is indicative not only of the high level of asymmetric induction at the acceptor carbonyl but also of stereospecific deprotonation of the donor. Aldehydes up to a chain length of four non-hydrogen atoms are tolerated as acceptors (Table 5.7). 2-Hydroxyaldehydes **50** are relatively good acceptors, and the D isomers are preferred over the L isomers [276]. Reactions that lead to thermodynamically unfavorable structures can proceed with low stereoselectivity at the reaction center [277]. A single-point mutant aldolase was recently found to be 2.5 times more effective than the wild type in accepting unphosphorylated glyceraldehyde [278].

E. coli RibA has been used in a multi-enzymatic commercial process for production of different purine- or pyrimidine-containing deoxyribonucleosides, for example **130**, in good yield (Figure 5.58) [279]. Similarly, 2-deoxy-D-ribose, ^{13}C-labeled at different positions, has been prepared from labeled acetaldehyde and **41** to serve as the precursor to isotopically labeled thymidine [280].

Tab. 5.7
Substrate tolerance of deoxy-D-ribose-5-phosphate aldolase.

R	Yield [%]	Rel. Rate [%]	Ref.
$CH_2OPO_3^{=}$	78	100	276
H	20	–	301
CH_2OH	65	0.4	276
CH_3	32	0.4	276
CH_2F	33	0.4	276
CH_2Cl	37	0.3	276
CH_2Br	30	–	283
CH_2SH	33	–	283
CH_2N_3	76	0.3	276
C_2H_5	18	0.3	276
$CH{=}CH_2$	12	–	283
$CHOH{-}CH_2OH$	62	0.3	276
$CHN_3{-}CH_2OH$	46	–	283
$CHOH{-}CH_3$	51	–	283
$CHOH{-}CH_2{-}C_6H_5$	46	–	283
$CH_2SCH_2{-}CHOH{-}CH_2OH$	27	–	283

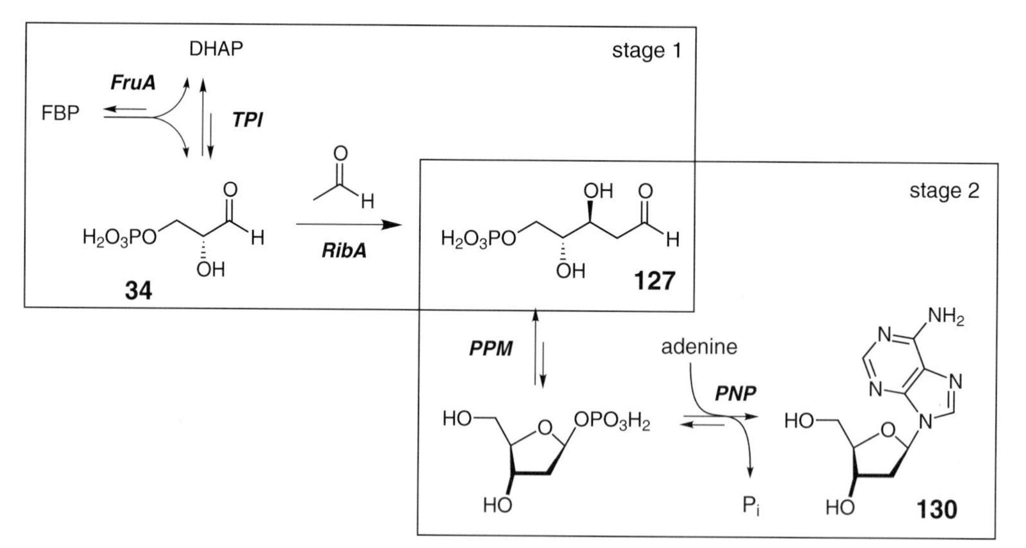

Fig. 5.58
Two-stage aldolase-based technical process for deoxyribonucleoside synthesis.

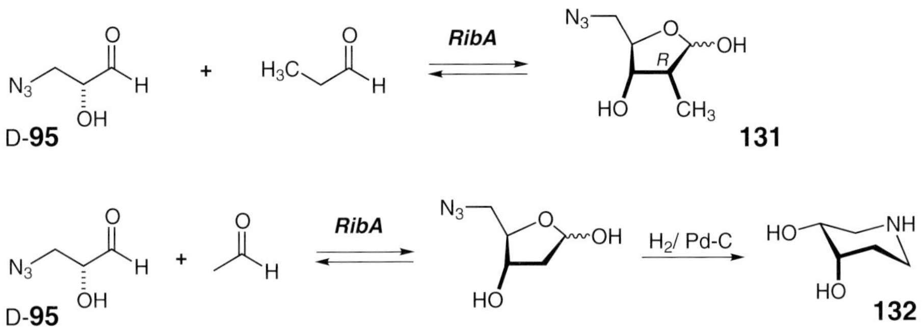

Fig. 5.59
Azasugar precursors prepared by RibA catalysis.

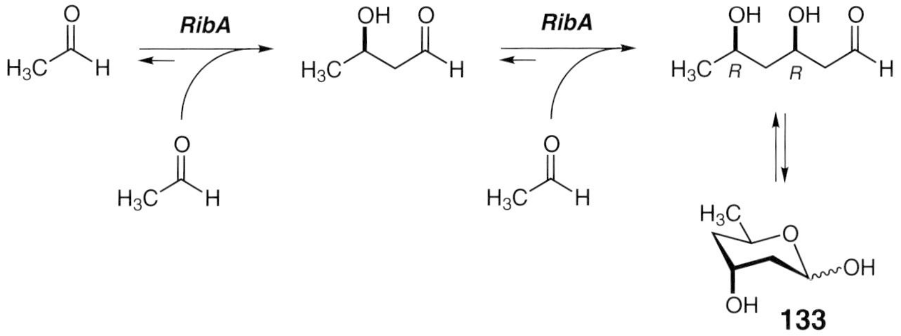

Fig. 5.60
RibA-catalyzed sequential aldol addition.

Starting from azidoaldehydes, several azasugars containing a lower density of substituents (for example **132**; Figure 5.59) have been prepared by sequential aldolization–hydrogenation [276]. More recently, pyranose synthons have been prepared as key intermediates in the synthesis of epothilones by using RibA catalysis (Chapter 7 in Part I of this book) [281].

When acetaldehyde is used as the only substrate, the initial aldol product can serve again as a suitable acceptor for sequential addition of a second donor molecule to give (3*R*,5*R*)-2,4,6-trideoxyhexose **133** (Figure 5.60) [282, 283]. Cyclization to stable hemiacetals masks the free aldehyde and thus effectively precludes formation of higher-order adducts. When the first acceptor is an α-substituted acetaldehyde, related aldol products from twofold donor additions can be prepared that are structurally related to mevinolactone. Combination of a RibA-catalyzed initial addition to other aldolases such as FruA or NeuA in a consecutive addition reaction has also been studied for synthesis of non-natural sugars [277, 282].

Fig. 5.61
Aldol reactions catalyzed in vivo by serine hydroxymethyl transferase and by threonine aldolases.

5.7
Glycine Aldolases

The metabolism of β-hydroxy-α-amino acids involves pyridoxal phosphate-dependent enzymes, classified as serine hydroxymethyltransferase or threonine aldolases, that catalyze reversible aldol-type cleavage to aldehydes and glycine (**134**) [284].

In vivo serine hydroxymethyltransferases (SHMT; EC 2.1.2.1) catalyze the formation of L-serine **135** from addition of **134** to formaldehyde (Figure 5.61) in the presence of the cofactor tetrahydrofolate. SHMT have been purified and cloned from a large variety of organisms [285]. Eukaryotic enzymes have been shown to accept a range of other aldehydes (Table 5.8) to furnish the corresponding β-hydroxy-α-amino acid adducts [286, 287]. The enzymes are highly selective for the L-configuration but will usually produce diastereomeric mixtures with low *erythro*/*threo* selectivity. In preparative reactions excess **134** can compensate for the unfavorable equilibrium constant [56], and economically viable yields of **135** have thus been obtained on a multi-molar scale at high product concentrations [288, 289]. Recombinant rabbit SHMT has been used for highly stereoselective synthesis of L-*erythro*-2-amino-3-hydroxy-1,6-hexanedicarboxylic acid **136** as a potential precursor to carbocyclic β-lactams and nucleosides [290].

Threonine aldolases catalyze the reversible addition of **134** to acetaldehyde to give threonine (**137**), but the distinction between L-threonine aldolase (ThrA; EC 4.1.2.5) and L-*allo*-threonine aldolase (EC 4.1.2.6) is vague, because many catalysts have only poor capacity for *threo*/*erythro* discrimination (**137**/**138**; Figure 5.61) [284, 291, 292]. In fact, both L- and D-selective enzymes have been purified and cloned from a variety of organisms [284]. The first stereospecific L-*allo*-threonine aldolase from *Aeromonas jandaei* has recently been cloned and characterized [293]. X-ray structures of *Thermotoga*

Tab. 5.8
Substrate tolerance of threonine-dependent aldolases.

R	SHMT		L-ThrA		D-ThrA		Ref.
	Selectivity (threo:erythro)	Yield [%]	Selectivity (threo:erythro)	Yield [%]	Selectivity (threo:erythro)	Yield [%]	
H	–	94					288, 289
CH_3	2:98	–	9:91	40	53:47	60	292
C_3H_9			24:76	21	67:33	37	292
C_5H_{11}	60:40	25	37:63	16	61:39	31	287, 292
$CH(CH_3)_2$			46:53	15	84:16	26	292
CH_2OBn			2:98	88			297
CH_2CH_2OBn			47:53	53			297
$(CH_2)_2C_6H_5$			28:72	10	87:13	16	292
C_6H_5	60:40	22	60:40	9	74:26	11	287, 292
p-$C_6H_4NO_2$			47:53	53	55:45	88	292
o-$C_6H_4NO_2$			58:42	93	72:28	89	292
m-C_6H_4OH			73:27	43	74:26	54	292
p-$C_6H_4CH_3$			55:45	17	43:57	25	292
2-imidazolyl	33:67	10	66:34	40	61:39	60	287, 292
2-furanyl	50:50	20					287
2-thienyl	56:44	11					287

maritima ThrA as apoenzyme and in complex with substrate and product have recently been determined [294].

Many ThrA enzymes have broad substrate tolerance for the aldehyde acceptor, notably including variously substituted aliphatic and aromatic aldehydes (Figure 5.63); however, α,β-unsaturated aldehydes are not accepted [291]. The *erythro*-selective L-threonine aldolase from the yeast *Candida humicola* has been used to catalyze the addition of **134** to D-glyceraldehyde acetonide **139** (Figure 5.62) [295]. The diastereomeric products **140** and **141**, formed in a 1:2 ratio, were separated and further processed to furnish (*S*,*S*,*R*)- and (*S*,*S*,*S*)-3,4-dihydroxyprolines **143** and **144**. The same enzyme has also been applied to the synthesis of nucleobase-modified amino acids of type **142** [296], and of compound **146** as an intermediate en route to the immunosuppressive lipid mycestericin D [297]. For the latter, benzyl protection of the acceptor (**145**) was found to increase the stereoselectivity, and the *erythro* product was obtained preferentially under kinetic control at low conversion. With benzyloxyacetaldehyde a cloned ThrA from *E. coli* resulted in higher diastereoselectivity than other enzymes under conditions of kinetic control; this was applied in the synthesis of novel sialyl Lewis^x mimetics [298].

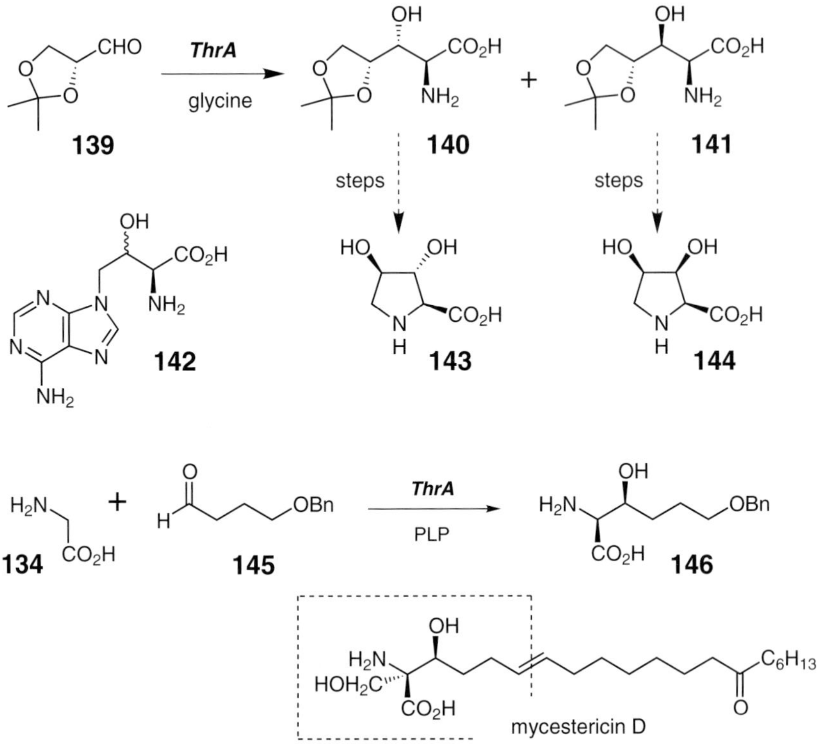

Fig. 5.62
Application of ThrA catalysis for stereoselective synthesis of dihydroxyprolines from glyceraldehyde, of an adenylamino acid in preparation of RNA mimics, and of a potential precursor to the immunosuppressive lipid mycestericin D.

More suitably, the specificity of ThrA enzymes can be used for kinetic resolution of diastereomer mixtures such as those produced by chemical synthesis (Figure 5.63). This is particularly promising for aryl analogs of threonine that are of interest as building blocks of pharmaceuticals, including vancomycin antibiotics. Thus, an L-ThrA from *Streptomyces amakusaensis* has been shown to be particularly useful for resolution of racemic *threo*-aryl serines **147** by retroaldolization under kinetic control to furnish enantiomerically pure D amino acids [290, 292, 299]. As a complementary example, the recombinant low-specificity D-ThrA from *Alcaligenes xyloxidans* has been used for resolution of DL-*threo*-β-(3,4-methylenedioxyphenyl)serine **148** by retro-aldol cleavage to furnish the desired L-*threo* isomer with a molar yield of 50% and almost 100% ee [300]. The latter compound serves as a synthetic intermediate en route to a parkinsonism drug.

DL-*threo*-**147** → (L-*ThrA*) → D-*threo*-**147** + 4-X-benzaldehyde + glycine **134**

DL-*threo*-**148** → (D-*ThrA*) → L-*threo*-**148** + piperonal + glycine

Fig. 5.63
Resolution of diastereomer mixtures by retro-aldolization under kinetic control for preparation of enantiomerically pure phenylserines, and racemate resolution process for a synthetic intermediate of an antiparkinsonism drug.

5.8 Recent Developments

As a result of a recent almost explosive increase in knowledge of the structures of all important classes of aldolases, including their liganded complexes, detailed understanding of substrate recognition and catalytic machinery for stereospecific carbon–carbon bond formation is advancing rapidly (Section 5.2.2). This, with progress fueled by whole-genome sequencing, means the enzymes are now amenable to rational engineering or directed molecular evolution toward improved physical properties and catalytic performance. Tailoring of enzyme properties such as improved activity and stability has been amply demonstrated to be feasible on demand [304, 305], and improved substrate tolerance and optimized stereoselectivity for asymmetric syntheses are immediate targets [306]. For identification of suitable mutant aldolases from engineering or evolutionary approaches, efficient assay techniques have been developed for identification of target catalysts in high-throughput mode based on chromogenic or fluorogenic substrates [307–309]. The same is true in the search for novel enzymes with interesting synthetic capabilities from mass screening of natural or man-made biodiversity [310]. Most recently, first attempts to redesign individual aldolase enzymes for an altered substrate spectrum have been reported [143, 144, 278], including a first example of modification of the stereospecificity of carbon–carbon bond creation by directed evolution [311]. Together with knowledge gained from the mechanistic profiles of catalytic antibodies as a complementary approach toward novel stereoselective aldolase biocatalysts (Chapter 6 in Part I of this book), such endeavors will probably also be instrumental in enabling better generic understanding of the mechanistic subtleties operating in the initiation and stereoelectronic steering of enzyme-catalyzed aldol reactions. Consequently, it is probable that current restrictions resulting from narrow donor specificity will soon be eliminated.

In addition to the major types of aldolases discussed above, several biosynthetically important enzymes promote aldol-related Claisen additions of acyl thioesters of coenzyme A (CoA) to ketones, often to α-oxoacids. This includes some of the key enzymes in central metabolism such as the citric acid cycle, and many others involved in the biosynthesis of fatty acids, steroids, terpenoids, macrolides and other secondary metabolites [14, 15]. Members of this class of C–C bond-forming enzymes are certainly attractive for asymmetric synthesis, because the equilibrium constants profit from a high driving force by ensuing thioester hydrolysis. High expense and limited CoA cofactor stability have so far precluded broader synthetic evaluation, but the technique might be more practical now that several cost-efficient in-situ cofactor recycling schemes for acyl-CoA substrates have been developed [312–316]. Particular progress is also advancing in the field of polyketides, a class of structurally diverse natural products with a broad range of biological activity. Polyketide biosynthesis is based on a fully modular scheme that involves distinct protein domains with ketosynthase (Claisen aldolization), or functional group processing activity, which iteratively build up the carbon backbone and an individually patterned polyketide structure. By choosing from known gene clusters of different polyketides from different microorganisms and plants that code for specific catalytic domains, new interspecies gene constructs can be assembled in a combinatorial fashion; this enables exploration of the feasibility of engineered biosynthesis of novel, non-natural polyketides with potentially useful bioactivity [317–320].

5.9 Summary and Conclusion

Progress in the synthetic use of aldol-active enzymes as efficient and highly stereoselective catalysts has been remarkable. Biocatalytic C–C bond formation is eminently useful for asymmetric synthesis of complex multifunctional molecules. Process design and development, however, requires an intimate knowledge base covering subjects such as substrate tolerance for donor and acceptor components or the influence of substrate structure on enantio- and diastereoselectivity of C–C bonding. This overview demonstrates that many enzymes for carbon–carbon bond formation are now readily available and predictably useful for synthetic applications. Clearly, such enzymes bear significant potential for application in asymmetric synthesis, particularly with regard to polyfunctionalized natural products and other biologically relevant classes of compounds. It is also evident that the technology is now well accepted in the chemical community as a powerful supplement to existing methodology of organic synthesis. Indeed, with the first processes already at the state of industrial commercialization on a large scale, enzymatic aldol reactions are maturing rapidly.

The scope of potential applications is broadening on all frontiers with increasing complexity of target structures. Limitations inherent in the properties of enzymes are appreciated and approaches are developed to resolve them. Hopefully, this compilation of the potential scope of the technique and its limitations will help to identify, and aid in the successful development of, future synthetic applications.

Examples of Experimental Procedures

General

Dihydroxyacetone phosphate (**41**) was prepared enzymatically from glycerol 1-phosphate (**74**) by the action of glycerol phosphate oxidase [177]. Analytical thin-layer chromatography was performed on silica gel plates using a 1:1 mixture of sat. ammonia–ethanol for development, and anisaldehyde stain for detection. The aldolases are commercially available or can be purified in accordance with published procedures [150]. Activity of aldolases (1 unit catalyzes cleavage of 1 μmol L-ketose 1-phosphates (**43**/**45**) per minute at 25 °C [150]) and amounts of **41** were determined photometrically by an assay coupled with glycerol phosphate dehydrogenase-catalyzed NADH oxidation [149].

L-Fructose 1-Phosphate from L-Glyceraldehyde by in-situ Formation of DHAP. GPO (70 U), catalase (1000 U), and RhuA (50 U) was added to a solution of L-glycerol 3-phosphate (**74**, 1.0 mmol) and L-glyceraldehyde (110 mg, 1.2 mmol) in 10 mL oxygen-saturated water at pH 6.8. The mixture was shaken at 20 °C under an oxygen atmosphere at 100 rpm. Conversion was monitored by enzymatic assay for equivalents of **41** produced, and by ^{1}H and ^{31}P NMR spectroscopy. After complete conversion and filtration through charcoal the pH was adjusted to 7.5 by addition of 1.0 M cyclohexylamine in ethanol and the solution was concentrated to dryness by rotary evaporation at ≤20 °C in vacuo. The solid residue was dissolved in 0.5 mL water and the resulting solution was filtered. Dry ethanol (2.5 mL) was added, then dry acetone until faint turbidity remained. Crystallization at 4 °C furnished L-fructose 1-phosphate bis(cyclohexylammonium) salt as colorless needles; yield 370 mg (85%).

L-Fuculose 1-Phosphate from DL-Lactaldehyde by Racemate Resolution [171]. Racemic lactaldehyde dimethylacetal (6.0 g, 50 mmol) in 50 mL water was hydrolyzed by treatment with cation-exchange resin (Dowex AG50W-X8, H^{+} form) at 60 °C for 8 h. After filtration, an aqueous solution of **41** (80 mL, 10 mmol; prepared by GPO oxidation of **74** as above) was added. The mixture was adjusted to pH 6.8 with 1 M NaOH and incubated with FucA (300 U) at room temperature. After complete conversion of **41**, as determined by TLC (R_F(**41**) = 0.39, R_F(**45**) = 0.30) and enzymatic assay, the mixture was filtered through charcoal and passed through an anion-exchange column (Dowex AG1-X8, HCO_3^- form, 100 mL). The column was washed with water (200 mL), and the product was eluted with 0.2 M triethylammonium hydrogen carbonate buffer. Repeated concentration from water (3 × 50 mL), ion exchange to the free acid (Dowex AG50W-X8, H^+ form, 100 mL), neutralization with cyclohexylamine, then crystallization from 90% aqueous ethanol provided the colorless bis(cyclohexylammonium) salt of **45** (3.76 g, 85%).

References

1 H. G. Davies, R. H. Green, D. R. Kelly, S. M. Roberts, *Biotransformations in Preparative Organic Chemistry*, Academic Press, London **1989**.

2 W. Gerhartz, *Enzymes in Industry: Production and Applications*, VCH, Weinheim 1991.

3 J. Halgas, *Biocatalysts in Organic Synthesis*, Elsevier, Amsterdam **1992**.

4 L. Poppe, L. Novák, *Selective Biocatalysis. A Synthetic Approach*, VCH, Weinheim **1992**.

5 C.-H. Wong, G. M. Whitesides, *Enzymes in Synthetic Organic Chemistry*, Pergamon, Oxford **1994**.

6 M. Sinnott, *Comprehensive Biological Catalysis*, Academic Press, San Diego **1998**.

7 W.-D. Fessner, *Biocatalysis – From Discovery to Application (Top. Curr. Chem., Vol. 200)*, Springer, Heidelberg **1998**.

8 U. Bornscheuer, R. Kazlauskas, *Hydrolases in Organic Synthesis – Regio- and Stereoselective Biotransformations*, Wiley–VCH, Weinheim **1999**.

9 K. Faber, *Biotransformations in Organic Chemistry*, 4th Ed., Springer, Heidelberg **2000**.

10 A. Liese, K. Seelbach, C. Wandrey, *Industrial Biotransformations*, Wiley–VCH, Weinheim **2000**.

11 K. Drauz, H. Waldmann, *Enzyme Catalysis in Organic Synthesis*, 2nd Ed., Wiley–VCH, Weinheim **2002**.

12 T. D. Machajewski, C.-H. Wong, *Angew. Chem. Int. Ed.* **2000**, *39*, 1352–1374.

13 H. Waldmann, D. Sebastian, *Chem. Rev.* **1994**, *94*, 911–937.

14 D. F. Henderson, E. J. Toone, in *Comprehensive Natural Product Chemistry, Vol. 3* (Ed.: B. M. Pinto), Elsevier Science, Amsterdam **1999**, p. 367–440.

15 D. Schomburg, M. Salzmann, *Enzyme Handbook*, Springer, Berlin **1990–1995**.

16 V. Schellenberger, H. D. Jakubke, *Angew. Chem. Int. Ed. Engl.* **1991**, *30*, 1437–1449.

17 D. Rozzell, F. Wagner, *Biocatalytic Production of Amino Acids and Derivatives*, Hanser, Munich **1992**.
18 E. J. Toone, E. S. Simon, M. D. Bednarski, G. M. Whitesides, *Tetrahedron* **1989**, *45*, 5365–5422.
19 S. David, C. Augé, C. Gautheron, *Adv. Carbohydr. Chem. Biochem.* **1991**, *49*, 175–237.
20 H. J. M. Gijsen, L. Qiao, W. Fitz, C.-H. Wong, *Chem. Rev.* **1996**, *96*, 443–473.
21 G. M. Watt, P. A. Lowden, S. L. Flitsch, *Curr. Opin. Struct. Biol.* **1997**, *7*, 652–660.
22 S. Takayama, G. J. McGarvey, C.-H. Wong, *Ann. Rev. Microbiol.* **1997**, *51*, 285–310.
23 N. Wymer, E. J. Toone, *Curr Opin Chem Biol* **2000**, *4*, 110–119.
24 W.-D. Fessner, V. Helaine, *Curr Opin. Biotechnol.* **2001**, *12*, 574–586.
25 W.-D. Fessner, C. Walter, *Top. Curr. Chem.* **1996**, *184*, 97–194.
26 W.-D. Fessner, in *Microbial Reagents in Organic Synthesis, Vol. 381* (Ed.: S. Servi), Kluwer Academic Publishers, Dordrecht, **1992**, pp. 43–55.
27 R. Kluger, *Chem. Rev.* **1990**, *90*, 1151–1169.
28 E. C. Webb, *Enzyme Nomenclature*, Academic Press, San Diego **1992**.
29 B. L. Horecker, O. Tsolas, C. Y. Lai, in *The Enzymes, Vol. VII* (Ed.: P. D. Boyer), Academic Press, New York, **1972**, pp. 213–258.
30 G. Trombetta, G. Balboni, A. di Iasio, E. Grazi, *Biochem. Biophys. Res. Commun.* **1977**, *74*, 1297–1301.
31 D. J. Kuo, I. A. Rose, *Biochemistry* **1985**, *24*, 3947–3952.
32 W.-D. Fessner, A. Schneider, H. Held, G. Sinerius, C. Walter, M. Hixon, J. V. Schloss, *Angew. Chem. Int. Ed. Engl.* **1996**, *35*, 2219–2221.
33 J. Sygusch, D. Beaudry, M. Allaire, *Proc. Natl. Acad. Sci. USA* **1987**, *84*, 7846–7850.
34 S. J. Gamblin, G. J. Davies, J. M. Grimes, R. M. Jackson, J. A. Littlechild, H. C. Watson, *J. Mol. Biol.* **1991**, *219*, 573–576.
35 G. Hester, O. Brenner-Holzach, F. A. Rossi, M. Struck-Donatz, K. H. Winterhalter, J. D. G. Smit, K. Piontek, *FEBS Lett.* **1991**, *292*, 237–242.
36 H. Kim, U. Certa, H. Dobeli, P. Jakob, W. G. J. Hol, *Biochemistry* **1998**, *37*, 4388–4396.
37 A. R. Dalby, D. R. Tolan, J. A. Littlechild, *Acta Cryst.* **2001**, *D 57*, 1526–1533.
38 N. Blom, J. Sygusch, *Nature Struct. Biol.* **1997**, *4*, 36–39.
39 K. H. Choi, J. Shi, C. E. Hopkins, D. R. Tolan, K. N. Allen, *Biochemistry* **2001**, *40*, 13868–13875.
40 A. Maurady, A. Zdanov, D. de Moissac, D. Beaudry, J. Sygusch, *J. Biol. Chem.* **2002**, *277*, 9474–9483.
41 J. Jia, W. J. Huang, U. Schörken, H. Sahm, G. A. Sprenger, Y. Lindqvist, G. Schneider, *Structure* **1996**, *4*, 715–724.

42 J. Jia, U. Schörken, Y. Lindqvist, G. A. Sprenger, G. Schneider, *Protein Sci.* **1997**, *6*, 119–124.
43 T. Izard, M. C. Lawrence, R. L. Malby, G. G. Lilley, P. M. Colman, *Structure* **1994**, *2*, 361–369.
44 M. C. Lawrence, J. A. R. G. Barbosa, B. J. Smith, N. E. Hall, P. A. Pilling, H. C. Ooi, S. M. Marcuccio, *J. Mol. Biol.* **1997**, *266*, 381–399.
45 J. A. Barbosa, B. J. Smith, R. DeGori, H. C. Ooi, S. M. Marcuccio, E. M. Campi, W. R. Jackson, R. Brossmer, M. Sommer, M. C. Lawrence, *J Mol Biol* **2000**, *303*, 405–421.
46 J. Allard, P. Grochulski, J. Sygusch, *Proc. Natl. Acad. Sci. USA* **2001**, *98*, 3679–3684.
47 M. K. Dreyer, G. E. Schulz, *J. Mol. Biol.* **1996**, *259*, 458–466.
48 M. K. Dreyer, G. E. Schulz, *J. Mol. Biol.* **1993**, *231*, 549–553.
49 M. Kroemer, G. E. Schulz, *Acta Cryst.* **2002**, *D 58*, 824–832.
50 A. C. Joerger, C. Gosse, W.-D. Fessner, G. E. Schulz, *Biochemistry* **2000**, *39*, 6033–6041.
51 N. S. Blom, S. Tetreault, R. Coulombe, J. Sygusch, *Nature Struct. Biol.* **1996**, *3*, 856–862.
52 S. J. Cooper, G. A. Leonard, S. M. McSweeney, A. W. Thompson, J. H. Naismith, S. Qamar, A. Plater, A. Berry, W. N. Hunter, *Structure* **1996**, *4*, 1303–1315.
53 S. M. Zgiby, G. J. Thomson, S. Qamar, A. Berry, *Eur. J. Biochem.* **2000**, *267*, 1858–1869.
54 D. R. Hall, C. S. Bond, G. A. Leonard, C. I. Watt, A. Berry, W. N. Hunter, *J. Biol. Chem.* **2002**, *277*, 22018–22024.
55 D. R. Hall, L. E. Kemp, G. A. Leonard, K. Marshall, A. Berry, W. N. Hunter, *Acta Cryst.* **2003**, *D 59*, 611–614.
56 R. N. Goldberg, Y. B. Tewari, *J. Phys. Chem. Ref. Data* **1995**, *24*, 1669–1698.
57 A. Pollak, H. Blumenfeld, M. Wax, R. L. Baughn, G. M. Whitesides, *J. Am. Chem. Soc.* **1980**, *102*, 6324–6336.
58 B. Bossow, W. Berke, C. Wandrey, *BioEngineering* **1992**, *8*, 12–19.
59 S. B. Sobolov, A. Bartoszkomalik, T. R. Oeschger, M. M. Montelbano, *Tetrahedron Lett.* **1994**, *35*, 7751–7754.
60 U. Kragl, U. Niedermeyer, M.-R. Kula, C. Wandrey, in *Ann. N. Y. Acad. Sci., Vol. 613* (Eds.: H. Okada, A. Tanaka, H. W. Blanch), The New York Academy of Sciences, New York, **1990**, pp. 167–175.
61 U. Kragl, D. Gygax, O. Ghisalba, C. Wandrey, *Angew. Chem. Int. Ed. Engl.* **1991**, *30*, 827–828.
62 U. Kragl, A. Gödde, C. Wandrey, N. Lubin, C. Augé, *J. Chem. Soc. Perkin Trans. I* **1994**, 119–124.
63 L. Espelt, P. Clapes, J. Esquena, A. Manich, C. Solans, *Langmuir* **2003**, *19*, 1337–1346.
64 W.-D. Fessner, M. Knorst, DE 10034586, **2002**; *Chem. Abstr. 136*:166160.
65 Y. Ohta, M. Shimosaka, K. Murata, Y. Tsukada, A. Kimura, *Appl. Microbiol. Biotechnol.* **1986**, *24*, 386–391.
66 K. Aisaka, S. Tamura, Y. Arai, T. Uwajima, *Biotechnol. Lett.* **1987**, *9*, 633–637.
67 K. Yamamoto, B. Kawakami, Y. Kawamura, K. Kawai, *Anal. Biochem.* **1997**, *246*, 171–175.

68 M. J. Kim, W. J. Hennen, H. M. Sweers, C.-H. Wong, *J. Am. Chem. Soc.* **1988**, *110*, 6481–6486.
69 C. Augé, S. David, C. Gautheron, *Tetrahedron Lett.* **1984**, *25*, 4663–4664.
70 C. Augé, S. David, C. Gautheron, A. Veyrières, *Tetrahedron Lett.* **1985**, *26*, 2439–2440.
71 C. Augé, C. Gautheron, *J. Chem. Soc. Chem. Commun.* **1987**, 859–860.
72 C. Augé, S. David, C. Gautheron, A. Malleron, B. Cavayé, *New. J. Chem.* **1988**, *12*, 733–744.
73 M. D. Bednarski, H. K. Chenault, E. S. Simon, G. M. Whitesides, *J. Am. Chem. Soc.* **1987**, *109*, 1283–1285.
74 E. S. Simon, M. D. Bednarski, G. M. Whitesides, *J. Am. Chem. Soc.* **1988**, *110*, 7159–7163.
75 T. Sugai, A. Kuboki, S. Hiramatsu, H. Okazaki, H. Ohta, *Bull. Chem. Soc. Jpn.* **1995**, *68*, 3581–3589.
76 M. Mahmoudian, D. Noble, C. S. Drake, R. F. Middleton, D. S. Montgomery, J. E. Piercey, D. Ramlakhan, M. Todd, M. J. Dawson, *Enzyme Microb. Technol.* **1997**, *20*, 393–400.
77 S. Blayer, J. M. Woodley, M. D. Lilly, M. J. Dawson, *Biotechnol. Prog.* **1996**, *12*, 758–763.
78 I. Maru, J. Ohnishi, Y. Ohta, Y. Tsukada, *Carbohydrate Res.* **1998**, *306*, 575–578.
79 C. H. Lin, T. Sugai, R. L. Halcomb, Y. Ichikawa, C.-H. Wong, *J. Am. Chem. Soc.* **1992**, *114*, 10138–10145.
80 Y. Ichikawa, J. L. C. Liu, G. J. Shen, C.-H. Wong, *J. Am. Chem. Soc.* **1991**, *113*, 6300–6302.
81 O. Blixt, J. C. Paulson, *Adv. Synth. Catal.* **2003**, *345*, 687–690.
82 J. Beliczey, U. Kragl, A. Liese, C. Wandrey, K. Hamacher, H. H. Coenen, T. Tierling, US 6355453, **2002**; *Chem. Abstr. 136*:231340.
83 W. Fitz, J. R. Schwark, C.-H. Wong, *J. Org. Chem.* **1995**, *60*, 3663–3670.
84 C. Augé, B. Bouxom, B. Cavayé, C. Gautheron, *Tetrahedron Lett.* **1989**, *30*, 2217–2220.
85 J. L. C. Liu, G. J. Shen, Y. Ichikawa, J. F. Rutan, G. Zapata, W. F. Vann, C.-H. Wong, *J. Am. Chem. Soc.* **1992**, *114*, 3901–3910.
86 C. Salagnad, A. Godde, B. Ernst, U. Kragl, *Biotechnol. Prog.* **1997**, *13*, 810–813.
87 R. J. Lins, S. L. Flitsch, N. J. Turner, E. Irving, S. A. Brown, *Angew. Chem. Int. Ed.* **2002**, *41*, 3405–3407.
88 W.-Y. Wu, B. Jin, D. C. M. Kong, M. von Itzstein, *Carbohydrate Res.* **1997**, *300*, 171–174.
89 A. Kuboki, H. Okazaki, T. Sugai, H. Ohta, *Tetrahedron* **1997**, *53*, 2387–2400.
90 A. J. Humphrey, C. Fremann, P. Critchley, Y. Malykh, R. Schauer, T. D. H. Bugg, *Bioorg. Med. Chem.* **2002**, *10*, 3175–3185.
91 C.-C. Lin, C.-H. Lin, C.-H. Wong, *Tetrahedron Lett.* **1997**, *38*, 2649–2652.
92 M. Murakami, K. Ikeda, K. Achiwa, *Carbohydrate Res.* **1996**, *280*, 101–110.

93 D. C. M. Kong, M. Vonitzstein, *Carbohydrate Res.* **1997**, *305*, 323–329.
94 M. J. Kiefel, J. C. Wilson, S. Bennett, M. Gredley, M. von Itzstein, *Bioorg. Med. Chem.* **2000**, *8*, 657–664.
95 M. A. Sparks, K. W. Williams, C. Lukacs, A. Schrell, G. Priebe, A. Spaltenstein, G. M. Whitesides, *Tetrahedron* **1993**, *49*, 1–12.
96 P. Z. Zhou, H. M. Salleh, J. F. Honek, *J. Org. Chem.* **1993**, *58*, 264–266.
97 W. Fitz, C.-H. Wong, *J. Org. Chem.* **1994**, *59*, 8279–8280.
98 G. B. Kok, M. Campbell, B. L. Mackey, M. von Itzstein, *Carbohydr. Res.* **2001**, *332*, 133–139.
99 S. David, A. Malleron, B. Cavaye, *New J. Chem* **1992**, *16*, 751–755.
100 C. Augé, C. Gautheron, S. David, A. Malleron, B. Cavayé, B. Bouxom, *Tetrahedron* **1990**, *46*, 201–214.
101 T. Sugai, G. J. Shen, Y. Ichikawa, C.-H. Wong, *J. Am. Chem. Soc.* **1993**, *115*, 413–421.
102 S. Ladisch, A. Hasegawa, R. X. Li, M. Kiso, *Biochemistry* **1995**, *34*, 1197–1202.
103 K. Koppert, R. Brossmer, *Tetrahedron Lett.* **1992**, *33*, 8031–8034.
104 A. Malleron, S. David, *New J. Chem.* **1996**, *20*, 153–159.
105 T. Miyazaki, H. Sato, T. Sakakibara, Y. Kajihara, *J. Am. Chem. Soc.* **2000**, *122*, 5678–5694.
106 L. K. Mahal, K. J. Yarema, C. R. Bertozzi, *Science* **1997**, *276*, 1125–1128.
107 K. J. Yarema, L. K. Mahal, R. E. Bruehl, E. C. Rodriguez, C. R. Bertozzi, *J. Biol. Chem.* **1998**, *273*, 31168–31179.
108 C. R. H. Raetz, *J. Bacteriol.* **1993**, *175*, 5745–5753.
109 F. M. Unger, *Adv. Carbohydr. Chem. Biochem.* **1981**, *38*, 323–388.
110 B. R. Knappmann, M. R. Kula, *Appl. Microbiol. Biotechnol.* **1990**, *33*, 324–329.
111 M. A. Ghalambor, E. C. Heath, *J. Biol. Chem.* **1966**, *241*, 3222–3227.
112 P. R. Srinivasan, D. B. Sprinson, *J. Biol. Chem.* **1959**, *234*, 716–722.
113 J. Hurwitz, A. Weissbach, *J. Biol. Chem.* **1959**, *234*, 710–712.
114 P. H. Ray, *J. Bacteriol.* **1980**, *141*, 635–644.
115 F. Stuart, I. S. Hunter, *Biochim. Biophys. Acta* **1993**, *1161*, 209–215.
116 J. M. Ray, R. Bauerle, *J. Bacteriol.* **1991**, *173*, 1894–1901.
117 L. M. Weaver, J. E. B. P. Pinto, K. M. Herrmann, *Bioorg. Med. Chem. Lett.* **1993**, *3*, 1421–1428.
118 I. A. Shumilin, C. Zhao, R. Bauerle, R. H. Kretsinger, *J. Mol. Biol.* **2002**, *320*, 1147–1156.
119 R. L. Doong, J. E. Gander, R. J. Ganson, R. A. Jensen, *Physiol. Plant.* **1992**, *84*, 351–360.
120 L. M. Reimer, D. L. Conley, D. L. Pompliano, J. W. Frost, *J. Am. Chem. Soc.* **1986**, *108*, 8010–8015.
121 K. M. Draths, J. W. Frost, *J. Am. Chem. Soc.* **1990**, *112*, 1657–1659.

122 K. M. Draths, T. L. Ward, J. W. Frost, *J. Am. Chem. Soc.* **1992**, *114*, 9725–9726.
123 D. R. Knop, K. M. Draths, S. S. Chandran, J. L. Barker, R. von Daeniken, W. Weber, J. W. Frost, *J. Am. Chem. Soc.* **2001**, *123*, 10173–10182.
124 K. M. Draths, J. W. Frost, *J. Am. Chem. Soc.* **1995**, *117*, 2395–2400.
125 J. L. Barker, J. W. Frost, *Biotechnol. Bioeng.* **2001**, *76*, 376–390.
126 N. Ran, D. R. Knop, K. M. Draths, J. W. Frost, *J. Am. Chem. Soc.* **2001**, *123*, 10927–10934.
127 A. K. Sundaram, R. W. Woodard, *J Org Chem* **2000**, *65*, 5891–5897.
128 W. A. Wood, in *The Enzymes, 3rd Ed, Vol. 7* (Eds.: P. D. Boyer), Academic, New York, **1972**, pp. 281–302.
129 R. H. Hammerstedt, H. Möhler, K. A. Decker, D. Ersfeld, W. A. Wood, *Methods Enzymol.* **1975**, *42*, 258–264.
130 R. K. Scopes, *Anal. Biochem.* **1984**, *136*, 525–529.
131 M. C. Shelton, E. J. Toone, *Tetrahedron Asymmetry* **1995**, *6*, 207–211.
132 T. Conway, R. Fliege, D. Jones-Kilpatrick, J. Liu, W. O. Barnell, S. E. Egan, *Mol. Microbiol.* **1991**, *5*, 2901–2911.
133 A. T. Carter, B. M. Pearson, J. R. Dickinson, W. E. Lancashire, *Gene* **1993**, *130*, 155–156.
134 N. Hugouvieux-Cotte-Pattat, J. Robert-Baudouy, *Mol. Microbiol.* **1994**, *11*, 67–75.
135 J. S. Griffiths, N. J. Wymer, E. Njolito, S. Niranjanakumari, C. A. Fierke, E. J. Toone, *Bioorg. Med. Chem.* **2002**, *10*, 545–550.
136 S. E. Egan, R. Fliege, S. Tong, A. Shibata, R. E. Wolf, T. Conway, *J. Bacteriol.* **1992**, *174*, 4638–4646.
137 T. S. M. Taha, T. L. Deits, *Biochem. Biophys. Res. Commun.* **1994**, *200*, 459–466.
138 N. C. Floyd, M. H. Liebster, N. J. Turner, *J. Chem. Soc. Perkin Trans. I* **1992**, 1085–1086.
139 S. T. Allen, G. R. Heintzelman, E. J. Toone, *J. Org. Chem.* **1992**, *57*, 426–427.
140 M. C. Shelton, I. C. Cotterill, S. T. A. Novak, R. M. Poonawala, S. Sudarshan, E. J. Toone, *J. Am. Chem. Soc.* **1996**, *118*, 2117–2125.
141 I. C. Cotterill, M. C. Shelton, D. E. W. Machemer, D. P. Henderson, E. J. Toone, *J. Chem. Soc. Perkin Trans. 1* **1998**, 1335–1341.
142 D. P. Henderson, M. C. Shelton, I. C. Cotterill, E. J. Toone, *J. Org. Chem.* **1997**, *62*, 7910–7911.
143 S. Fong, T. D. Machajewski, C. C. Mak, C.-H. Wong, *Chem Biol* **2000**, *7*, 873–883.
144 N. Wymer, L. V. Buchanan, D. Henderson, N. Mehta, C. H. Botting, L. Pocivavsek, C. A. Fierke, E. J. Toone, J. H. Naismith, *Structure* **2001**, *9*, 1–10.
145 H. P. Meloche, E. L. O'Connell, *Methods Enzymol* **1982**, *90*, 263–269.

146 A. M. Elshafei, O. M. Abdel-Fatah, *AFS, Adv. Food Sci.* **2002**, *24*, 30–36.
147 D. P. Henderson, I. C. Cotterill, M. C. Shelton, E. J. Toone, *J. Org. Chem.* **1998**, *63*, 906–907.
148 W.-D. Fessner, G. Sinerius, *Bioorg. Med. Chem.* **1994**, *2*, 639–645.
149 M. D. Bednarski, E. S. Simon, N. Bischofberger, W.-D. Fessner, M. J. Kim, W. Lees, T. Saito, H. Waldmann, G. M. Whitesides, *J. Am. Chem. Soc.* **1989**, *111*, 627–635.
150 W.-D. Fessner, G. Sinerius, A. Schneider, M. Dreyer, G. E. Schulz, J. Badia, J. Aguilar, *Angew. Chem. Int. Ed. Engl.* **1991**, *30*, 555–558.
151 C. H. von der Osten, A. J. Sinskey, C. F. Barbas, R. L. Pederson, Y. F. Wang, C.-H. Wong, *J. Am. Chem. Soc.* **1989**, *111*, 3924–3927.
152 M. T. Zannetti, C. Walter, M. Knorst, W.-D. Fessner, *Chemistry Eur. J.* **1999**, *5*, 1882–1890.
153 C. de Montigny, J. Sygusch, *Eur. J. Biochem.* **1996**, *241*, 243–248.
154 F. Götz, S. Fischer, K.-H. Schleifer, *Eur. J. Biochem.* **1980**, *108*, 295–301.
155 W. J. Lees, G. M. Whitesides, *J. Org. Chem.* **1993**, *58*, 1887–1894.
156 A. Straub, F. Effenberger, P. Fischer, *J. Org. Chem.* **1990**, *55*, 3926–3932.
157 D. L. Bissett, R. L. Anderson, *J. Biol. Chem.* **1980**, *255*, 8750–8755.
158 W.-D. Fessner, O. Eyrisch, *Angew. Chem. Int Ed. Engl.* **1992**, *31*, 56–58.
159 O. Eyrisch, G. Sinerius, W.-D. Fessner, *Carbohydrate Res.* **1993**, *238*, 287–306.
160 E. Garcia-Junceda, G. J. Shen, T. Sugai, C.-H. Wong, *Bioorg Med Chem* **1995**, *3*, 945–953.
161 A. Brinkkotter, A. Shakeri Garakani, J. W. Lengeler, *Arch. Microbiol.* **2002**, *177*, 410–419.
162 G. J. Williams, S. Domann, A. Nelson, A. Berry, *Proc. Natl. Acad. Sci. USA* **2003**, *100*, 3143–3148.
163 Y. Takagi, *Methods Enzymol.* **1966**, *9*, 542–545.
164 T. H. Chiu, K. L. Evans, D. S. Feingold, *Methods Enzymol* **1975**, *42*, 264–269.
165 M. A. Ghalambor, E. C. Heath, *Methods Enzymol.* **1966**, *9*, 538–542.
166 P. Moralejo, S. M. Egan, E. Hidalgo, J. Aguilar, *J. Bacteriol* **1993**, *175*, 5585–5594.
167 J. Nishitani, G. Wilcox, *Gene* **1991**, *105*, 37–42.
168 Z. Lu, E. C. C. Lin, *Nucleic Acids Res* **1989**, *17*, 4883–4884.
169 A. Ozaki, E. J. Toone, C. H. von der Osten, A. J. Sinskey, G. M. Whitesides, *J. Am. Chem. Soc.* **1990**, *112*, 4970–4971.
170 W.-D. Fessner, J. Badia, O. Eyrisch, A. Schneider, G. Sinerius, *Tetrahedron Lett.* **1992**, *33*, 5231–5234.
171 W.-D. Fessner, A. Schneider, O. Eyrisch, G. Sinerius, J. Badia, *Tetrahedron Asymmetry* **1993**, *4*, 1183–1192.
172 S. A. Phillips, P. J. Thornalley, *Eur. J. Biochem.* **1993**, *212*, 101–105.

173 J. P. Richard, *Biochem. Soc. Trans.* **1993**, *21*, 549–553.
174 W.-D. Fessner, C. Walter, *Angew. Chem. Int. Ed. Engl.* **1992**, *31*, 614–616.
175 D. C. Crans, G. M. Whitesides, *J. Am. Chem. Soc.* **1985**, *107*, 7019–7027.
176 P. D'Arrigo, V. Piergianni, G. Pedrocchi-Fantoni, S. Servi, *J. Chem. Soc. Chem. Commun.* **1995**, 2505–2506.
177 W.-D. Fessner, G. Sinerius, *Angew. Chem. Int. Ed. Engl.* **1994**, *33*, 209–212.
178 R. Schoevaart, F. van Rantwijk, R. A. Sheldon, *J Org Chem* **2000**, *65*, 6940–6943.
179 H. L. Arth, G. Sinerius, W.-D. Fessner, *Liebigs Ann.* **1995**, 2037–2042.
180 H. L. Arth, W.-D. Fessner, *Carbohydrate Res.* **1997**, *305*, 313–321.
181 D. G. Drueckhammer, J. R. Durrwachter, R. L. Pederson, D. C. Crans, L. Daniels, C.-H. Wong, *J. Org. Chem.* **1989**, *54*, 70–77.
182 R. Schoevaart, F. van Rantwijk, R. A. Sheldon, *J. Org. Chem.* **2001**, *66*, 4559–4562.
183 R. A. Periana, R. Motiu-DeGrood, Y. Chiang, D. J. Hupe, *J. Am. Chem. Soc.* **1980**, *102*, 3923–3927.
184 F. Effenberger, A. Straub, *Tetrahedron Lett.* **1987**, *28*, 1641–1644.
185 S. H. Jung, J. H. Jeong, P. Miller, C.-H. Wong, *J. Org. Chem.* **1994**, *59*, 7182–7184.
186 R. L. Pederson, J. Esker, C.-H. Wong, *Tetrahedron* **1991**, *47*, 2643–2648.
187 M. L. Valentin, J. Bolte, *Bull. Soc. Chimi. Fr.* **1995**, *132*, 1167–1171.
188 T. Gefflaut, M. Lemaire, M. L. Valentin, J. Bolte, *J. Org. Chem.* **1997**, *62*, 5920–5922.
189 E. L. Ferroni, V. DiTella, N. Ghanayem, R. Jeske, C. Jodlowski, M. O'Connell, J. Styrsky, R. Svoboda, A. Venkataraman, B. M. Winkler, *J. Org. Chem.* **1999**, *64*, 4943–4945.
190 W.-D. Fessner, *Curr Opin Chem Biol* **1998**, *2*, 85–97.
191 I. Henderson, K. B. Sharpless, C.-H. Wong, *J. Am. Chem. Soc.* **1994**, *116*, 558–561.
192 M.-J. Kim, I. T. Lim, *Synlett* **1996**, 138–140.
193 W.-D. Fessner, C. Gosse, G. Jaeschke, O. Eyrisch, *Eur. J. Org. Chem.* **2000**, 125–132.
194 C.-H. Wong, G. M. Whitesides, *J. Org. Chem.* **1983**, *48*, 3199–3205.
195 K. K. C. Liu, C.-H. Wong, *J. Org. Chem.* **1992**, *57*, 4789–4791.
196 T. Ziegler, A. Straub, F. Effenberger, *Angew. Chem., Int. Ed. Engl.* **1988**, *27*, 716–717.
197 L. Azema, F. Bringaud, C. Blonski, J. Perie, *Bioorg. Med. Chem.* **2000**, *8*, 717–722.
198 W. Zhu, Z. Li, *Perkin Trans. 1* **2000**, 1105–1108.
199 M. D. Bednarski, H. J. Waldmann, G. M. Whitesides, *Tetrahedron Lett* **1986**, *27*, 5807–5810.
200 N. J. Turner, G. M. Whitesides, *J. Am. Chem. Soc.* **1989**, *111*, 624–627.

201 D. Crestia, C. Guerard, J. Bolte, C. Demuynck, *J. Mol. Catal. B: Enzym* **2001**, *11*, 207–212.

202 C.-H. Wong, F. Moris-Varas, S.-C. Hung, T. G. Marron, C.-C. Lin, K. W. Gong, G. Weitz-Schmidt, *J. Am. Chem. Soc.* **1997**, *119*, 8152–8158.

203 O. Eyrisch, M. Keller, W.-D. Fessner, *Tetrahedron Lett.* **1994**, *35*, 9013–9016.

204 M. Petersen, M. T. Zannetti, W.-D. Fessner, *Top. Curr. Chem.* **1997**, *186*, 87–117.

205 J. R. Durrwachter, C.-H. Wong, *J. Org. Chem.* **1988**, *53*, 4175–4181.

206 K. K. C. Liu, R. L. Pederson, C.-H. Wong, *J. Chem. Soc., Perkin Trans* **1991**, *1*, 2669–2673.

207 J. R. Durrwachter, D. G. Drueckhammer, K. Nozaki, H. M. Sweers, C.-H. Wong, *J. Am. Chem. Soc.* **1986**, *108*, 7812–7818.

208 D. Franke, V. Lorbach, S. Esser, C. Dose, G. A. Sprenger, M. Halfar, J. Thömmes, R. Müller, R. Takors, M. Müller, *Chem. Eur. J.* **2003**, *9*, 4188–4196.

209 C. W. Borysenko, A. Spaltenstein, J. A. Straub, G. M. Whitesides, *J. Am. Chem. Soc.* **1989**, *111*, 9275–9276.

210 C.-H. Wong, R. Alajarin, F. Moris-Varas, O. Blanco, E. Garcia-Junceda, *J. Org. Chem.* **1995**, *60*, 7360–7363.

211 W.-D. Fessner, *GIT Fachz. Lab.* **1993**, *37*, 951–956.

212 W. Schmid, G. M. Whitesides, *J. Am. Chem. Soc.* **1990**, *112*, 9670–9671.

213 W.-C. Chou, C. Fotsch, C.-H. Wong, *J. Org. Chem.* **1995**, *60*, 2916–2917.

214 H. J. M. Gijsen, C.-H. Wong, *Tetrahedron Lett.* **1995**, *36*, 7057–7060.

215 B. P. Maliakel, W. Schmid, *J. Carbohydr. Chem* **1993**, *12*, 415–424.

216 F. Nicotra, L. Panza, G. Russo, A. Verani, *Tetrahedron Asymmetry* **1993**, *4*, 1203–1204.

217 G. Guanti, L. Banfi, M. T. Zannetti, *Tetrahedron Lett* **2000**, *41*, 3181–3185.

218 G. Guanti, M. T. Zannetti, L. Banfi, R. Riva, *Adv. Synth. Catal.* **2001**, *343*, 682–691.

219 R. Duncan, D. G. Drueckhammer, *J. Org. Chem.* **1996**, *61*, 438–439.

220 F. Effenberger, A. Straub, V. Null, *Liebigs Ann.* **1992**, 1297–1301.

221 W. C. Chou, L. H. Chen, J. M. Fang, C.-H. Wong, *J. Am. Chem. Soc.* **1994**, *116*, 6191–6194.

222 G. C. Look, C. H. Fotsch, C.-H. Wong, *Acc. Chem. Res.* **1993**, *26*, 182–190.

223 C.-H. Wong, R. L. Halcomb, Y. Ichikawa, T. Kajimoto, *Angew. Chem. Int. Ed. Engl.* **1995**, *34*, 412–432.

224 R. L. Pederson, M.-J. Kim, C.-H. Wong, *Tetrahedron Lett* **1988**, *29*, 4645–4648.

225 K. K. C. Liu, T. Kajimoto, L. Chen, Z. Zhong, Y. Ichikawa, C.-H. Wong, *J. Org. Chem.* **1991**, *56*, 6280–6289.

226 P. Z. Zhou, H. M. Salleh, P. C. M. Chan, G. Lajoie, J. F.

Honek, P. T. C. Nambiar, O. P. Ward, *Carbohydrate Res.* **1993**, *239*, 155–166.
227 W. J. Lees, G. M. Whitesides, *Bioorg. Chem.* **1992**, *20*, 173–179.
228 R. R. Hung, J. A. Straub, G. M. Whitesides, *J. Org. Chem.* **1991**, *56*, 3849–3855.
229 F. Moris-Varas, X.-H. Qian, C.-H. Wong, *J. Am. Chem. Soc.* **1996**, *118*, 7647–7652.
230 M. L. Mitchell, L. V. Lee, C.-H. Wong, *Tetrahedron Lett.* **2002**, *43*, 5691–5693.
231 M. Schuster, W. F. He, S. Blechert, *Tetrahedron Lett* **2001**, *42*, 2289–2291.
232 M. Mitchell, L. Qaio, C.-H. Wong, *Adv. Synth. Catal.* **2001**, *343*, 596–599.
233 A. Romero, C.-H. Wong, *J. Org. Chem.* **2000**, *65*, 8264–8268.
234 O. Eyrisch, W.-D. Fessner, *Angew. Chem. Int. Ed. Engl.* **1995**, *34*, 1639–1641.
235 M. Schultz, H. Waldmann, H. Kunz, W. Vogt, *Liebigs Ann.* **1990**, 1019–1024.
236 D. C. Myles, P. J. I. Andrulis, G. M. Whitesides, *Tetrahedron Lett* **1991**, *32*, 4835–4838.
237 R. Chenevert, M. Dasser, *J Org Chem* **2000**, *65*, 4529–4531.
238 K. Matsumoto, M. Shimagaki, T. Nakata, T. Oishi, *Tetrahedron Lett* **1993**, *34*, 4935–4938.
239 M. Shimagaki, H. Muneshima, M. Kubota, T. Oishi, *Chem. Pharm. Bull.* **1993**, *41*, 282–286.
240 R. Chenevert, M. Lavoie, M. Dasser, *Can. J. Chem.* **1997**, *75*, 68–73.
241 E. Grazi, M. Mangiarotti, S. Pontremoli, *Biochemistry* **1962**, *1*, 628–631.
242 J. F. Williams, P. F. Blackmore, *Int. J. Biochem.* **1983**, *15*, 797–816.
243 K. K. Arora, J. G. Collins, J. K. MacLeod, J. F. Williams, *Biol. Chem. Hoppe Seyler* **1988**, *369*, 549–557.
244 A. Moradian, S. A. Benner, *J. Am. Chem. Soc.* **1992**, *114*, 6980–6987.
245 M. Schürmann, G. A. Sprenger, *J Biol Chem* **2001**, *276*, 11055–11061.
246 S. Thorell, M. Schürmann, G. A. Sprenger, G. Schneider, *J. Mol. Biol.* **2002**, *319*, 161–171.
247 M. Schürmann, M. Schürmann, G. A. Sprenger, *J. Mol. Catal. B: Enzym.* **2002**, *19–20*, 247–252.
248 E. Racker, in *The Enzymes, Vol. 5* (Eds.: P. Boyer, H. Lardy, K. Myrbäck), Academic Press, New York, **1961**, pp. 397–406.
249 G. A. Kotchetov, *Methods Enzymol.* **1982**, *90*, 209–223.
250 U. Schörken, G. A. Sprenger, *Biochim. Biophys. Acta* **1998**, *1385*, 229–243.
251 U. Nilsson, L. Meshalkina, Y. Lindqvist, G. Schneider, *J. Biol. Chem.* **1997**, *272*, 1864–1869.
252 E. Fiedler, S. Thorell, T. Sandalova, R. Golbik, S. Konig, G. Schneider, *Proc. Natl. Acad. Sci. USA* **2002**, *99*, 591–595.
253 C. French, J. M. Ward, *Biotechnol. Lett.* **1995**, *17*, 247–252.

254 G. R. Hobbs, R. K. Mitra, R. P. Chauhan, J. M. Woodley, M. D. Lilly, *J. Biotechnol.* **1996**, *45*, 173–179.
255 J. Bongs, D. Hahn, U. Schörken, G. A. Sprenger, U. Kragl, C. Wandrey, *Biotechnol. Lett.* **1997**, *19*, 213–215.
256 S. Brocklebank, J. M. Woodley, M. D. Lilly, *J. Mol. Catal. B: Enzym.* **1999**, *7*, 223–231.
257 R. P. Chauhan, J. M. Woodley, L. W. Powell, *Ann. N. Y. Acad. Sci.* **1996**, *799*, 545–554.
258 N. J. Turner, *Curr Opin Biotechnol* **2000**, *11*, 527–531.
259 K. G. Morris, M. E. B. Smith, N. J. Turner, M. D. Lilly, R. K. Mitra, J. M. Woodley, *Tetrahedron Asymmetry* **1996**, *7*, 2185–2188.
260 F. Effenberger, V. Null, T. Ziegler, *Tetrahedron Lett* **1992**, *33*, 5157–5160.
261 Y. Kobori, D. C. Myles, G. M. Whitesides, *J. Org. Chem.* **1992**, *57*, 5899–5907.
262 J. Bolte, C. Demuynck, H. Samaki, *Tetrahedron Lett* **1987**, *28*, 5525–5528.
263 L. Hecquet, C. Demuynck, G. Schneider, J. Bolte, *J. Mol. Catal. B: Enzym* **2001**, *11*, 771–776.
264 F. T. Zimmermann, A. Schneider, U. Schörken, G. A. Sprenger, W.-D. Fessner, *Tetrahedron: Asymmetry* **1999**, *10*, 1643–1646.
265 A. J. Humphrey, S. F. Parsons, M. E. B. Smith, N. J. Turner, *Tetrahedron Lett* **2000**, *41*, 4481–4485.
266 S. V. Taylor, L. D. Vu, T. P. Begley, U. Schörken, S. Grolle, G. A. Sprenger, S. Bringer-Meyer, H. Sahm, *J. Org. Chem.* **1998**, *63*, 2375–2377.
267 F. Rohdich, C. A. Schuhr, S. Hecht, S. Herz, J. Wungsintaweekul, W. Eisenreich, M. H. Zenk, A. Bacher, *J. Am. Chem. Soc.* **2000**, *122*, 9571–9574.
268 S. Hecht, K. Kis, W. Eisenreich, S. Amslinger, J. Wungsintaweekul, S. Hertz, F. Rohdich, A. Bacher, *J. Org. Chem.* **2001**, *66*, 3948–3952.
269 E. Racker, *J. Biol. Chem.* **1952**, *196*, 347–365.
270 P. Hoffee, O. M. Rosen, B. L. Horecker, *Methods Enzymol.* **1966**, *9*, 545–549.
271 P. A. Hoffee, *Arch. Biochem. Biophys.* **1968**, *126*, 795–802.
272 P. Valentin-Hansen, F. Boetius, K. Hammer-Jespersen, I. Svendsen, *Eur. J. Biochem.* **1982**, *125*, 561–566.
273 C. F. Barbas, Y. F. Wang, C.-H. Wong, *J. Am. Chem. Soc.* **1990**, *112*, 2013–2014.
274 A. Heine, G. DeSantis, J. G. Luz, M. Mitchell, C.-H. Wong, I. A. Wilson, *Science* **2001**, *294*, 369–374.
275 H. Sakuraba, H. Tsuge, I. Shimoya, R. Kawakami, S. Goda, Y. Kawarabayasi, N. Katunuma, H. Ago, M. Miyano, T. Ohshima, *J. Biol. Chem.* **2003**, *278*, 10799–10806.
276 L. Chen, D. P. Dumas, C.-H. Wong, *J. Am. Chem. Soc.* **1992**, *114*, 741–748.
277 H. J. M. Gijsen, C.-H. Wong, *J. Am. Chem. Soc.* **1995**, *117*, 2947–2948.
278 J. Liu, G. DeSantis, C.-H. Wong, *Can. J. Chem.* **2002**, *80*, 643–645.

279 W. Tischer, H.-G. Ihlenfeldt, O. Barzu, H. Sakamoto, E. Pistotnik, P. Marliere, S. Pochet, WO 0114566, **2001**; *Chem. Abstr. 134*:208062.

280 N. Ouwerkerk, J. H. Van Boom, J. Lugtenburg, J. Raap, *Eur. J. Org. Chem.* **2000**, 861–866.

281 J. Liu, C.-H. Wong, *Angew. Chem. Int. Ed.* **2002**, *41*, 1404–1407.

282 H. J. M. Gijsen, C.-H. Wong, *J. Am. Chem. Soc.* **1995**, *117*, 7585–7591.

283 C.-H. Wong, E. Garcia-Junceda, L. R. Chen, O. Blanco, H. J. M. Gijsen, D. H. Steensma, *J. Am. Chem. Soc.* **1995**, *117*, 3333–3339.

284 J. Q. Liu, T. Dairi, N. Itoh, M. Kataoka, S. Shimizu, H. Yamada, *J. Mol. Catal. B: Enzym* **2000**, *10*, 107–115.

285 R. Contestabile, A. Paiardini, S. Pascarella, M. L. Di Salvo, S. D'Aguanno, F. Bossa, *Eur. J. Biochem.* **2001**, *268*, 6508–6525.

286 B. T. Lotz, C. M. Gasparski, K. Peterson, M. J. Miller, *J. Chem. Soc. Chem. Commun.* **1990**, *16*, 1107–1109.

287 A. Saeed, D. W. Young, *Tetrahedron* **1992**, *48*, 2507–2514.

288 D. Ura, T. Hashimukai, T. Matsumoto, N. Fukuhara, EP 421477 **1991**; *Chem. Abstr. 115*:90657.

289 D. M. Anderson, H.-H. Hsiao, in *Biocatalytic Production of Amino Acids & Derivatives* (Eds.: D. Rozzell, F. Wagner), Hanser, Munich, **1992**, pp. 23–41.

290 M. Bycroft, R. B. Herbert, G. J. Ellames, *J. Chem. Soc. Perkin Trans. 1* **1996**, 2439–2442.

291 V. P. Vassilev, T. Uchiyama, T. Kajimoto, C.-H. Wong, *Tetrahedron Lett.* **1995**, *36*, 4081–4084.

292 T. Kimura, V. P. Vassilev, G. J. Shen, C.-H. Wong, *J. Am. Chem. Soc.* **1997**, *119*, 11734–11742.

293 J. Q. Liu, T. Dairi, M. Kataoka, S. Shimizu, H. Yamada, *J. Bacteriol.* **1997**, *179*, 3555–3560.

294 C. L. Kielkopf, S. K. Burley, *Biochemistry* **2002**, *41*, 11711–11720.

295 M. Fujii, T. Miura, T. Kajimoto, Y. Ida, *Synlett* **2000**, 1046–1048.

296 T. Miura, M. Fujii, K. Shingu, I. Koshimizu, J. Naganoma, T. Kajimoto, Y. Ida, *Tetrahedron Lett.* **1998**, *39*, 7313–7316.

297 K. Shibata, K. Shingu, V. P. Vassilev, K. Nishide, T. Fujita, M. Node, T. Kajimoto, C.-H. Wong, *Tetrahedron Lett.* **1996**, *37*, 2791–2794.

298 S. H. Wu, M. Shimazaki, C. C. Lin, L. Qiao, W. J. Moree, G. Weitz-Schmidt, C.-H. Wong, *Angew. Chem. Int. Ed. Engl.* **1996**, *35*, 88–90.

299 R. B. Herbert, B. Wilkinson, G. J. Ellames, *Can. J. Chem.* **1994**, *72*, 114–117.

300 J. Q. Liu, M. Odani, T. Yasuoka, T. Dairi, N. Itoh, M. Kataoka, S. Shimizu, H. Yamada, *Appl Microbiol Biotechnol* **2000**, *54*, 44–51.

301 H. J. M. Gijsen, C.-H. Wong, *J. Am. Chem. Soc.* **1994**, *116*, 8422–8423.

302 V. Dalmas, C. Demuynck, *Tetrahedron Asymmetry* **1993**, *4*, 2383–2388.

303 L. Hecquet, J. Bolte, C. Demuynck, *Tetrahedron* **1994**, *50*, 8677–8684.
304 U. T. Bornscheuer, M. Pohl, *Curr. Opin. Chem. Biol.* **2001**, *5*, 137–143.
305 K. A. Powell, S. W. Ramer, S. B. del Cardayré, W. P. C. Stemmer, M. B. Tobin, P. F. Longchamp, G. W. Huisman, *Angew. Chem. Int. Ed.* **2002**, *41*, 3949–3959.
306 K. E. Jaeger, T. Eggert, A. Eipper, M. T. Reetz, *Appl. Microbiol. Biotechnol.* **2001**, *55*, 519–530.
307 R. Perez Carlon, N. Jourdain, J. L. Reymond, *Chem. Eur. J.* **2000**, *6*, 4154–4162.
308 E. Gonzalez-Garcia, V. Helaine, G. Klein, M. Schuermann, G. A. Sprenger, W.-D. Fessner, J.-L. Reymond, *Chem. Eur. J.* **2003**, *9*, 893–899.
309 A. Sevestre, V. Helaine, G. Guyot, C. Martin, L. Hecquet, *Tetrahedron Lett.* **2003**, *44*, 827–830.
310 D. Wahler, J.-L. Reymond, *Curr. Opin. Chem. Biol.* **2001**, *5*, 152–158.
311 G. J. Williams, S. Domann, A. Nelson, A. Berry, *Proc. Natl. Acad. Sci. USA* **2003**, *100*, 3143–3148.
312 S. S. Patel, H. D. Conlon, D. R. Walt, *J. Org. Chem.* **1986**, *51*, 2842–2844.
313 T. Ouyang, D. R. Walt, S. S. Patel, *Bioorg. Chem.* **1990**, *18*, 131–135.
314 T. Ouyang, D. R. Walt, *J. Org. Chem.* **1991**, *56*, 3752–3755.
315 R. Jossek, A. Steinbuchel, *FEMS Microbiol. Lett.* **1998**, *168*, 319–324.
316 Y. Satoh, K. Tajima, H. Tannai, M. Munekata, *J. Biosci. Bioeng.* **2003**, *95*, 335–341.
317 C. Khosla, R. J. X. Zawada, *Trends Biotechnol.* **1996**, *14*, 335–341.
318 P. F. Leadlay, *Curr. Opin. Chem. Biol.* **1997**, *1*, 162–168.
319 C. R. Hutchinson, *Proc. Natl. Acad. Sci. USA* **1999**, *96*, 3336–3338.
320 T. A. Cropp, B. S. Kim, B. J. Beck, Y. J. Yoon, D. H. Sherman, K. A. Reynolds, *Biotechnol. Gen. Eng. Rev.* **2002**, *19*, 159–172.

6
Antibody-catalyzed Aldol Reactions

Fujie Tanaka and Carlos F. Barbas, III

6.1
Introduction

Antibody catalysts have been designed to process a wide range of aldol reactions. Although aldol reactions can be catalyzed chemically and enzymatically, catalytic antibodies significantly extend the range of aldol reactions. Natural aldolase enzymes have substrate specificity that are predetermined by nature (although they can be engineered); antibody catalysts designed by synthetic chemists can have substrate specificity different from those of natural enzymes. In this chapter, we describe antibodies that catalyze aldol reactions (aldolase antibodies), their features, reactions and applications ranging from synthetic chemistry to the treatment of cancer.

6.2
Generation of Aldolase Antibodies

6.2.1
Antibody as Catalyst Scaffold

An antibody immunoglobulin G (IgG) molecule is represented schematically in Figure 6.1. One IgG is composed of four polypeptide chains, two identical light chains, and two identical heavy chains. The molecular weight of an IgG is ~150,000 daltons and each IgG has two identical antigen-binding sites. The fragment antigen binding (Fab) portion of the IgG molecules have identical antigen binding sites, however; the molecular weight of this antibody fragment is ~50,000 daltons. One antigen binding site is constructed by display of the six complementarity-determining regions (CDR) on a structurally conserved protein framework. The CDR regions are designated LCDR1, LCDR2, and LCDR3 in the light chain variable domain (VL) and HCDR1, HCDR2, and HCDR3 in the heavy variable domain (VL). The immune system has the potential to provide an almost

Modern Aldol Reactions. Vol. 1: Enolates, Organocatalysis, Biocatalysis and Natural Product Synthesis.
Edited by Rainer Mahrwald

ISBN: 3-527-30714-1

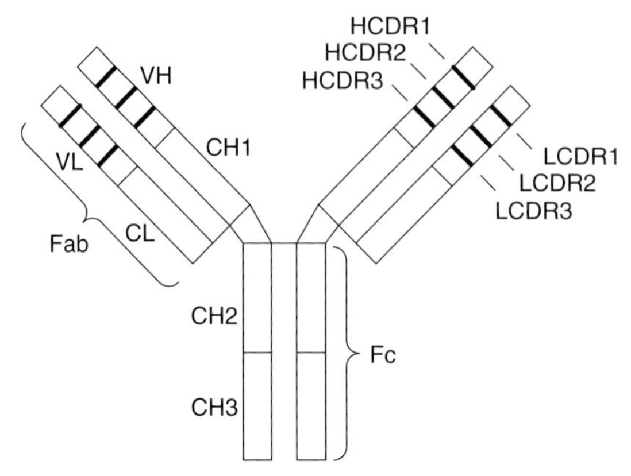

Fig. 6.1
Schematic representation of an antibody (IgG) molecule. One antibody molecule has two identical active sites. A light chain is built up from one variable domain (VL) and one constant domain (CL), and a heavy chain from one variable domain (VH), followed by constant domains (CH1, CH2, and CH3). Complementarity determining regions, CDR1–CDR3, vary most significantly among different immunoglobulins and determine the specificity of the antigen–antibody interactions.

limitless array of diverse antibodies that can bind to their antigens with high specificity. The amino acid residues in VL and VH, especially in the six CDR, vary among different immunoglobulins, and determine the specificity of antigen–antibody interactions, whereas constant domains support the structural stability of the antibody and provide connectivity to the rest of the immune system via conserved interactions. By taking advantage of the immune system and the highly specific binding of antibodies, antibodies that process chemical transformations have been developed [1]. Antigen binding sites of antibodies can serve as the catalytic active sites for catalysis of a wide variety of reactions.

6.2.2
Generation of Aldolase Antibodies that Operate via an Enamine Mechanism

Together with our colleagues, we have developed antibodies that catalyze aldol reactions via the covalent enamine mechanism of natural class I aldolases. Natural class I aldolases utilize the ε-amino group of a lysine in their active site to form an enamine, a carbon nucleophile, in their catalyzed aldol reactions [2]. One of the most important issues for development of such antibody catalysts is the pK_a of the active site lysine ε-amino group. The pK_a of the ε-amino group of lysine in aqueous solution is 10.7 [3], and the ε-amino group is protonated at neutral pH and thus is not nucleophilic under physiological conditions. For the ε-amino group to be nucleophilic it must be in its uncharged form, i.e. the pK_a must be perturbed as observed in class I aldolase enzymes.

Although many types of antibody catalyst have been generated, selection of antibody catalysts has typically been based on binding to transition state analogs of the reactants or charged compounds designed using information from the reaction coordinate of a given chemical transformation [1]. Although this strategy has occasionally provided antibody catalysts that use a covalent mechanism, expecting the fortuitous is obviously not a reasonable way to gain access to aldolase antibodies that operate via a covalent enamine mechanism. Thus, together with our colleagues we have developed catalytic function and residue-based selections by using 1,3-diketones for immunization, a strategy later termed reactive immunization [4]. Reactive immunization provides a means of selecting antibody catalysts in vivo on the basis of their capacity to perform a chemical reaction [4, 5]. In this approach, a designed reactive immunogen is used for immunization and chemical reaction(s), for example formation of a covalent bond, occurs in the binding pocket of the antibodies during their induction. The chemical reactivity and mechanism integrated into the antibody by the covalent trap with the reactive immunogen are used in catalytic reactions with substrate molecules. The most highly proficient and efficient catalytic antibodies, aldolase antibodies, have a nucleophilic lysine residue in their active site and were generated by this reactive immunization strategy.

6.2.2.1 Reactive Immunization with the Simple Diketone Derivative

Historically, Frank Westheimer used 1,3-diketones as mechanistic probes of the enzyme acetoacetate decarboxylase [6]. The active site of acetoacetate decarboxylase has a nucleophilic lysine ε-amino group and forms a Schiff base in its catalytic cycle. In his studies, the active site lysine ε-amino group of this enzyme reacted covalently with the 1,3-diketones. Later 1,3-diketones were used as modification reagents for the nucleophilic lysine ε-amino groups of many enzymes [7].

We used 1,3-diketone hapten **1**-carrier protein conjugate for immunization in our experiments to generate aldolase antibodies (Scheme 6.1) [4]. Class I aldolases use the ε-amino group of a lysine in their active site to form a Schiff base with one of their substrates and this substrate becomes the aldol donor substrate. Schiff-base formation reduces the activation energy for proton abstraction from the Cα atom and for subsequent enamine formation. The enamine, a carbon nucleophile, then reacts with an aldehyde substrate, the aldol acceptor, to form a new C–C bond. The Schiff base is then hydrolyzed and the product is released [2]. The 1,3-diketone hapten acts as a mechanism-based trap of the requisite lysine residue in the active site and this lysine is necessary for formation of the essential enamine intermediate. The molecular steps involved in trapping the requisite lysine residue are essentially the same chemical steps as are involved in activating

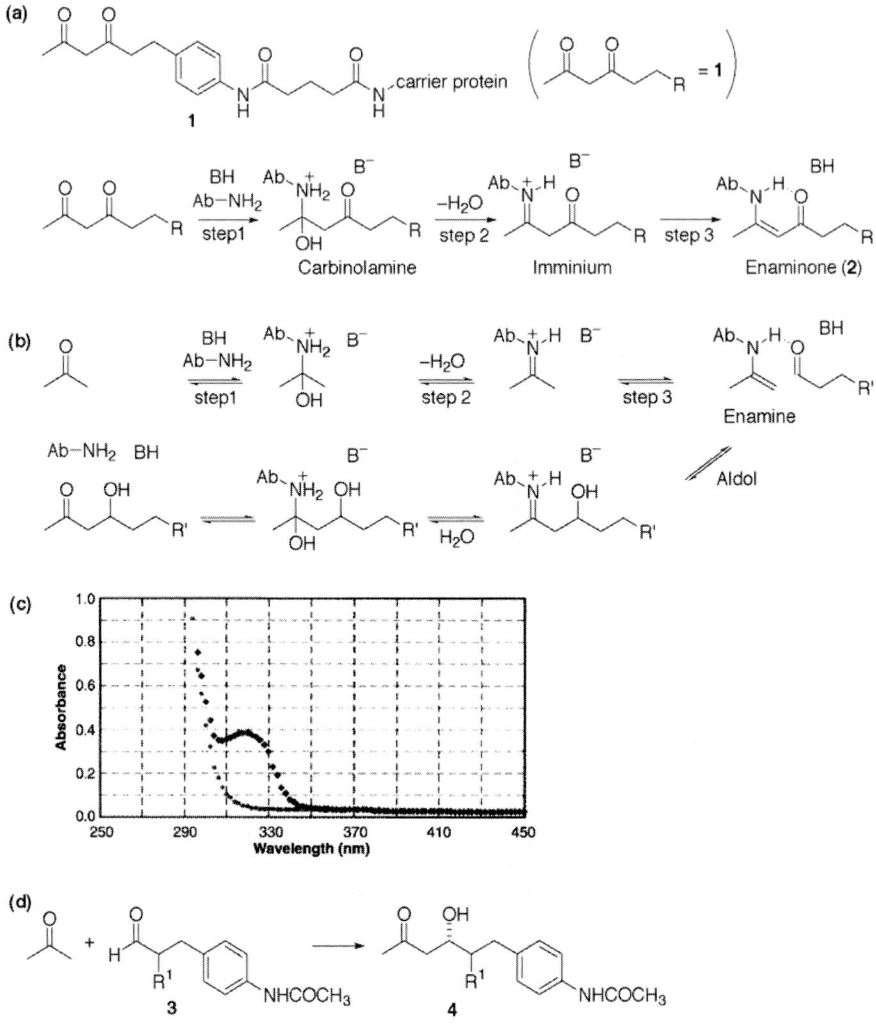

Name of aldolase Ab	Substrate **3** R^1	k_{cat} (min^{-1})	K_m (μM)
38C2	H	6.7×10^{-3}	17
38C2	CH_3	4.0×10^{-2}	48
33F12	CH_3	8.3×10^{-2}	125

Scheme 6.1
(a) Diketone **1** used for aldolase antibody induction and selection, and mechanism of trapping of the essential ε-amino group of a Lys residue in the antibody (Ab–NH_2) binding pocket by using **1**. (b) Antibody-catalyzed aldol-addition reaction (Ab–NH_2 = aldolase antibody). (c) UV spectra of non-catalytic and aldolase antibodies with **1** [1]. The aldolase antibody forms enaminone **2**, readily detected at $\lambda = 316$ nm. (d) Kinetic data for aldolase antibody-catalyzed reactions.

a substrate ketone to an enamine. These include: (1) catalysis of carbinolamine formation, (2) dehydration, and (3) α-deprotonation of the imminium intermediate to form enamine. This mechanistic mimicry enables selection of an active site that can adapt to the chemical and steric changes that occur during the course of the reaction. Further, the diketone structure provides appropriate binding sites for the two substrates of the intermolecular reaction and facilitates crossing of the entropic barrier intrinsic to this bimolecular reaction. The driving force for reaction of the 1,3-diketone hapten with the antibody is the formation of a stable covalent conjugated enaminone (**2**) between hapten **1** and ε-amino group of lysine (Scheme 6.1).

Antibodies prepared from hybridomas derived after immunization with 1,3-diketone hapten **1**-carrier protein conjugate were screened for their capacity to form the stable enaminone (absorption maximum $\lambda_{max} = 316$ nm) with hapten **1**. Of the 20 antibodies screened two, 38C2 and 33F12, reacted to form a covalent bond with the 1,3-diketone and generated the enaminone absorption band in the UV. These two antibodies catalyzed aldol addition of acetone to aldehyde **3** to give β-hydroxyketone **4**. The saturation kinetics of the antibody-catalyzed reactions were described by the Michaelis–Menten equation. Both antibodies catalyzed the addition of acetone to the *si* face of **3**, irrespective of the stereochemistry at α-position of this aldehyde. These antibodies also catalyzed the retro-aldol reaction of **4**. The other antibodies were unable to catalyze the aldol and retro-aldol reactions, indicating that only those that formed the critical enaminone intermediate were active. The catalytic activity of 38C2 and 33F12 was completely inhibited when either hapten **1** or 2,4-pentanedione was added before addition of acetone or aldehyde **3**, indicating the antigen-binding site is also the catalytically active site.

For the aldol reaction of acetone and aldehyde **3** at pH 7.5, aldolase antibodies 38C2 and 33F12 had $(k_{cat}/K_m)/k_{uncat} \approx 10^9$. The efficiency of catalysis is largely because of an entropic advantage in the antibody-catalyzed reaction, which is reflected as a high effective molarity, $k_{cat}/k_{uncat} > 10^5$ M. With the best substrates identified to date for the retro-aldol reaction [8–10], the catalytic proficiency $(k_{cat}/K_m)/k_{uncat} \approx 6 \times 10^{10}$. The catalytic efficiency (k_{cat}/K_m) of these aldolase antibodies with these substrates is only ~20–40-fold below that of the most studied aldolase enzyme, fructose-1,6-bisphosphate aldolase 4.9×10^4 s^{-1} M^{-1} [2b]. The pK_a of the active site lysine was determined to be 5.5 and 6.0 for antibodies 33F12 and 38C2, respectively, from the pH dependence of enaminone formation with 3-methyl-2,4-pentanedione [11]. These perturbed pK_a values are consistent with the catalysis and mechanism of these antibodies.

6.2.2.2 Combining Reactive Immunization with Transition-state Analogs

To improve on the initial concepts of reactive immunization, hapten **5** (Scheme 6.2) was used for induction of aldolase antibodies. Although 1,3-diketone hapten **1** enabled generation of the efficient aldolase antibodies 38C2 and 33F12, the hapten does not address the tetrahedral geometry of

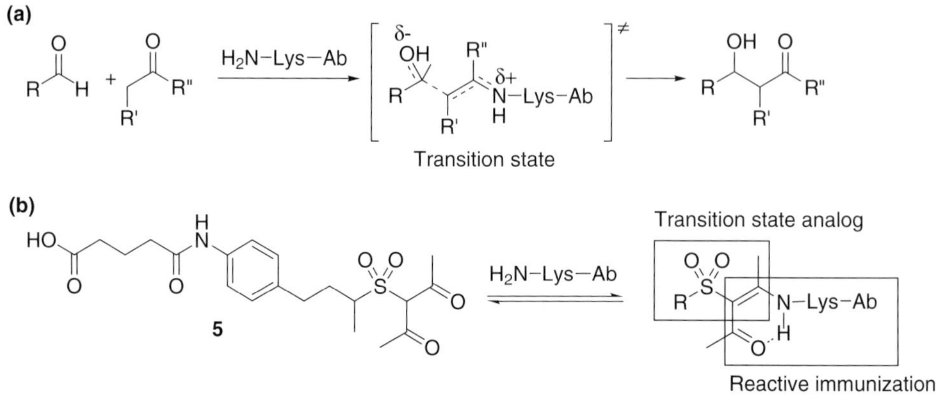

Scheme 6.2
Diketone **5**, used for generation of aldolase antibodies, includes conceptual elements derived from reactive immunization and transition state analog design. Ab-Lys-NH_2 = aldolase antibody.

the rate-determining transition state of the C–C bond-forming step. Hapten **5** contains features common to the transition state analog of the enzyme-catalyzed reaction. The use of a transition state analog approach has been successfully used to generate antibodies that catalyze a diverse set of reactions whereas the 1,3-diketone functionality proved to be key for the reactive immunization strategy [12]. The tetrahedral geometry of the sulfone moiety in hapten **5** mimics the tetrahedral transition state of the C–C bond-forming step and should therefore facilitate nucleophilic attack of the enamine intermediate on the acceptor aldehyde substrate.

Nine antibodies out of 17 screened reacted with 2,4-pentanedione to form a stable enaminone, and all nine antibodies were catalytic. These results are consistent with a covalent catalytic mechanism in which a reactive amine is programmed in these antibodies. It is of practical interest that immunization with hapten **5** generated the two families of catalysts that operate on opposite optical isomers. Antibodies 93F3 and 84G3 generated with hapten **5** have reactivity antipodal to that of antibodies 38C2 and 33F12 generated with hapten **1**, and antibodies 40F12 and 42F1 generated with hapten **5** have the same enantio preference as antibodies 38C2 and 33F12. The rate enhancement, k_{cat}/k_{uncat}, of the best of the nine antibodies (84G3) was 2.3×10^8; its catalytic proficiency $[(k_{cat}/K_m)/k_{uncat}]$ was 7.7×10^{13}, approximately 1000-fold higher than that reported for any other catalytic antibody [1c]. In this instance the catalytic efficiency (k_{cat}/K_m) of 84G3, 3.3×10^5 s^{-1} M^{-1}, is slightly greater than that of the best characterized aldolase enzyme, fructose-1,6-biphosphate aldolase [2b].

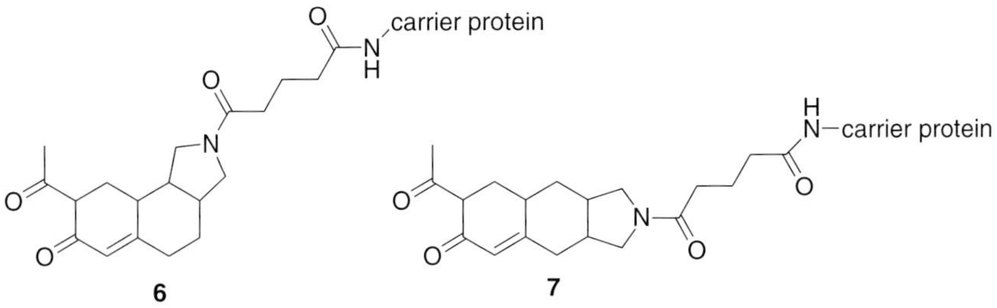

Fig. 6.2
Alternative diketones **6** and **7**, used for generation of aldolase antibodies.

6.2.2.3 Reactive Immunization with other Diketones

To expand the concept of reactive immunization with diketones, a mixture of diketones **6** and **7** (Figure 6.2) were used for generation of the aldolase antibody 24H6 that operates via an enamine mechanism to catalyze a distinct set of aldol reactions [13].

6.3 Aldolase Antibody-catalyzed Aldol and Retro-aldol Reactions

Aldolase antibodies generated with any of the diketone haptens are typically very broad in scope, accepting a wide variety of substrates. The broad substrate specificity of aldolase antibodies are unprecedented in the field of catalytic antibodies. Traditionally, antibody catalysts are very specific and only catalyze the reaction of a single substrate (or substrate combination). This is because, in normal immunization, a series of somatic mutations usually leads to highly specific binding of the inducing antigen [1, 14].

For aldolase antibodies selected with the diketones the usual process of somatic refinement might be aborted because any clone that carries an antibody that has made a covalent bond with the antigen will be selected above clones containing only non-covalent antibody–antigen interactions. This is because no matter how many productive non-covalent interactions are generated by competing clones, they cannot equal the binding energy achieved by a formation of a single covalent bond. When a covalent bond is formed early in the process of antibody evolution, any selective pressure on the refinement process may cease. Antibodies selected in this way are efficient because, like Nature's enzymes, they were selected on the basis of a chemical reaction. Unlike enzymes, however, they are broad in scope because the usual requirement for refinement of the binding pocket has been circumvented.

6.3.1
Antibody 38C2-catalyzed Aldol Reactions

Aldolase antibodies 38C2 and 33F12 generated by immunization with diketone **1** are capable of accelerating more than 100 different aldol reactions [4, 8, 11, 15, 16]. Some examples of cross-aldol reactions are shown in Table 6.1. For cross-aldol reactions, a variety of ketones are accepted as donors, including aliphatic open-chain ketones (for example acetone to pentanone), aliphatic cyclic ketones (cyclopentanone to cycloheptanone), functionalized open-chain ketones (hydroxyacetone, dihydroxyacetone, fluoroacetone), and functionalized cyclic ketones (2-hydroxycyclohexanone). As with the donors, the antibodies also accept different kinds of aldehyde substrate, for example benzaldehyde derivatives **8–10**, α,β-unsaturated aldehyde **11**, and aliphatic aldehydes **12** and **13** with products as indicated in Table 6.1.

These aldolase antibody-catalyzed reactions are often highly enantioselective. The enantioselectivity rules for 38C2-catalyzed aldol reactions are simple and general, although this selectivity was not directly programmed by the diketone hapten used for the immunization. Asymmetric induction is a consequence of the asymmetry of the active site that directs the attack of the enamine intermediate while stereochemically fixing the face of the acceptor aldehyde. With acetone as the aldol donor substrate a new stereogenic center is formed by attack of the *si* face of the aldehyde, usually with $>95\%$ enantiomeric excess (ee). With hydroxyacetone as the donor substrate, the *Z*-enamine of hydroxyacetone reacts on the *re* face of the aldehyde. The antibody-catalyzed reactions of hydroxyacetone as the donor provided only the single regioisomer as shown in the formation of **16**, **18**, **20**, **24**, and **25** (Table 6.1). Lower enantioselectivity was achieved with acceptor aldehydes containing an sp^3 center in the α-position, for example the reaction with **12**, although reactions with hydroxyacetone as the donor increased the enantioselectivity (for example, formation of **24** and **25** in Table 6.1). In the antibody-catalyzed reactions of 2-butanone with aldehydes, a mixture of both possible regioisomers of aldol products was formed with moderate selectivity by preferential bond formation at the most substituted carbon atom of the ketone.

Although a variety of secondary aldols can be prepared by aldolase antibody 38C2-catalyzed cross-aldol reactions, tertiary aldols are typically not accessible via intermolecular cross-aldol reactions. For preparation of enantiomerically enriched tertiary aldols, aldolase antibody 38C2-catalyzed retro-aldol reactions can be used (Section 6.3.2).

Antibodies 38C2 and 33F12 also catalyzed self-aldol condensations of propionaldehyde and provided the aldol-elimination product **26**, and the antibodies did not catalyze the consecutive aldol reaction of **26** with propionaldehyde (Scheme 6.3). The antibodies did not catalyze the self-aldol reactions of aldehydes bearing a longer alkyl chain ($\geq$ valeraldehyde, for example, aldehydes **3**, **12**, and **13**). It might not be possible for the antibody to accept two

Tab. 6.1
Antibody 38C2-catalyzed cross-aldol reactions.

Acceptor	*Donor*	*Product*	*ee (%)*
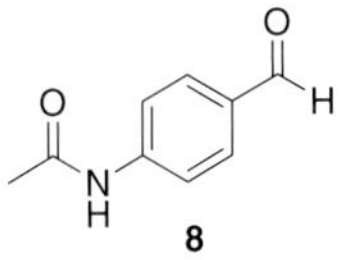 **8**		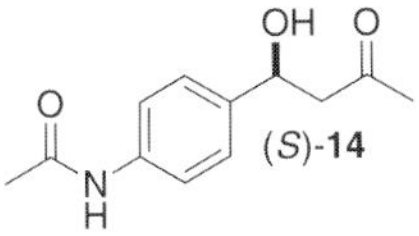 (*S*)-**14**	98
		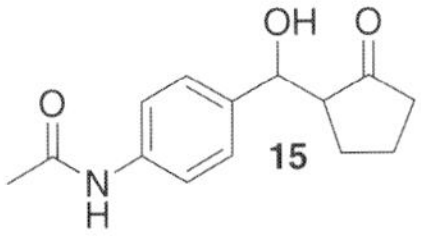 **15**	
	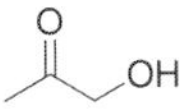	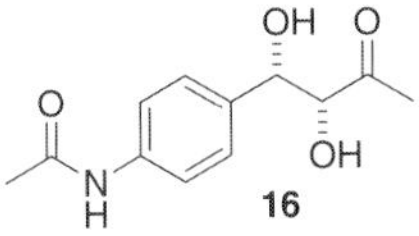 **16**	
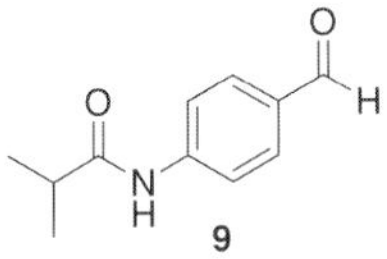 **9**		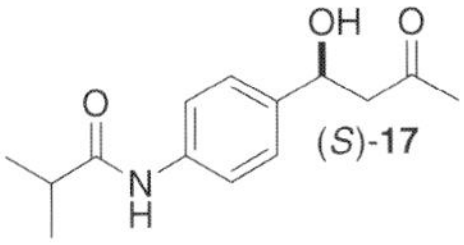 (*S*)-**17**	>99
	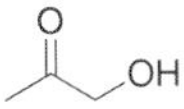	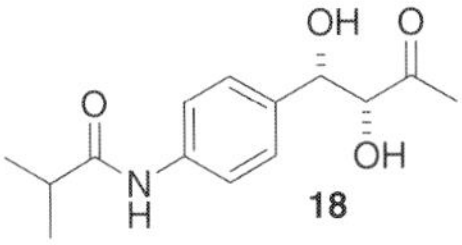 **18**	
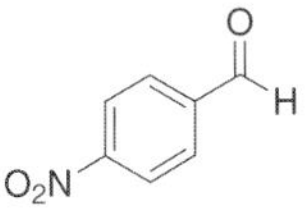 **10**		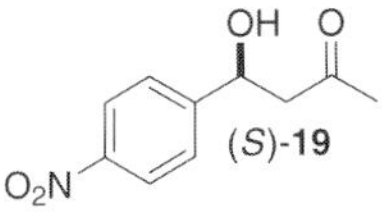 (*S*)-**19**	98
	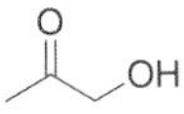	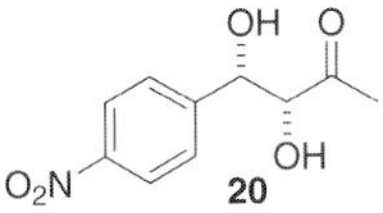 **20**	
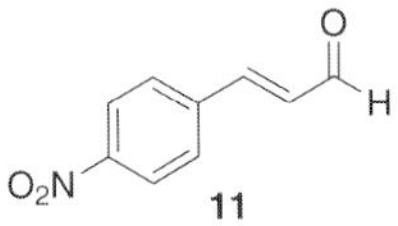 **11**		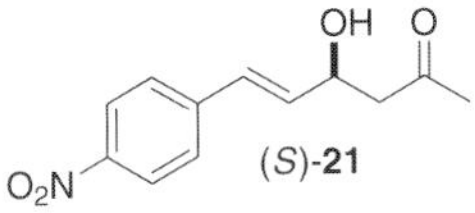 (*S*)-**21**	99
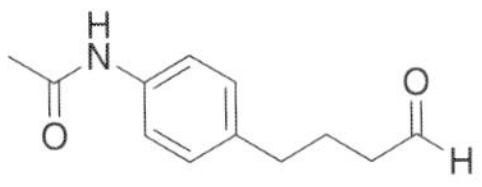 **12**		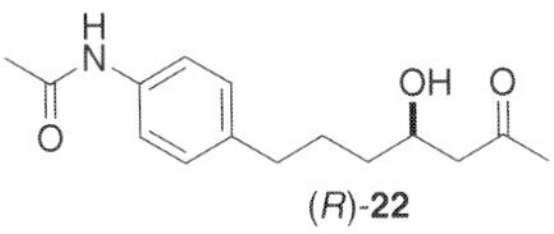 (*R*)-**22**	20
	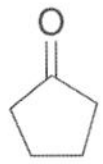	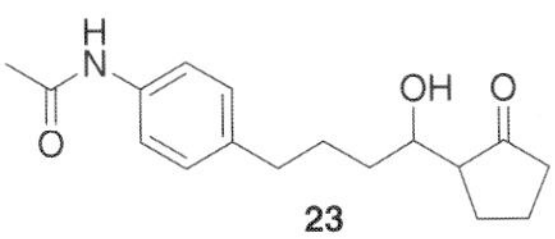 **23**	

Tab. 6.1 *(continued)*

Acceptor	*Donor*	*Product*	*ee (%)*
	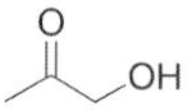	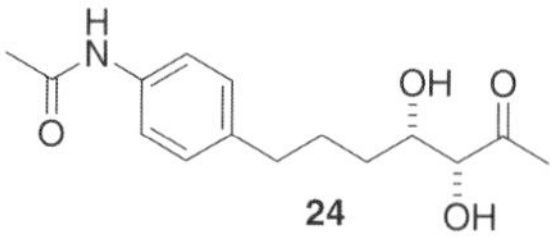 24	77 (de >99)
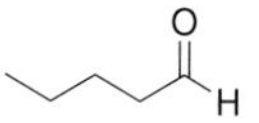 13	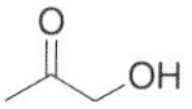	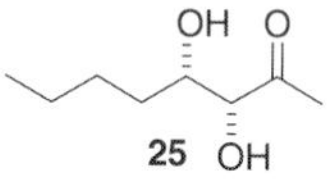 25	98

Substrates	*Product*	k_{cat} *(min⁻¹)*	K_m *(μM)*
8 + acetone	14	0.048	204
8 + hydroxyacetone	16	0.11	348
9 + acetone	17	0.21	123
9 + hydroxyacetone	18	1.1	184
10 + acetone	19	0.04	27
12 + acetone	22	0.11	256

molecules of aldehyde with a larger substituent in the active-site cavity. Self-aldol reactions of acetone or cyclopentanone were also catalyzed by the antibodies in the absence of acceptor aldehyde for a cross-aldol reaction. In this instance the aldol product was also converted to the elimination product by the antibodies (for example, formation of **27**).

Antibodies 38C2 and 33F12 also catalyzed intramolecular aldol reactions and Robinson annulation (Scheme 6.4); 1,5-diketones **28** and **30** were converted into **29** and **31**, respectively [15]. To explore the scope of Baldwin's rules in antibody-catalyzed intramolecular aldol reactions, 38C2 was incubated with three different aliphatic diketones – 2,4-hexanedione, 2,5-heptanedione and 2,6-octanedione. No catalysis was observed in the

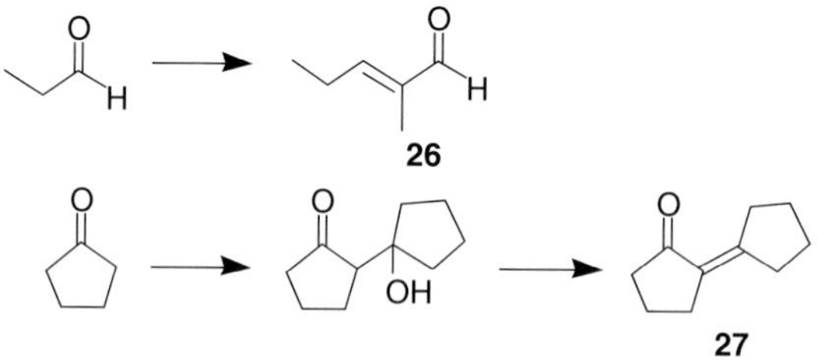

26

27

Scheme 6.3
Antibody 38C2-catalyzed self-aldol reactions.

(S)-**28** → (S)-**29**

(R)-**28** → (R)-**29**

30 → (S)-**31** >95%ee

Substrate	Product	k_{cat} (min^{-1})	K_m (mM)
(S)-**28**	(S)-**29**	0.186	12.4
(R)-**28**	(R)-**29**	0.126	2.45
30	**31**	0.086	2.34

Scheme 6.4
Antibody 38C2-catalyzed intramolecular aldol reactions.

reaction pathway from 2,4-hexanedione to 3-methylcyclopent-2-enone, presumably because of the Baldwin disfavored *5(enolendo-trig)* process involved in the attack of the enamine at C2 in 2,4-hexanedione. Although the corresponding ring closure (followed by water elimination) of 2,6-octanedione to give 3-methylcyclohept-2-enone is Baldwin favored (a *7(enolendo-trig)* process) in this instance also no product formation was observed. In contrast, the Baldwin favored ring closure reaction of 2,5-heptanedione, (a *6(enolendo-trig)* process) followed by elimination of water and giving 3-methylcyclohex-2-enone was catalyzed by antibody 38C2 [8].

6.3.2
Antibody 38C2-Catalyzed Retro-aldol Reactions and their Application to Kinetic Resolution

The antibodies also catalyzed retro-aldol reactions of secondary [9, 11, 16] and tertiary aldols [10]. In these retro-aldol reactions antibodies 38C2 and 33F12 processed hydroxyketones whose stereochemistry was the same as that of the aldol reaction product. Kinetic resolution by the retro-aldol reaction therefore provided the opposite enantiomer from the forward aldol reaction (Scheme 6.5 and Table 6.2). For example, (*R*)-**14** (> 99% ee) was obtained by the 38C2-catalyzed kinetic resolution of (±)-**14** (Table 6.2) whereas

38C2 or 33F12
Enantioselective
aldol reaction

Racemic

38C2 or 33F12
Enantioselective
retro-aldol reaction
(Kinetic resolution)

Scheme 6.5
Stereochemistry of antibody 38C2-catalyzed aldol reactions and kinetic resolution.

(*S*)-**14** (98% ee) was formed in the 38C2-catalyzed forward-aldol reaction (Table 6.1). The recovered aldols were highly enantiomerically enriched by the aldolase antibody-catalyzed kinetic resolution; examples include secondary aldols **14**, **17**, **21**, and **32–35**, and tertiary aldols **36–40** (Tables 6.2 and 6.3). Kinetic resolution with antibody 38C2 usually afforded higher enantiomeric excess for the recovered aldols than the forward aldol reactions, because one enantiomer was completely consumed in the resolution. For example, (*S*)-**32** was obtained in >99% ee by 38C2-catalyzed kinetic resolution after 67% conversion whereas in the 38C2-catalyzed forward aldol reaction (*R*)-**32** was obtained in 58% ee.

In the aldol reactions to form secondary aldols the equilibrium constants favor the aldol product. For example, the aldol reaction of acetone with benzaldehyde has an equilibrium constant of 12 M^{-1} [17]. Aqueous solutions containing 1 M acetone and 1 mM benzaldehyde reached equilibrium when 92% of the benzaldehyde had reacted to form the aldol. In the retro-aldol reaction of a 1 mM solution of the aldol, equilibrium was reached at 99% conversion to benzaldehyde. Aldolase antibodies 38C2 and 33F12 efficiently provided enantiomerically enriched secondary aldols via aldol reactions or retro-aldol reactions, i.e. kinetic resolution of racemic aldols. The concentration of the substrates determined the reaction direction, aldol or retro-aldol, with these aldolase antibodies.

In the aldol reaction of acetone and acetophenone, the equilibrium constant is 0.002 M^{-1} [18]. A 1 mM solution of acetophenone would require the concentration of acetone to be 10,000 M (neat acetone = 13.6 M) to reach 95% conversion. In the retro-aldol reaction, a 1 mM solution of this tertiary aldol would be converted almost completely to its constituent ketones at equilibrium. This is consistent with the fact that aldolase antibodies 38C2 and 33F12 can be used for kinetic resolution of tertiary aldols, but not for the forward aldol reactions to form tertiary aldols.

While general access to optically active tertiary aldols is not available by traditional synthetic methods, a variety of enantiomerically enriched tertiary aldols can be prepared by aldolase antibody-catalyzed kinetic resolution. Significantly, the aldolase antibodies process not only keto-aldols but also aldehyde-aldols for example, the reaction to provide (*S*)-**40**.

Tab. 6.2
Secondary aldols prepared by antibody 38C2-catalyzed kinetic resolution.

Product	*Conversion*	*ee (%)*
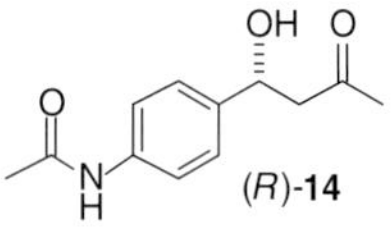 (*R*)-**14**	52%	>99
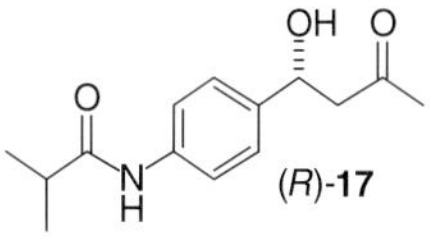 (*R*)-**17**	52%	>99
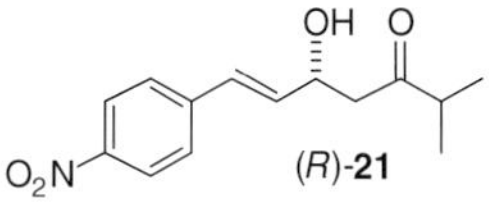 (*R*)-**21**	51%	>99
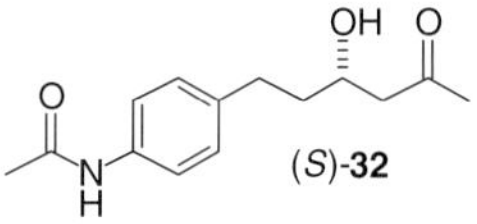 (*S*)-**32**	67%	>99
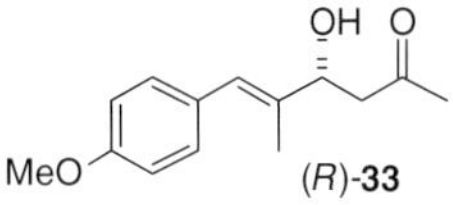 (*R*)-**33**	50%	95
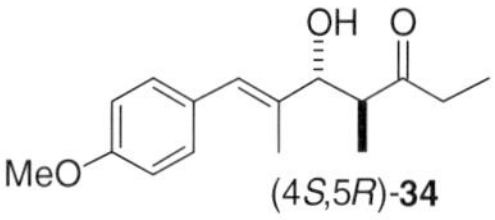 (4*S*,5*R*)-**34**	54%	98
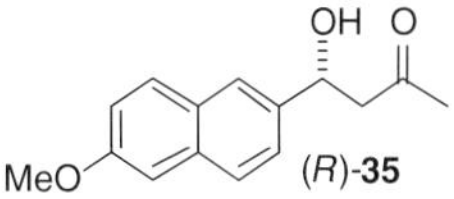 (*R*)-**35**		

Substrate	k_{cat} (min^{-1})	K_m (μM)	k_{cat}/k_{uncat}	$k_{cat}/K_m/k_{uncat}$ (M^{-1})
(*S*)-**17**	1.4	270	1.7×10^7	6.2×10^{10}
(±)-**33**	2.2	16	1.0×10^5	6.3×10^9
(±)-*anti*-**34**	1.4	93	1.9×10^4	2.1×10^8
(±)-**35**	1.0	14	1.0×10^6	7.1×10^{10}

6.3.3 Aldol and Retro-aldol Reactions Catalyzed by Antibodies 93F3 and 84G3

Aldolase antibodies 93F3 and 84G3 also have promiscuous active sites and catalyze reactions of a variety of substrates [12, 19]. These antibodies pro-

Tab. 6.3
Tertiary aldols prepared by antibody 38C2-catalyzed kinetic resolution.

Product	*Conversion (%)*	*ee (%)*
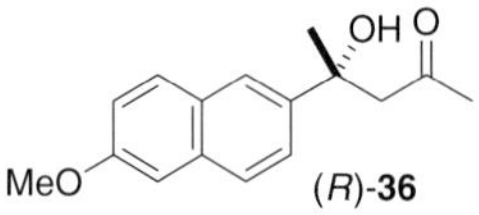 (*R*)-**36**	50	>99
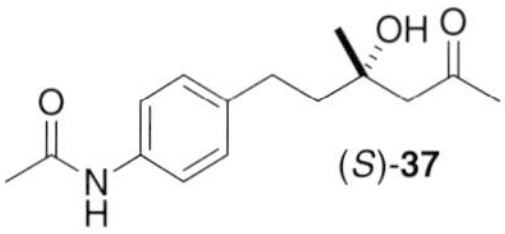 (*S*)-**37**	52	80
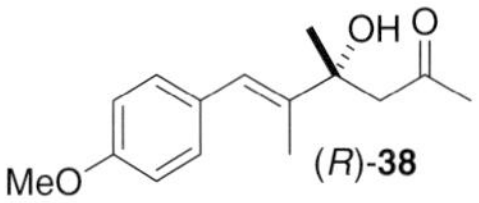 (*R*)-**38**	50	94
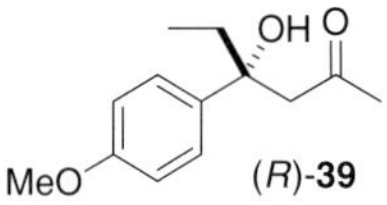 (*R*)-**39**	50	>99
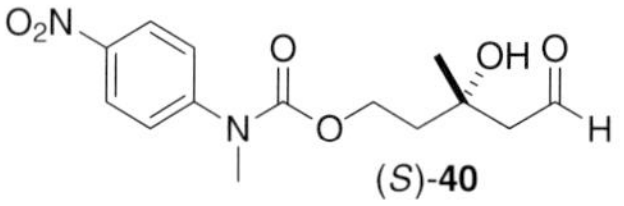 (*S*)-**40**	50	95

Substrate	k_{cat} (min^{-1})	K_m (mM)	k_{cat}/k_{uncat}	$k_{cat}/K_m/k_{uncat}$ (M^{-1})
(*S*)-**36**	1.8	0.12	1.2×10^6	1.3×10^{10}
(±)-**37**	4.6	0.11	8.4×10^5	7.7×10^9
(±)-**38**	0.15	1.62		
(±)-**40**	0.02	0.13		

vide the opposite enantioselectivity to the 38C2 and 33F12-catalyzed reactions. Examples of antibody 93F3-catalyzed aldol reactions are shown in Table 6.4. Antibodies 93F3 and 84G3 provided (*R*)-**17**, (*R*)-**19**, and (*R*)-**21** in the aldol reactions of acetone and aldehydes **9**, **10**, and **11**, respectively, whereas antibodies 38C2 and 33F12 provided the corresponding (*S*) enantiomers in the aldol reactions (Table 6.1). When unsymmetrical ketones were used in antibody 84G3-catalyzed cross-aldol reactions with aldehydes the reactions occurred exclusively at the less substituted carbon atom of the ketones, irrespective of the presence of heteroatoms in the ketones: Aldol products (*R*)-**41–44** were regioselectively and enantioselectively obtained. This is a notable feature, because antibodies 28C2 and 33F12 gave the corresponding regioisomer mixtures and because the background reaction fa-

Tab. 6.4
Aldol reactions catalyzed by antibodies 93F3 and 84G3.

Acceptor	*Donor*	*Product*	*Antibody*	*ee (%)*
9	acetone	(*R*)-**17**	93F3	>99
10	acetone	(*R*)-**19**	93F3	95
11	acetone	(*R*)-**21**	93F3	98
10	2-butanone	(*R*)-**41**	84G3	95 (dr 100:1)
	2-pentanone	(*R*)-**42**	84G3	95 (dr 99:1)
	methoxyacetone (OMe)	(*R*)-**43**	84G3	95 (dr 100:1)
	(methylthio)acetone (SMe)	(*R*)-**44**	84G3	95 (dr 98:2)

vored the formation of the other regioisomers. In contrast with antibodies 38C2 and 33F12, aldolase antibodies 93F3 and 84G3 were very poor catalysts for aldol reactions involving hydroxyacetone as donor with aldehydes as acceptors.

When kinetic resolution of aldol (±)-**17** was performed with antibody 93F3, (*S*)-**17** (>99% ee) was obtained at 52% conversion (Table 6.5), whereas antibody 38C2 provided (*R*)-**17** in the kinetic resolution (Table 6.2). The catalytic proficiency of antibodies 93F3 and 84G3 usually exceeded that of antibody 38C2, especially for retro-aldol reactions of 3-keto-5-hydroxy-type substrates (for example, **45** and **46**). In contrast with antibodies 38C2 and

Tab. 6.5
Examples of kinetic resolution catalyzed by antibodies 93F3 and 84G3.

Product	*Conversion*	*ee (%)*
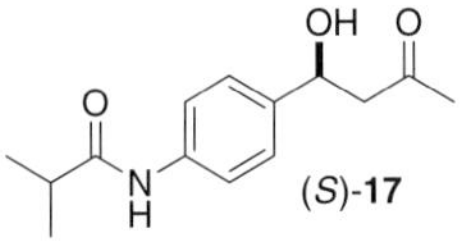 (*S*)-**17**	52%	>99
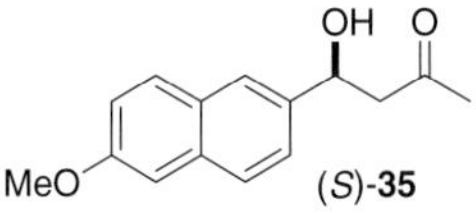 (*S*)-**35**	50%	99
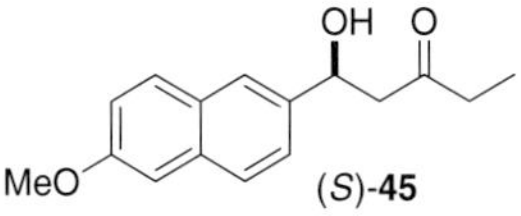 (*S*)-**45**	50%	>99
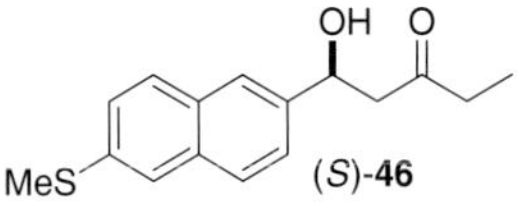 (*S*)-**46**		
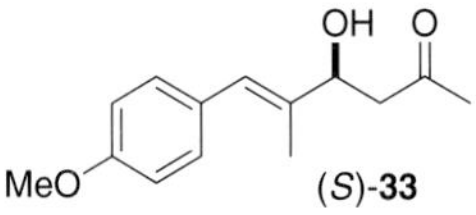 (*S*)-**33**	50%	96

Substrate	*Antibody*	k_{cat} *(min^{-1})*	K_m *(μM)*	k_{cat}/k_{uncat}	$k_{cat}/K_m/k_{uncat}$ *(M^{-1})*
(±)-**35**	93F3	2.63	15	2.7×10^6	1.8×10^{11}
	84G3	3.5	23	3.6×10^6	1.6×10^{11}
(±)-**45**	93F3	43.3	6.5	4.9×10^7	7.4×10^{12}
	84G3	46.8	10.3	5.2×10^7	5.0×10^{12}
(*R*)-**46**	93F3	69.6	2.6	1.9×10^8	7.4×10^{13}
	84G3	81.4	4.2	2.3×10^8	5.4×10^{13}

33F12, antibodies 93F3 and 84G3 cannot process retro-aldol reactions of tertiary aldols.

6.3.4
Preparative-scale Kinetic Resolution Using Aldolase Antibodies in a Biphasic Aqueous–Organic Solvent System

Antibody-catalyzed reactions are typically performed in aqueous buffer, because catalytic antibodies function ideally in an aqueous environment. Many organic molecules of interest are, however, poorly soluble in water. For the transformation of such molecules by aldolase antibodies, a biphasic

Tab. 6.6
Preparative scale kinetic resolution in biphasic system.

Product	*Antibody*	*Time*	*Recovery (%)*	*ee (%)*
(*R*)-**47**	38C2 (255 mg, 0.025 mol%)	88 h	1.55 g (49)	>97
(*S*)-**47**	84G3 (16 mg, 0.015 mol%)	340 h	154 mg (48)	95
(*R*)-**35**	38C2 (15.4 mg, 0.10 mol%)	144 h	25 mg (50)	97
(*S*)-**35**	84G3 (210 mg, 0.065 mol%)	91 h	469 mg (47)	97
	reuse (2nd round)	172 h	441 mg (42)	97
	reuse (3rd round)	259 h	458 mg (43)	97
(*R*)-**36**	38C2 (18 mg, 0.12 mol%)	193 h	22 mg (44)	99
(*S*)-**45**	84G3 (500 mg, 0.0086 mol%)	65 h	10 g (50)	>99

aqueous–organic solvent system was especially useful in a large-scale reaction [20]. Although water-miscible solvents can be used to increase the solubility of the substrates, antibody 38C2-catalyzed kinetic resolution of (±)-**47** in 20% CH_3CN–buffer resulted in reduced reactivity and a lower enantioselectivity than the reaction in 2–5% CH_3CN–buffer. The same reaction in toluene–phosphate buffer yielded (*R*)-**47** in >97% ee with 49% recovery. Examples of kinetic resolution in biphasic systems are given in Table 6.6. In the biphasic system, racemic substrate (50–100 mM) in toluene or chlorobenzene is mixed with an antibody solution in buffer, for example, a 20-g-scale reaction is performed in a reaction volume of ~700 mL. After completion of the reaction (determined by monitoring the ee of the substrate by HPLC) the mixture is cooled (−20 °C) and the organic phase is easily separated from the frozen aqueous antibody solution. The product is purified by conventional column chromatography. The antibody solution is thawed and can be reused. Although the activity of the recycled catalysts was lower than on first use and longer reaction time was necessary for reactions using recycling catalysts, the enantioselectivity of the reaction with the recycling catalysts was retained.

6.3.5
Aldolase Antibody-catalyzed Reactions in Natural Product Synthesis

Aldolase antibody-catalyzed aldol reactions and kinetic resolution are an efficient means of synthesis of highly enantiomerically pure aldols. These processes have been used for the total synthesis of cytotoxic natural products epothilone A (**48**) and C (**49**) (Scheme 6.6) [21]. These compounds have

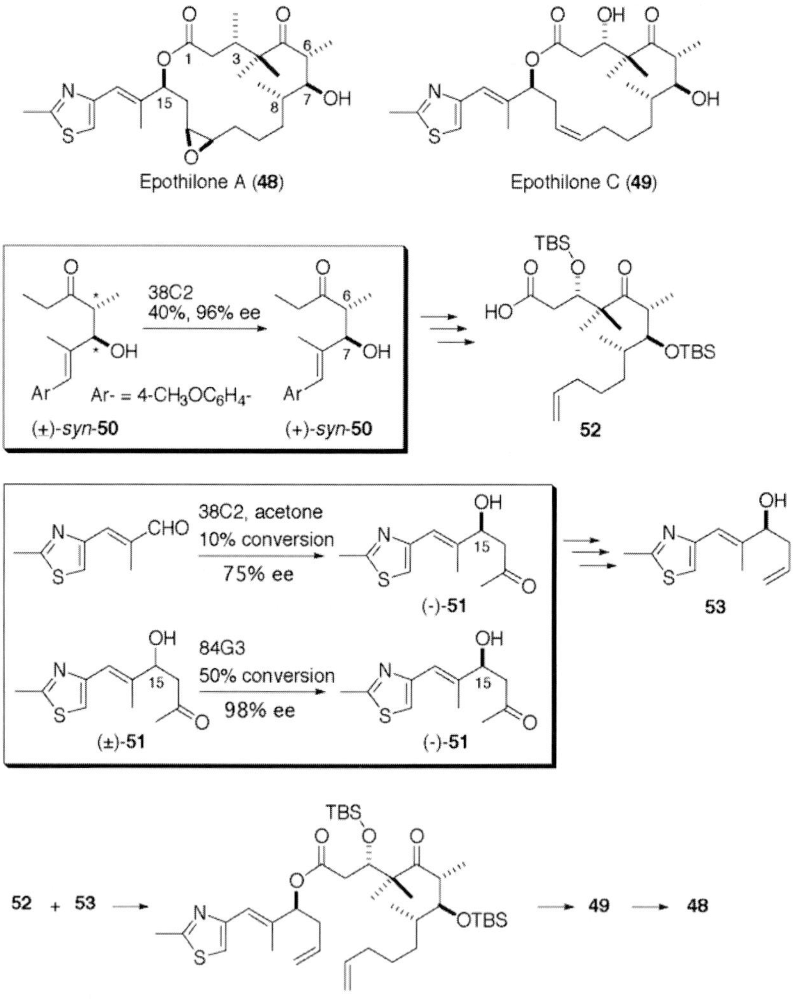

Scheme 6.6
Syntheses of epothilones using aldolase antibody-catalyzed reactions.

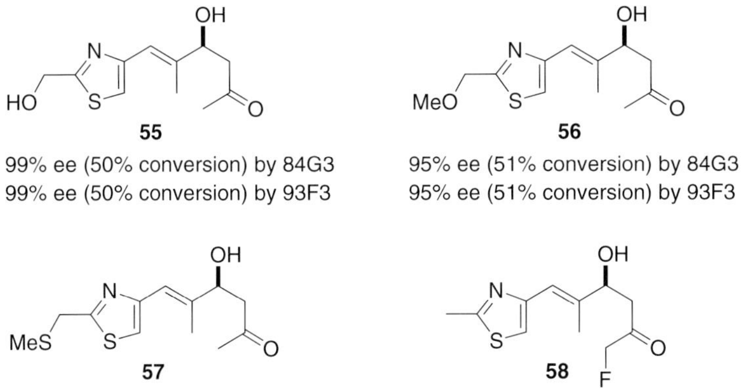

55: 99% ee (50% conversion) by 84G3; 99% ee (50% conversion) by 93F3

56: 95% ee (51% conversion) by 84G3; 95% ee (51% conversion) by 93F3

57: 99% ee (55% conversion) by 84G3; 99% ee (54% conversion) by 93F3

58: 96% ee (54% conversion) by 84G3; 98% ee (52% conversion) by 93F3

Fig. 6.3
Chiral precursors for syntheses of epothilones and their derivatives, which were prepared by aldolase antibody-catalyzed reactions.

a taxol-like mode of action, functioning by stabilization of cellular microtubules. The structural moieties (+)-*syn*-**50** and (−)-**51** were prepared by antibody 38C2-catalyzed reactions and converted to the intermediates **52** and **53**, respectively. The key compound (−)-**51** (98% ee) was also obtained by kinetic resolution using antibodies 93F3 and 84G3 [22]. Chiral precursors **55**–**58** (Figure 6.3) for the syntheses of other epothilones and their derivatives were also prepared by the aldolase antibody-catalyzed kinetic resolution [22].

Other syntheses of natural products have capitalized on the 38C2-catalyzed aldol addition of hydroxyacetone to install 1,2-*syn*-diol functionality. Aldolase antibody 38C2 was also used for synthesis of brevicomins **59**–**61** (Scheme 6.7) [23] and 1-deoxy-L-xylulose (**62**) (Scheme 6.8) [24]. Antibody 38C2-catalyzed kinetic resolution of a tertiary aldol was used for the synthesis of (+)-frontalin (**63**) (Scheme 6.9) [10].

6.3.6 Retro-aldol Reactions in Human Therapy: Prodrug Activation by Aldolase Antibody

An important application of aldolase antibodies is prodrug activation in chemotherapeutic strategies. Activation of a prodrug into an active drug at the tumor site enables selective destruction of those tumor cells. This type of site-specific targeting is known as antibody-directed enzyme prodrug therapy (ADEPT) [25]. The ADEPT complex serves two functions. The antibody portion of the complex enables delivery of the drug directly to the tu-

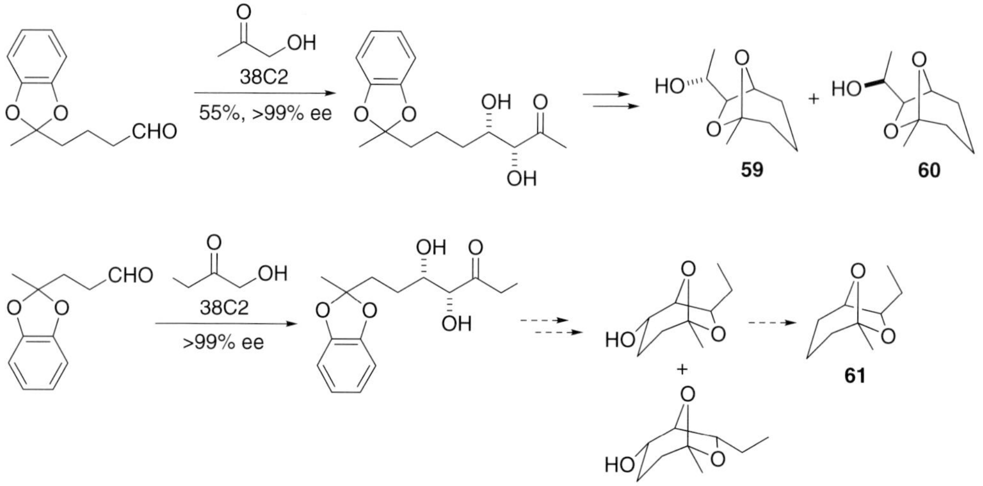

Scheme 6.7
Syntheses of brevicomins using aldolase antibody-catalyzed reactions.

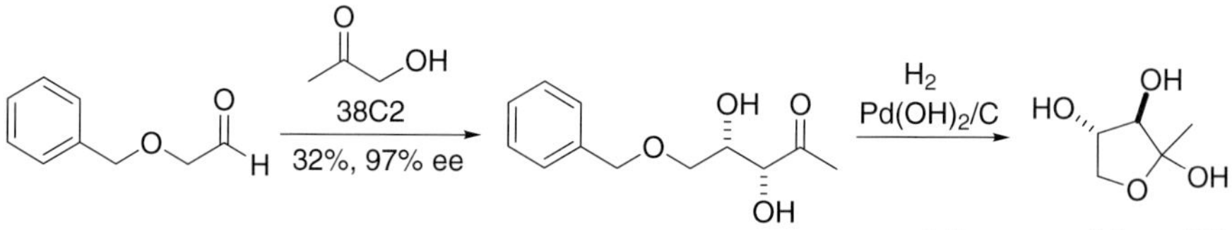

Scheme 6.8
Synthesis of 1-deoxy-L-xylulose using aldolase antibody-catalyzed reaction.

mor by recognition of antigen on the tumor cell surfaces. The enzyme part of the conjugate catalyzes the prodrug activation reaction at the tumor site. This system enhances the efficiency of anti-cancer drugs and reduces peripheral cytotoxicity because of the prodrug's low toxicity. Most ADEPT sys-

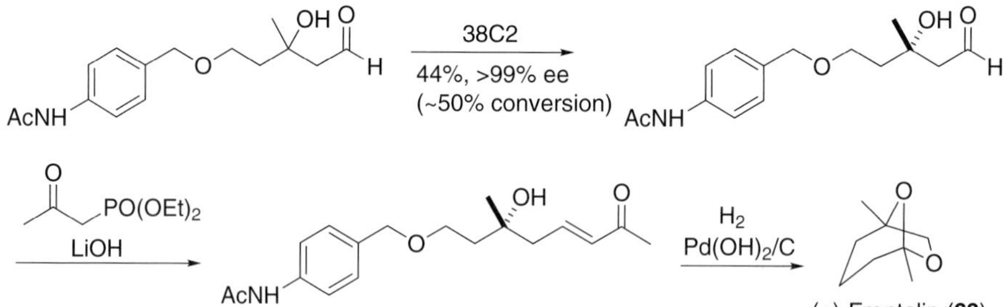

Scheme 6.9
Synthesis of (+)-frontalin using aldolase antibody-catalyzed reaction.

tems incorporate a bacterial enzyme, and the problems of this system are: (1) identification of an enzyme not already present in humans, and (2) the immunogenicity of such a bacterial enzyme. These problems can be circumvented by use of a humanized catalytic antibody which catalyzes the activation reaction of the prodrug selectively, in place of the foreign enzyme [26]. Immunogenicity of the enzyme component of ADEPT can therefore be solved using humanized catalytic antibodies.

Antibody 38C2 catalyzes the activation reactions of prodrugs that incorporate a trigger portion designed to be released by sequential retro-aldol-retro-Michael reactions (Scheme 6.10) [27, 28]. The retro-Michael step is also catalyzed by antibody 38C2, although this step occurs spontaneously in buffer. Because the retro-aldol reaction of the tertiary aldol in this reaction cascade is not catalyzed by any known natural enzymes, this masking of the anti-cancer drugs substantially reduces their toxicity. Combination of doxorubicin-prodrug **64** and antibody 38C2 strongly inhibited cell growth of cancer cell lines, whereas the same concentration of **64** alone was far less potent [27]. Camptothecin- and etoposide-prodrugs **65** and **66** were also activated by 38C2. The aldolase antibody-prodrug system has also proven to be efficient in an animal model of cancer [28].

Incorporation of an additional 8.4-Å linker, as shown in prodrugs **65** and **66**, enables a diverse group of drugs to be used for application of aldolase antibody-catalyzed prodrug activation. Aldolase antibody 33F12 has an active site lysine ε-amino group at ~10 Å depth in a narrow pocket (Section 6.6). The reaction sites of the masking linkers of **65** and **66** reach into the active site whereas the bulky drug molecules remain outside the active-site cavity.

The prodrug activation strategy has also been demonstrated in the context of protein activation. Native insulin modified with aldol-terminated linkers at the primary amines crippled the biological activity of insulin (Scheme 6.11). The modified insulin was defective with regard to receptor binding and stimulation of glucose transport. Antibody 38C2 cleaved the linker, released insulin, and restored insulin activity in an animal model [29]. Antibody 38C2 was also used for the retro-aldol or retro-aldol-retro-Michael reactions of aldol functionality on dendrimers and polymers that have potential for drug encapsulation and delivery [30, 31].

6.4 Aldolase Antibodies for Reactions Related to an Enamine Mechanism and the Nucleophilic Lysine ε-Amino Group

Because aldolase antibodies operate by an enamine mechanism, they also catalyze other reactions that proceed by a similar mechanism. For example, as described in the section on prodrug activation reactions, antibodies 38C2 and 33F12 catalyze β-elimination (retro-Michael) reactions (Section 6.3.6). These antibodies also catalyze decarboxylation of β-keto acids (Scheme 6.12)

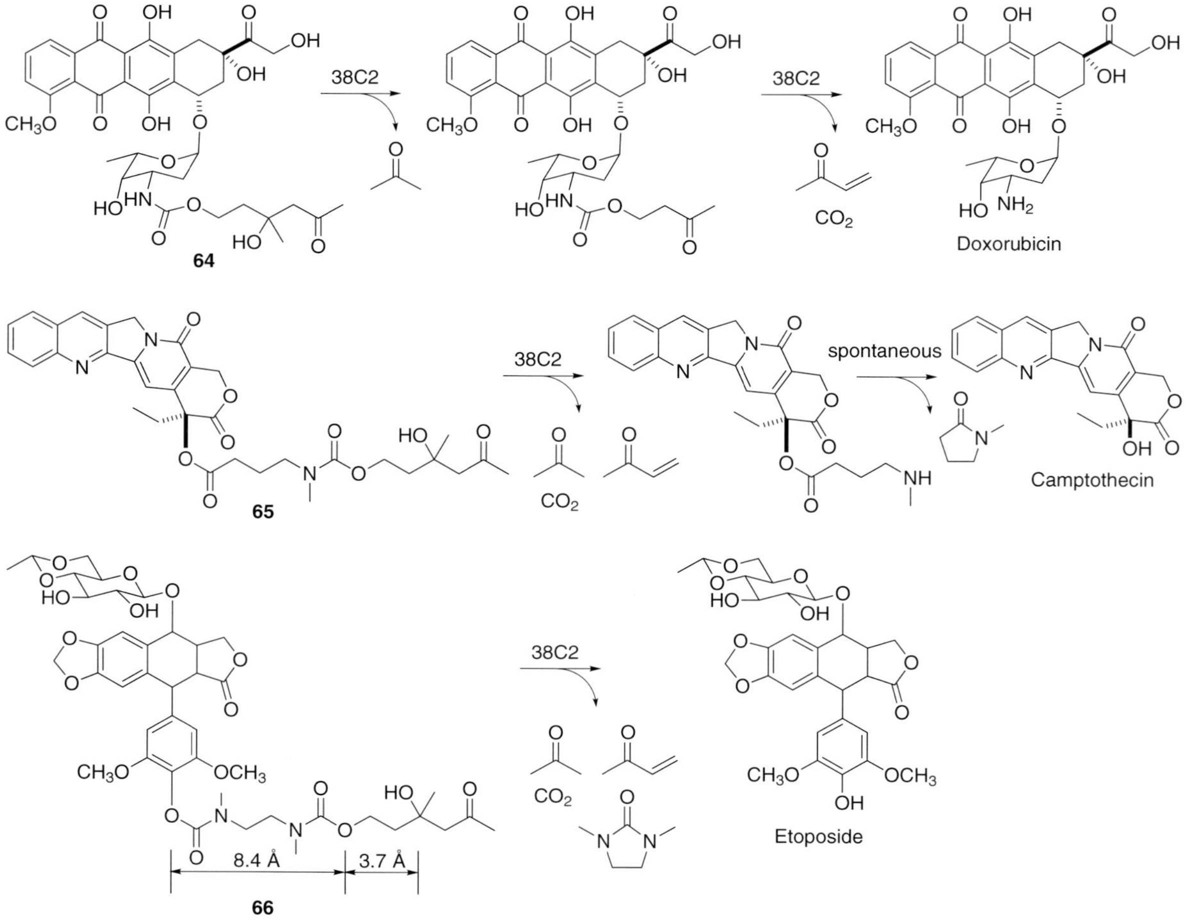

Scheme 6.10 Prodrug activation via a tandem retro-aldol-retro-Michael reaction catalyzed by antibody 38C2. Prodrugs **65** and **66** incorporate an additional linker arm to enable efficient catalysis of drug release.

NH_2
H_2N
Insulin
NH_2
38C2
O_2N
O
OH O
O
O
O
OH
O
O
OH O
NH
O
O
N
H
Insulin
O
OH O
HN
O

Scheme 6.11
Modification of insulin with retro-aldol-retro-Michael linkers and its reactivation by aldolase antibody 38C2.

[32], allylic rearrangement of steroids (Scheme 6.13) [33], and deuterium-exchange reactions (Table 6.7) [34].

In the antibody 38C2-catalyzed decarboxylation of **67**, incorporation of ^{18}O into product **68** was observed in the presence of $H_2{}^{18}O$, consistent with decarboxylation proceeding via an enamine intermediate [32].

Antibody 38C2 catalyzed deuterium-exchange reactions at the α-position of a variety of ketones and aldehydes (Table 6.7) [34]. Because aldehydes bearing a longer alkyl chain (≥valeraldehyde) were not substrates for the

O
O
N
H
67
O
OH
H
Lys—Ab
N^+
H_2N—Lys—Ab
O
N
H
O
O^-
CO_2
H
Lys—Ab
N
O
N
H
^{18}O
$H_2{}^{18}O$
O
N
H
68

Scheme 6.12
Antibody 38C2-catalyzed decarboxylation.

38C2

38C2

38C2

Scheme 6.13
Antibody 38C2-catalyzed allylic rearrangements.

38C2-catalyzed self-aldol reactions (Section 6.3.1), deuterium-exchanged aldehydes accumulated in the presence of 38C2 in D_2O.

Aldolase antibodies can be covalently modified with cofactor derivatives at the active site lysine residue to enable catalysis of cofactor-dependent reactions [35, 36]. A variety of 1-acyl β-lactam derivatives formed the stable amide linkages to active site lysine, whereas 1,3-diketones bound reversibly (Scheme 6.14) [35]. Antibody 38C2 was modified with lactam **69**, and the modified antibody catalyzed thiazolium-dependent decarboxylation of PhCOCOOH. Antibody 38C2 was also covalently modified with the succinic anhydride derivative bearing a bis-imidazole functionality that chelates Cu(II), and the modified antibody was used for Cu(II)-dependent ester hydrolysis [36]. The reactions of the active site lysine ε-amino group of aldolase antibodies depend on the substrates – the catalytic lysine residue of antibody 38C2 did not react with polymer *p*-nitrophenyl ester [37]. An effect of metal-cofactors on antibody 38C2-catalyzed aldol reactions has also been reported [38]. The active site lysine ε-amino group of antibody 38C2 also functioned as general base for the Kemp elimination [39].

Over the years, we have studied a wide-variety of enamine based chemistries using aldolase antibodies. Several key synthetic reactions have been explored using aldolase antibodies. These include investigation of antibody-catalyzed additions to imines, i.e. antibody-based Mannich reactions that operate through an enamine intermediate, and the analogous reaction with nitrostyrene-derived electrophiles (Michael reactions) and Diels–Alder reactions wherein the antibody either generates an enamine derived diene or activates a dienophile through an iminium intermediate. With the exception of the Michael reaction, only recently shown to proceed exclusively with a maleimide electrophile [40], all other reactions failed, presumably because of steric constraints imposed by what is otherwise a promiscuous active site. Significantly, these exploratory studies with catalytic antibodies set the stage

Tab. 6.7
Antibody 38C2-catalyzed deuterium-exchange at the α-position.

Substrate	k_{cat} *(min^{-1})*	K_m *(mM)*	k_{cat}/k_{uncat}
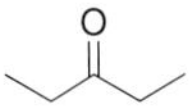	10.7	79	1.1×10^8
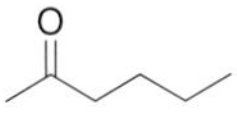	9.7 $\begin{bmatrix} 3.5\ (CH_2) \\ 0.6\ (CH_3) \end{bmatrix}$	105	9.7×10^7
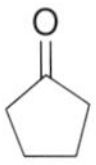	29.9	25	3.0×10^8
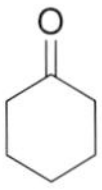	84.9	5	5.9×10^8
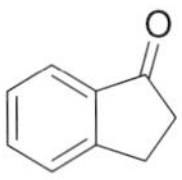	0.95	13.5	9.5×10^6
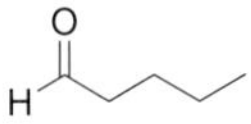	19.0	0.7	
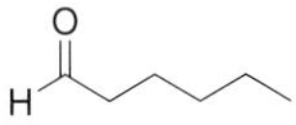	27.6	0.6	
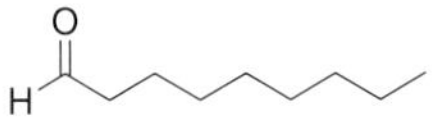	127	1	
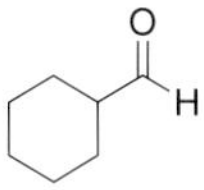	5.8	0.5	

for application of these principles in the area of organocatalysis by our laboratory and former members of the laboratory familiar with these studies [41].

6.5
Concise Catalytic Assays for Aldolase Antibody-catalyzed Reactions

Spectroscopic or visible detection of antibody-catalyzed reactions enhances rapid characterization of catalysts on a small scale. Such detection systems are also useful for high-throughput screening for new aldolase antibody

H2N–Lys–Ab

H2N–Lys–Ab

Lys–Ab

69

Scheme 6.14
Cofactor introduction at a unique active site lysine of aldolase antibodies.

catalysts and for evolution of aldolase antibodies in vitro. Examples of substrates for spectroscopic or visible detection of the antibody-catalyzed reactions are shown in Scheme 6.15. Substrate **47** was the first UV–visible-active aldol substrate designed for following retro-aldol reactions. Reaction results in liberation of the yellow product **70** [9] and this reagent served as the basis for the development of fluorescent versions when fluorescent aldehydes and ketones were later identified [42]. Substrates **35**, **36**, **71**, and **72** liberate the fluorescent products **73**, **74**, **75**, and **76**, respectively, by aldolase antibody-catalyzed retro-aldol reaction [42]. Thus the progress of the reactions with these fluorogenic substrates can be followed by fluorescence. Substrates **77**–**79** are used for analysis of retro-aldol-retro-Michael reactions. Substrates **77** and **78** generate fluorescent compounds resolufin (**80**) and umbelliferone (**81**), respectively, after the reactions [42, 43]. Substrate **79** generates a 2-naphthol derivative **82** that forms a visible colored azo dye with diazonium salts [44]. Maleimide derivative **83** is useful for detection of the carbon–carbon bond-formation catalyzed by aldolase antibodies [40]. It should be noted that all the other systems monitor carbon–carbon bond cleavage, not formation. The reaction of **83** with acetone provides **84**, the fluorescence of which is much greater than that of **83**.

6.6
Structures of Aldolase Antibodies and Reaction Mechanism of Nucleophilic Lysine ε-Amino Group

Antibodies 38C2 and 33F12, generated with diketone **1**, are highly homologous with regard to sequence and have an essential lysine catalytic residue at the same position (H93) [11, 45]. Aldolase antibodies 40F12 and 42F1, generated with hapten **5**, have the same enantio-preference as antibodies 38C2 and 33F12. They are also highly similar to 38C2 and 33F12 in their

47 → 70 UV/VIS λmax 400 nm

35 → 73 fluorescence λex 330 nm, λem 452 nm

36 → 74 fluorescence λex 330 nm, λem 452 nm

71 → 75 fluorescence λex 364 nm, λem 531 nm

72 → 76 fluorescence λex 364 nm, λem 531 nm

77 → → 80 fluorescence λex 544 nm, λem 590 nm

78 → 81 fluorescence λex 360 nm, λem 460 nm

79 → 82 → bright red dye precipitate

(reagents: CH_3, $N_2^+Cl^-$, Cl, 1/2 $ZnCl_2$)

83 + acetone → 84 fluorescence λex 315 nm, λem 365 nm

Scheme 6.15
Substrates for fluorescent and visible detection of aldolase antibody-catalyzed reactions.

amino acid sequences and have an essential lysine catalytic residue at the same position (H93) [16]. On the other hand, aldolase antibodies 93F3 and 84G3, generated with hapten **5**, have antipodal reactivity to antibodies 38C2 and 33F12 and have different amino acid sequences from antibodies 38C2, 33F12, 40F12, and 42F12. The essential lysine residue is at position L89. X-ray crystal structures of the Fab fragments of 33F12 [11] and 93F3 [1d] are shown for comparison in Figures 6.4 and 6.5.

(a)

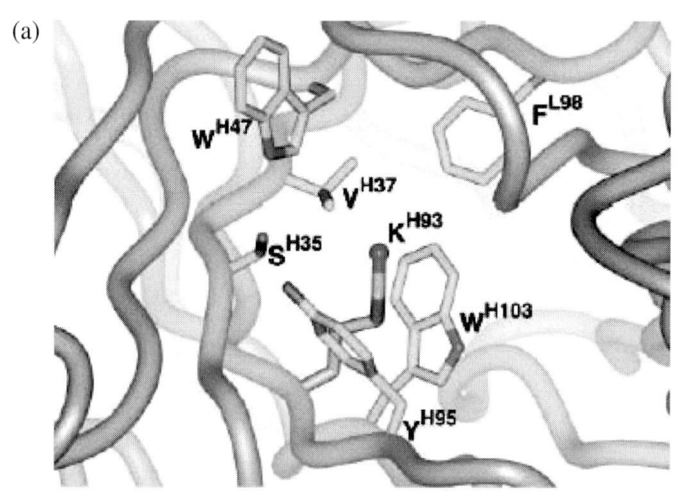

(b)

Fig. 6.4
X-ray structure of aldolase antibody 33F12. (a) Top view. Residues within 5 Å of the ε-amino group of the catalytic lysine (H93) are indicated. (b) Side view.

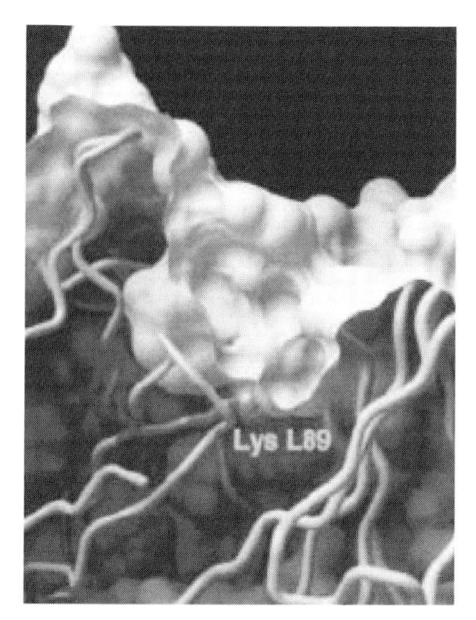

Fig. 6.5
X-ray structure of aldolase antibody 93F3. The essential lysine residue (L89) is indicated.

The structure of 33F12 shows that the entrance of the antigen binding site of 33F12 is a narrow elongated cleft. The binding pocket is more than 11 Å deep and is comparable with combining sites of antibodies raised against other small haptenic molecules. At the bottom of the pocket LysH93 is found within a hydrophobic environment. Within this pocket, only one charged residue is within an 8-Å radius of the nitrogen of ε-amino group of LysH93. No salt bridges or hydrogen bonds can be formed between LysH93 and any other residues in 33F12. A Hansch plot, a study of the free energy relationship for substrate partitioning into *n*-octanol and k_{cat}/K_m of the retro-aldol reactions, shows that the active site in 33F12 is 1.1 times more hydrophobic than *n*-octanol [11]. This environment perturbs the pK_a of LysH93 and enables it to exist in its uncharged form, facilitating its function as a strong nucleophile. The hydrophobic environment used by these aldolase antibodies to tune the pK_a of the ε-amino group of the active site lysine residue is in contrast with the mechanism used by most natural class I aldolase enzymes; in the natural aldolases an electrostatic mechanism perturbs the pK_a of the ε-amino group of the lysine [2, 16, 46].

Notably, with the exception of a single residue, the residues in van der Waals contact with LysH93 are conserved in both antibodies and are encoded in the germline gene segments used by these antibodies. This conservation suggests that LysH93 appeared early in the process of antibody evolution in a germline antibody. The insertion of this residue into this hydrophobic microenvironment resulted in chemical reactivity that was efficient enough to be selected. Once this covalent process appeared the binding pocket did not further evolve toward high specificity as would be indicated by the selection of somatic variants of the germline sequence. The broad substrate specificity of aldolase antibodies prepared by reactive immunization is likely to be the result of the special ontogeny of antibodies induced by immunogens that form covalent bonds within the binding pocket during the induction.

Although all aldolase antibodies generated with 1,3-diketones use an enamine mechanism in their aldol and retro-aldol reactions, the micromechanisms of the reactions might differ among these aldolase antibodies as they do among aldolase enzymes [2, 47]. In the 38C2-catalyzed retro-aldol reactions of a series of aldols prepared from acetone and *p*-substituted cinnamaldehyde, correlation between log k_{cat} and the Hammett substituent σ is not linear [13, 48]. A positive linear correlation between log k_{cat} and σ was, on the other hand, obtained for aldolase antibody 24H6, generated with a mixture of haptens **6** and **7**; this indicates 38C2 and 24H6 differ in their micromechanisms, including the rate-limiting steps. However, similar amino acid sequences of the aldolase antibodies reflect the broad similarity in substrate specificity, enantioselectivity, and micromechanism.

6.7
Evolution of Aldolase Antibodies In Vitro

Although aldolase antibodies are broad in scope, the efficiency with which any given aldol is processed can vary significantly. To create aldolase antibodies with altered substrate specificity and turnover, phage libraries were screened using different diketone derivatives [49]. In this approach, libraries were prepared by recombining the catalytic machinery of well-characterized aldolase antibodies with a naive V gene repertoire. In vitro selection systems enabled the use of multiple haptens without animal re-immunization and enabled the experimenter to combine insights gained by the study of existing catalytic antibodies with the diversity of the immune repertoire.

This strategy was used to prepare catalysts that would efficiently process cyclohexanone-aldols **85**, because retro-aldol reactions of **85** were relatively slow compared with those involving acetone-aldols using existing aldolase antibodies 38C2 and 33F12 (Scheme 6.16). The phage libraries [50] prepared by combination of active site residues of 38C2 and 33F12 with a naive V gene repertoire were selected with **86** and **1** as reactants. Fab 28 obtained from this selection catalyzed the retro-aldol reactions of *anti*-**87**, *syn*-**87**, and **35**. The k_{cat} values of Fab 28 were superior to those of parental antibodies for cyclohexanone-aldols *anti*-**87** and *syn*-**87** by approximately three- to tenfold [49]. In addition, Fab 28 catalyzed the reaction of acetone-aldol **35** with a k_{cat} value similar to that of antibody 33F12. The stereochemistries of the preferred substrate enantiomers of Fab 28 were the same as those of the parental antibodies 38C2 and 33F12. On the basis of on the design of

Scheme 6.16
Reactions and compounds for in vitro evolution of aldolase antibodies.

the library, Fab 28 retained specific sequence elements of the parental antibodies and the essential LysH93 of the catalytic mechanism. The remaining primary sequence of Fab 28 is not related to the parental antibodies. In addition, because a naive V gene library was generated using human bone marrow cDNA, Fab 28 is a human aldolase antibody. This strategy will be useful for providing human antibodies for catalytic antibody-mediated prodrug activation described in Section 6.3.6 without the need for re-immunization and selection.

A correlation was observed between the k_{cat} of the antibody-catalyzed retro-aldol reaction and the apparent K_d of the corresponding diketones (i.e. the reactivity to the diketones) within the family of aldolase antibodies 38C2, 33F12, Fab 28, and antibodies selected with Fab 28 in vitro. Stronger binding (lower K_d value) to acetone-diketone **88** (i.e. higher reactivity with acetone-diketone **88**) correlated with a higher k_{cat} value for the reaction of acetone-aldol **35**, and stronger binding to cyclohexanone-diketone **89** correlated with a higher k_{cat} value for the reaction of cyclohexanone-aldol *anti*-**87** (Figure 6.6) [51]. Selection using a structure-altered diketone provided catalytic antibodies that had altered substrate specificity as directed by the structure of the selecting diketone. The correlation indicates that antibodies 38C2, 33F12, and the in vitro evolved aldolase antibodies share a similar micromechanism including the rate-limiting step in their catalyzed reactions. 1,3-Diketones have also been used for phage selections of small peptides that catalyze aldol and retro-aldol reactions via an enamine mecha-

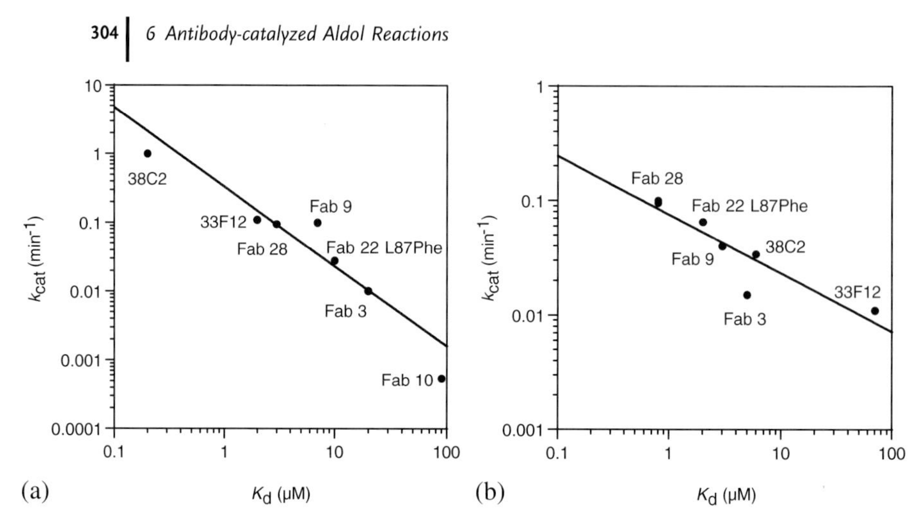

Fig. 6.6
(a) Plot of k_{cat} for the retro-aldol reaction of **35** against K_d of diketone **88**. (b) Plot of k_{cat} for the retro-aldol reaction of *anti*-**87** versus K_d of diketone **89**.

nism [52]. Selection with 1,3-diketones seems to be a general and effective route to catalysts that operate via an enamine mechanism.

Although in vitro selection strategies using 1,3-diketones can be used to obtain access to aldolase antibodies, direct selection of catalysis resulted in improved catalysts. Fluorogenic and chromogenic substrates described in Section 6.5 might be useful for the catalytic selection. Genetic selection is also useful for the selection of catalysts. Antibody-catalyzed reaction of substrate **90**, a prodrug, generates *p*-aminobenzoic acid (PABA), an essential metabolite for *E. coli*. When an *E. coli* strain that cannot synthesize PABA, because of a genetic defect was provided with the gene for an aldolase antibody, expression of the aldolase antibody provided the *E. coli* with the ability to metabolize **90** and survive. Such a genetic selection has been shown in preliminary experiments to provide a growth advantage to a strain expressing aldolase antibody 38C2 (Scheme 6.17) [53]. Ideally, such a system will be used in the future to rapidly evolve aldolase antibodies based on both turnover efficiency and substrate specificity.

Scheme 6.17
Reaction for genetic selection of aldolase antibodies.

91

92 + → 94, 72D4 → 93

Scheme 6.18
Antibody-catalyzed reaction using cofactor primary amine.

6.8 Cofactor-mediated Antibody-catalyzed Aldol and/or Retro-aldol Reactions

Antibody 72D4, generated by immunization with **91**, catalyzed the aldol reaction of acetone and aldehyde **92** and the retro-aldol reaction of **93**, with some enantioselectivity, in the presence of a primary amine cofactor **94** (Scheme 6.18) [54]. The antibody did not catalyze the reactions in the absence amine **94**, and evidence supported an enamine mechanism (enamine formation with **94**) for this cofactor amine-mediated antibody-catalyzed reaction.

An antibody that mimics threonine aldolase, which uses pyridoxal as the cofactor to catalyze the aldol reaction of glycine with aldehydes, has also been reported. Antibody 10H2 catalyzed the retro-aldol reaction of β-hydroxy-α-amino acid in the presence of pyridoxal [55].

6.9 Summary and Conclusion

Class I aldolases function as Nature's most fundamental carbon–carbon bond forming enzymes. Designer catalysts – aldolase antibodies – that mimic the aldolases have been created by using a reaction-based selection strategy with 1,3-diketones. The covalent reaction mechanism is a fundamental part of the selection. Broad scope, enhanced catalytic activity, and defined chemical mechanism are three features of these aldolase antibodies that distinguish them from traditional antibody catalysts. The substrate specificities of aldolase antibodies are different from those of naturally existing enzymes. Thus, applications of aldolase antibody-catalyzed reactions are very wide, from asymmetric synthesis to chemical transformations in cancer therapy [56, 57].

6.10 Experimental Procedures

Aldolase antibodies 38C2 and 84G3 are commercially available from Aldrich.

Example 1. Kinetic Resolution in a Biphasic System [20]. A solution of antibody 84G3 (500 mg, 6.67 μmol active site, 0.0086 mol%) in phosphate buffered saline (PBS; pH 7.4, 87.5 mL) at room temperature was added to a solution of (±)-**45** (20 g, 77.4 mmol) in toluene (600 mL) in a Teflon tube. The mixture was shaken at 250 rpm. under argon at the same temperature and the ee was monitored by chiral-phase HPLC. When ee > 99% was reached (65 h) the mixture was cooled to −20 °C for several hours to enable the aqueous phase to freeze. The organic phase was decanted. The frozen aqueous phase was left to thaw and extracted with toluene (3 × 3 vol. equiv.) and 1:1 EtOAc–toluene (3 × 3 vol. equiv.). The combined organic phase was dried over Na_2SO_4, concentrated in vacuo, and purified by silica gel column chromatography (1:3 EtOAc–hexane) to afford (*S*)-**45** (10 g, 50%).

Example 2. Kinetic Resolution [22]. A solution of antibody 84G3 (0.34 g, 0.00227 mmol) in PBS (27.2 mL) was added to a degassed solution of (±)-**51** (16.8 g, 75 mmol) in PBS (1.55 L)–CH_3CN (140 mL). The mixture was incubated under an argon atmosphere at 37 °C for 5 days. After consumption of more than 98% of one enantiomer, as judged by HPLC analysis, the mixture was dialyzed using Amicon membranes to recover the antibody and the filtrate was passed through a reversed-phase column (C_{18}). The column was first washed with water and then the desired compounds were eluted with MeOH. The solvents were removed under vacuum and the residue was purified by silica gel column chromatography (EtOAc–hexanes, 9:1 to 2:1) to afford (−)-**51** (7.6 g, 45%, >98% ee).

Example 3. Aldol Reaction [8]. Antibody 38C2 (120 μM in PBS, 8.0 mL) was added at room temperature to a mixture of **11** (110 mg, 0.61 mmol) in DMF (15 mL), acetone (31 mL), and degassed PBS (571 mL). The final concentrations were: **11**, 1.0 mM; 38C2, 1.9 μM; and acetone, 5% (*v*/*v*). The reaction mixture was kept under argon at room temperature in the dark for 7 days. The reaction mixture was saturated with NaCl and extracted with EtOAc (3 × 150 mL). The organic phase was dried over $MgSO_4$, concentrated in vacuo, and purified by silica gel column chromatography (EtOAc–hexane, 1:2) to afford (*S*)-**21** (96 mg, 67%, 91% ee).

Acknowledgments

We would like to thank Richard A. Lerner, our long-term collaborator in the aldolase antibody projects performed at the Scripps Research Institute, for his contributions.

References

1 (a) Schultz, P. G.; Lerner, R. A. *Science* **1995**, *269*, 1835. (b) Hilvert, D. *Annu. Rev. Biochem.* **2000**, *69*, 751. (c) Blackburn, G. M., Garcon, A., Catalytic antibodies. In Kelly, D. R. Ed. Biotechnology vol. 8b, *Biotechnology Biotransformations II*; Wiley–VCH: Weinheim, 2000; p. 403. (d) Schultz, P. G.; Yin, J.; Lerner, R. A. *Angew. Chem. Int. Ed.* **2002**, *41*, 4427. (e) Tanaka, F. *Chem. Rev.* **2002**, *102*, 4885.

2 (a) Lai, C. Y.; Nakai, N.; Chang, D. *Science* **1974**, *183*, 1204. (b) Morris, A. J.; Tolan, D. R. *Biochemistry* **1994**, *33*, 12291. (c) Choi, K. H.; Shi, J.; Hopkins, C. E.; Tolan, D. R.; Allen, K. N. *Biochemistry* **2001**, *40*, 13868. Allard, J.; Grochulski, P.; Sygusch, J. *Proc. Natl. Acad. Sci. U.S.A.* **2001**, *98*, 3679. Heine, A.; DeSantis, G.; Luz, J. G.; Mitchell, M.; Wong, C.-H.; Wilson, I. A. *Science* **2001**, *294*, 369.

3 Lide, D. R. Ed. CRC Handbook of Chemistry and Physics 83rd Edition; CRC Press: New York, 2002, p. 7-1.

4 Wagner, J.; Lerner, R. A.; Barbas, C. F., III *Science* **1995**, *270*, 1797.

5 Wirshing, P.; Ashley, J. A.; Lo, C.-H. L.; Janda, K. D.; Lerner, R. A. *Science* **1995**, *270*, 1775.

6 Tagaki, W.; Guthrie, J. P.; Westheimer, F. H. *Biochemistry* **1968**, *7*, 905.

7 Gilbert, H. F., III; O'Leary, M. H. *Biochemistry* **1975**, *14*, 5194. Gilbert, H. F., III; O'Leary, M. H. *Biochem. Biophys. Acta* **1977**, *483*, 79. Otwell, H. B.; Cipollo, K. L.; Dunlap, R. B. *Biochem. Biophys. Acta* **1979**, *568*, 297. Guidinger, P. F.; Nowak, T. *Biochemistry* **1991**, *30*, 8851.

8 Hoffmann, T.; Zhong, G.; List, B.; Shabat, D.; Anderson, J.; Gramatikova, S.; Lerner, R. A.; Barbas, C. F., III *J. Am. Chem. Soc.* **1998**, *120*, 2768.

9 Zhong, G.; Shabat, D.; List, B.; Anderson, J.; Sinha, R. A.; Lerner, R. A.; Barbas, C. F., III *Angew. Chem., Int. Ed.* **1998**, *37*, 2481.

10 List, B.; Shabat, D.; Zhong, G.; Turner, J. M.; Li, A.; Bui, T.; Anderson, J.; Lerner, R. A.; Barbas, C. F., III *J. Am. Chem. Soc.* **1999**, *121*, 7283.

11 Barbas, C. F., III; Heine, A.; Zhong, G.; Hoffmann, T.; Gramatikova, S.; Bjornestsdt, R.; List, B.; Anderson, J.; Stura, E. A.; Wilson, I. A.; Lerner, R. A. *Science* **1997**, *278*, 2085.

12 Zhong, G.; Lerner, R. A.; Barbas, C. F., III *Angew. Chem. Int. Ed.* **1999**, *38*, 3738.

13 Shulman, H.; Makarov, C.; Ogawa, A. K.; Romesberg, F.; Keinan, E. *J. Am. Chem. Soc.* **2000**, *122*, 10743.
14 Fujii, I.; Tanaka, F.; Miyashita, H.; Tanimura, R.; Kinoshita, K. *J. Am. Chem. Soc.* **1995**, *117*, 6199. Tanaka, F.; Kinoshita, K.; Tanimura, R.; Fujii, I. *J. Am. Chem. Soc.* **1996**, *118*, 2332. Patten, P. A.; Gray, N. S.; Yang, P. L.; Marks, C. B.; Wedemayer, G. J.; Boniface, J. J.; Stevens, R. C.; Schultz, P. G. *Science*, **1996**, *271*, 1086. Wedemayer, G. J.; Pattern, P. A.; Wang, L. H.; Schultz, P. G.; Stevens, R. C. *Science* **1997**, *276*, 1665. Yin, J.; Mundorff, E. C.; Yang, P. L.; Wendt, K. U.; Hanway, D.; Stevens, R. C.; Schultz, P. G. *Biochemistry* **2001**, *40*, 10764.
15 Zhong, G.; Hoffmann, T.; Lerner, R. A.; Danishefsky, S.; Barbas, C. F., III *J. Am. Chem. Soc.* **1997**, *119*, 8131. List, B.; Lerner, R. A.; Barbas, C. F., III *Org. Lett.* **1999**, 1, 59 and 353.
16 Karlstrom, A.; Zhong, G.; Rader, C.; Larsen, N. A.; Heine, A.; Fuller, R.; List, B.; Tanaka, F.; Wilson, I. A.; Barbas, C. F., III; Lerner, R. A. *Proc. Natl. Acad. Sci. U.S.A.* **2000**, *97*, 3878.
17 Guthrie, J. P.; Cossar, J.; Taylor, K. F. *Can. J. Chem.* **1984**, *62*, 1958.
18 Guthrie, J. P.; Wang, X.-P. *Can. J. Chem.* **1992**, *70*, 1055.
19 Maggiotti, V.; Resmini, M.; Gouverneur, V. *Angew. Chem. Int. Ed.* **2002**, *41*, 1012. Maggiotti, V.; Wong, J.-B.; Razet, R.; Cowley, A. R.; Gouverneur, V. *Tetrahedron: Asymmetry* **2002**, *13*, 1789.
20 Turner, J. M.; Bui, T.; Lerner, R. A.; Barbas, C. F., III; List, B. *Chem. Eur. J.* **2000**, *6*, 2772.
21 Sinha, S.; Barbas, C. F., III; Lerner, R. A. *Proc. Natl. Acad. Sci. U.S.A.* **1998**, *95*, 14603.
22 Sinha, S.; Sun, J.; Miller, G.; Barbas, C. F., III; Lerner, R. A. *Org. Lett.* **1999**, *1*, 1623. Shinha, S. C.; Sun, J.; Miller, G. P.; Wartmann, M.; Lerner, R. A. *Chem. Eur. J.* **2001**, *7*, 1691.
23 List, B., Shabat, D., Barbas III, C. F., and Lerner, R. A. *Chem. Eur. J.* **1998**, *4*, 881.
24 Shabat, D.; List, B.; Lerner, R. A.; Barbas, C. F., III *Tetrahedron Lett.* **1999**, *40*, 1437.
25 Niculesco-Duvanz, I.; Springer, C. J. *Adv. Drug Delivery Rev.* **1997**, *26*, 151.
26 Wenthworth, P.; Datta, A.; Blakey, D.; Boyle, T.; Partridge, L. J.; Blackburn, G. M. *Proc. Natl. Acad. Sci. U.S.A.* **1996**, *93*, 799.
27 Shabat, D.; Rader, C.; List, B.; Lerner, R. A.; Barbas, C. F., III *Proc. Natl. Acad. Sci. U.S.A.* **1999**, *96*, 6925.
28 Shabat, D.; Lode, H. N.; Pertl, U.; Reisfeld, R. A.; Rader, C.; Lerner, R. A.; Barbas, C. F., III *Proc. Natl. Acad. Sci. U.S.A.* **2001**, *98*, 7528.
29 Worrall, D. S.; McDunn, J. E.; List, B.; Reichart, D.; Hevener, A.; Gustafson, T.; Barbas, C. F., III; Lerner, R. A.; Olefsky, J. M. *Proc. Natl. Acad. Sci. U.S.A.* **2001**, *98*, 13514.
30 Cordova, A.; Janda, K. D. *J. Am. Chem. Soc.* **2001**, *123*, 8248.

31 Gopin, A.; Pessah, N.; Shamis, M.; Rader, C.; Shabat, D. *Angew. Chem. Int. Ed.* **2003**, *42*, 327.

32 Bjornestedt, R.; Zhong, G.; Lerner, R. A.; Barbas, C. F., III *J. Am. Chem. Soc.* **1996**, *118*, 11720.

33 Lin, C.-H.; Hoffman, T. Z.; Wirshing, P.; Barbas, C. F., III; Janda, K. D.; Lerner, R. A. *Proc. Natl. Acad. Sci. U.S.A.* **1997**, *94*, 11773.

34 Shulman, A.; Sitry, D.; Shulman, H.; Keinan, E. *Chem. Eur. J.* **2002**, *8*, 229.

35 Tanaka, F.; Lerner, R. A.; Barbas, C. F., III *Chem. Commun.* **1999**, 1383.

36 Nicholas, K. M.; Wentworth, P. Jr.; Harwig, C. W.; Wentworth, A. D.; Shafton, A.; Janda, K. D. *Proc. Natl. Acad. Sci. U.S.A.* **2002**, *99*, 2648.

37 Satchi-Fainaro, R.; Wrasidlo, W.; Lode, H. N.; Shabat, D. *Bioorg. Med. Chem.* **2002**, *10*, 3023.

38 Finn, M. G.; Lerner, R. A.; Barbas, C. F., III *J. Am. Chem. Soc.* **1998**, *120*, 2963.

39 James, L. C.; Tawfik, D. S. *Protein Science* **2001**, *10*, 2600.

40 Tanaka, F.; Thayumanavan, R.; Barbas, C. F., III *J. Am. Chem. Soc.* **2003**, *125*, 8523.

41 Sakthivel, K.; Notz, W.; Bui, T.; Barbas, C. F., III *J. Am. Chem. Soc.* **2001**, *123*, 5260. Cordova, A.; Watanabe, S.; Tanaka, F.; Notz, W.; Barbas, C. F., III *J. Am. Chem. Soc.* **2002**, *124*, 1866. Pidathala, C.; Hoang, L.; Vignola, N.; List, B. *Angew. Chem. Int. Ed.* **2003**, *42*, 2785.

42 List, B.; Barbas, C. F., III; Lerner, R. A. *Proc. Natl. Sci. Acad. U.S.A.* **1998**, *95*, 15351.

43 Jourdain, N.; Carlon, R. P.; Reymond, J.-L. *Tetrahedron Lett.* **1998**, *39*, 9415. Carlon, R. P.; Jourdain, N.; Reymond, J.-L. *Chem. Eur. J.* **2000**, *6*, 4154.

44 Tanaka, F.; Kerwin, L.; Kubitz, D.; Lerner, R. A.; Barbas, C. F., III *Bioorg. Med. Chem. Lett.* **2001**, *11*, 2983.

45 The GenBank accession numbers are following: 33F12 VL, AF242212; 33F12 VH, AF242213; 38C2 VL, AF242214; 38C2 VH, AF242215; 40F12 VL, AF242216; 40F12 VH, AF242217; 42F1 VL, AF242218; 42F1 VH, AF242219.

46 Westheimer, F. H. *Tetrahedron* **1995**, *51*, 3.

47 Littlechild, J. A.; Watson, H. C. *Trends Biochem. Sci.* **1993**, *18*, 36. Morris, A. J.; Davenport, R. C.; Tolan, D. R. *Protein Engineering* **1996**, *9*, 61.

48 Shulman, H.; Keinan, E. *Bioorg. Med. Chem. Lett.* **1999**, *9*, 1745.

49 Tanaka, F.; Lerner, R. A.; Barbas, C. F., III *J. Am. Chem. Soc.* **2000**, *122*, 4835.

50 Barbas III, C. F., Burton, D. R., Scott, J. K., and Silverman, G. J. Eds. Phage Display: A Laboratory Manual. Cold Spring Harbor Laboratory Press, Cold Spring Harbor, New York, 2001.

51 Tanaka, F.; Fuller, R.; Shim, H.; Lerner, R. A.; Barbas, C. F., III *J. Mol. Biol.* **2004**, *335*, 1007.

52 Tanaka, F.; Barbas, C. F., III *Chem Commun.* **2001**, 769. Tanaka, F.; Barbas, C. F., III *J. Am. Chem. Soc.* **2002**, *124*, 3510.

53 Gildersleeve, J.; Janes, J.; Ulrich, H.; Yang, P.; Turner, J.; Barbas, C.; Schultz, P. *Bioorg. Med. Chem. Lett.* **2002**, *12*, 1691 and 2789.
54 Reymond, J.-L.; Chen, Y. *Tetrahedron Lett.* **1995**, *36*, 2575. Reymond, J.-L.; Chen, Y. *J. Org. Chem.* **1995**, *60*, 6970. Reymond, J.-L. *Angew. Chem. Int. Ed.* **1995**, *34*, 2285.
55 Tanaka, F.; Oda, M.; Fujii, I. *Tetrahedron Lett.* **1998**, *39*, 5057.
56 Barbas, C. F., III; Rader, C.; Segal, D. J.; List, B.; Turner, J. M. *Advances in Protein Chemistry* **2001**, *55*, 317.
57 Tanaka, F.; Barbas, C. F., III *J. Immunol. Methods* **2002**, *269*, 67.

7
The Aldol Reaction in Natural Product Synthesis: The Epothilone Story

Dieter Schinzer

Dedicated to Clayton H. Heathcock, one of the pioneers of modern aldol reactions who figured out many of the import stereochemical principles of this fascinating chemistry

7.1
History of Epothilones: Biological Source, Isolation, and Structural Elucidation

The epothilones, a new class of macrocyclic compounds which show cytotoxic activity, have been isolated as secondary metabolites by Höfle and Reichenbach from myxo bacteria [1]. These are so-called gliding bacteria which cruise on slime tracks forming swarms. They grow preferentially on rodded material, like dung. Their morphogenetic potential is quite interesting, forming cylindrical vegetative cells, which under starvation conditions turn into fruiting bodies with solid walls, that further differentiate into myxo spores [2]. Besides the epothilones – which were isolated from a strain of *Sorangium cellulosum* – many other structural diverse natural products have been isolated from myxo bacteria [3].

7.2
History of Epothilones: The Total Synthesis Race

The publication of the structures of epothilone A and B in 1996 by Höfle *et al.* [4] in connection with the biological data of the tubulin assay was of great interest for the chemical community. The epothilones were identified by the Merck group in 1995 as tubulin-polymerizing natural products as a single active structural type out of a compound library of more than 60.000 molecules [5]. The cytotoxic effect of Taxol® via binding to tubulin and thus stabilizing the assembly, combined with the scope of its clinical application, prompted major activities around the world. Taxol® is far from being an ideal drug: low water solubility requires formulation vehicles, such as

Modern Aldol Reactions. Vol. 1: Enolates, Organocatalysis, Biocatalysis and Natural Product Synthesis.
Edited by Rainer Mahrwald

ISBN: 3-527-30714-1

Cremophore® which create allergic side effects and other risks [6]. Two different types of resistance arise during cancer therapy with Taxol®; an overexpression of the P-glycoprotein (P-gp) efflux system and tubulin mutation which both cause major drawbacks. Furthermore, the remarkable efficacy of epothilone against MDR (multiple drug resistante) cell lines makes these compounds superior in comparison to known anticancer agents [7]. This started a race for the first total synthesis in which many world leading chemists got involved. In addition, biologists, medicinal chemists, and clinicians started many efforts along these lines. Besides various total syntheses, many partial solutions, a large variety of synthetic analogues have been synthesized and have created a tremendous number of publications and patent applications around the world [8]. Finally, different large (and small) pharmaceutical companies took over the project and started clinical trials of epothilones or of their synthetic analogues in order to replace Taxol® as a therapeutic anticancer agent [9].

The first total synthesis was published by Danishefsky *et al.* [10], shortly followed by independent routes from Nicolaou *et al.* [11] and Schinzer *et al.* [12].

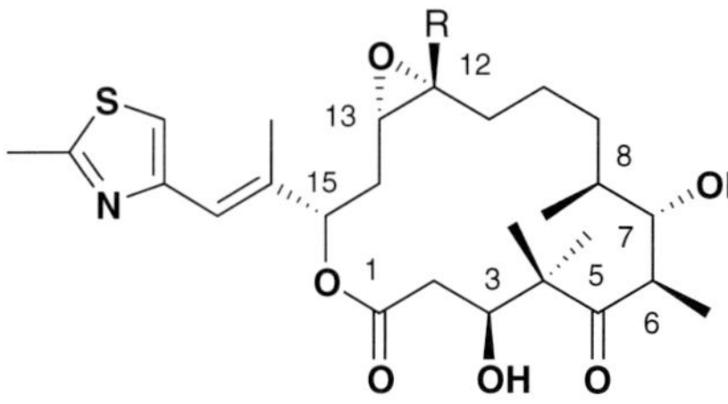

Scheme 7.1
Epothilone A: R=H; epothilone B: R=Me

7.2.1
Different Strategies with Aldol Reactions: The Danishefsky Synthesis of Epothilone A Relying on Intramolecular Aldol Reaction

The first total synthesis of epothilone A combines the two halves of the molecule via a B-alkyl Suzuki coupling [13] yielding the desired cyclization precursor. The critical cyclization was achieved by an intramolecular aldol reaction, utilizing KHMDS as base providing the natural configuration with a 6:1 ratio, 51% yield.

The selectivity of the macro condensation was best at −78 °C and work-up of the potassium alkoxide at 0 °C. Protonation at lower temperature yielded in more of (*R*)-configured material. Under special conditions, even the (*R*)-isomer predominated over the "natural" configuration. The reason for this surprising behavior is still unknown and under investigation.

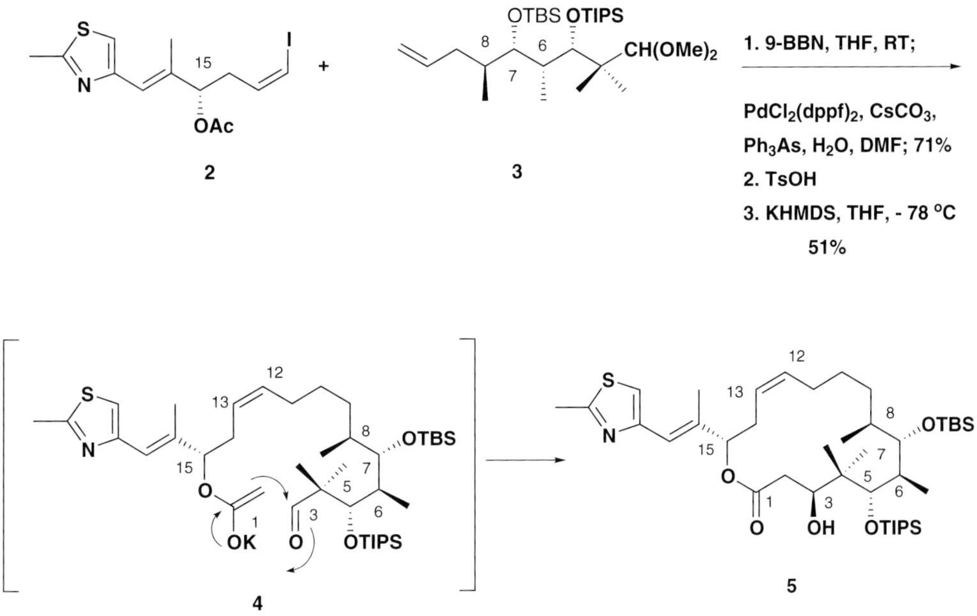

1. 9-BBN, THF, RT; $PdCl_2(dppf)_2$, $CsCO_3$, Ph_3As, H_2O, DMF; 71%
2. TsOH
3. KHMDS, THF, - 78 °C 51%

Scheme 7.2
First total synthesis of epothilone A by Danishefsky *et al.*

7.2.2
Different Strategies with Aldol Reactions: The Nicolaou Synthesis of Epothilone A Using an Unselective Aldol Reaction

The Nicolaou strategy is based on RCM [14] as the final ring closing step in order to establish the 16-membered macrolide. The linear subunit of C1–C12 was obtained by an intermolecular aldol reaction, in which an α-chiral aldehyde was coupled with an achiral enolate, yielding the desired aldol product as a 2:3 ratio of diastereomers **8** and **9**.

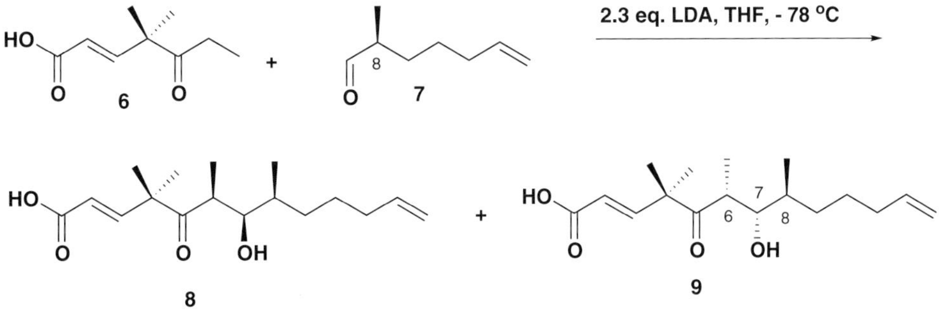

2.3 eq. LDA, THF, - 78 °C

Scheme 7.3
Model study for the total synthesis of epothilone A by Nicolaou *et al.*

2.3 eq. LDA, THF, - 78 °C

1. EDC, 4-DMAP, 12 h; 52%

2. $RuCl_2(=CHPh)(PCy_3)_2$, CH_2Cl_2, 20 h 79%

Scheme 7.4
Second total synthesis of epothilone A by Nicolaou *et al.*

This result is not surprising because only one chiral element was used in that reaction. A second chiral element should improve the selectivity. For that reason Nicolaou introduced a chiral enolate, which is shown in Scheme 7.4.

In his second synthesis, Nicolaou used a chiral enolate of type **10** which contained the (*S*)-configuration at C3 and the required oxidation level at C1. However, again almost no selectivity was observed in the aldol coupling [15].

7.2.3
Different Strategies with Aldol Reactions: The Schinzer Synthesis of Epothilone A with Complete Stereocontrol in the Aldol Reaction

The Schinzer synthesis of epothilone A is also based on a convergent strategy using three key fragments [16]. The 16-membered macrolide is closed via RCM to provide the key intermediate for the final epoxidation.

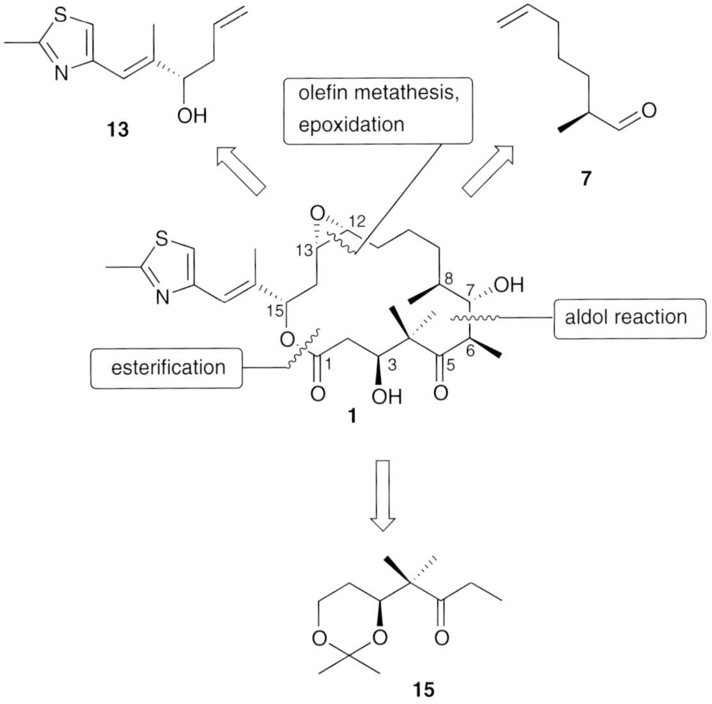

Scheme 7.5
Third total synthesis of epothilone A by Schinzer *et al.*

As shown before, the major stereochemical issue in epothilone chemistry is the absolute stereocontrol of the triad C6–C7–C8. In the Schinzer synthesis, a six-membered chiral acetonide of type **15** containing the (*S*)-configuration at C3 was used. Addition of α-chiral aldehyde **7** (X = C) to the preformed enolate (generated with LDA in THF at −78 °C) resulted in the formation of the (6*R*,7*S*)-diastereomer **16** as a the major isomer in ratio of 25:1 in 76% yield [17].

In a synthesis of oxa-epothilones, Schinzer used aldehyde **7** (X = O) in a diastereoselective aldol reaction which exclusively yielded aldol product **16** (X = O) [18].

Schinzer then performed the same reaction with α-chiral aldehyde **17**. Remarkably, this reaction provided again a single isomer **18** with the "natural" epothilone configuration in 49% yield [19]. The fully functionalized aldehyde **19** from Schinzer's epothilone B synthesis was also used in the same aldol process. However, a 10:1 mixture of diastereomers was obtained with the "correct" stereochemistry as the major product **20** [20]. In the meantime, Schinzer's chiral enolate **15** has been used extensively by other groups in academia [21] and in the pharmaceutical industry to synthesize a large number of analogues [22]. Most of the examples with highly func-

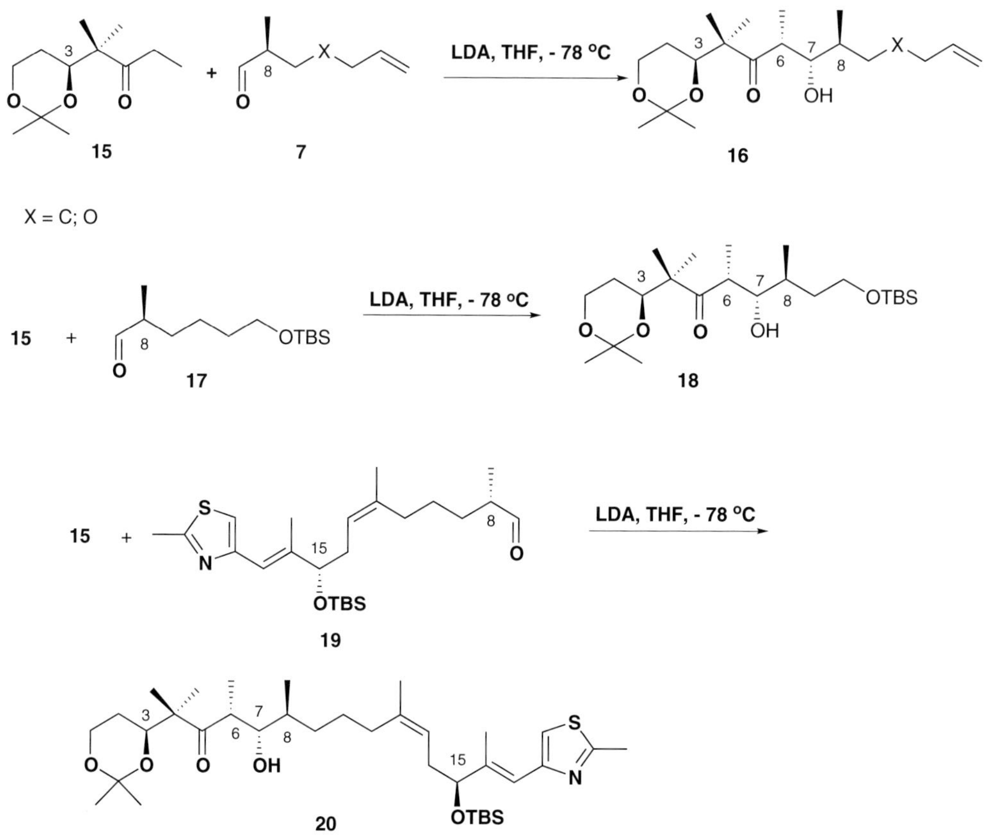

Scheme 7.6
Selective double stereodifferentiating aldol reactions

tionalized chiral aldehydes generated single isomers in high chemical yields.

These outstanding results are connected to the nature of the acetonide, because the lithium counter ion attached to the enolate oxygen can chelate to one oxygen of the acetonide, as shown in Scheme 7.7. The chelated structure **21** was calculated with force field (MM+) and quantum mechanics (PM3) methods, both methods showed minima in favor of the chelated enolate **21** [19].

Thus, with these results he was able to explain the stereochemical results. There are two aspects to understanding these remarkable observations. The simple diastereoselectivity of the aldol reaction can be explained based on Heathcock's studies [23], in which the enolate geometry is governed by the bulkiness of the group next to the ketone (in the Schinzer case a modified *t*-butyl group bearing an extra oxygen). This should result in the generation of a (*Z*)-lithium enolate, further stabilized by the oxygen at C3, forming a

Scheme 7.7
Chelated nucleophile **21** with Zimmerman-Traxler TS.

rigid bicyclic structure **21**. The diastereofacial selectivity to explain the stereochemical outcome in aldol condensations with α-chiral aldehydes was first proposed by Cram [24] and later on by Felkin [25]. Both models give the same products even though totally different reactive conformations are used. In Schinzer's case, the aldol addition occurred with the opposite sense of diastereofacial addition to the theoretical models predicted.

Intrigued by these results the Schinzer group decided to study this particular aldol reaction in detail. First, they conducted experiments with so-called simple aldehydes containing no chiral center to check the inherent selectivity of the chiral enolate. The reaction generated only the two *syn*-type diastereomeric aldol products **26** and **27**, as shown in table 1 and no *anti*-

Tab. 7.1
Aldol reactions with achiral aldehydes

R	*26*	*:*	*27*	*Yield [%]*
Ph	4	:	1	81
Pr	5	:	1	63
i-Pr	5	:	1	59
t-Bu	3	:	1	49
1-Heptenyl	5	:	1	74

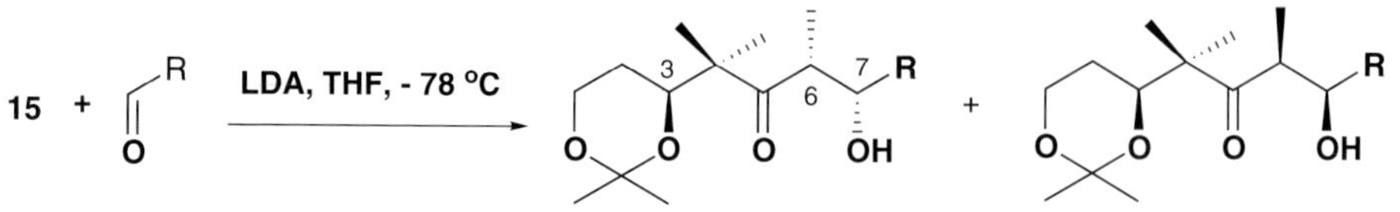

Scheme 7.8
Inherent selectivity of nucleophile **15**

aldol product was observed at all. In all cases studied, the 6(*R*),7(*S*)-isomer **26** was the predominant diastereomer. The inherent selectivity of the chiral enolate is in the range of about 4:1 [26].

Next, the double-stereodifferentiating [27] aldol process with α-chiral aldehydes was examined, seperate experiments using both enantiomeric forms of phenyl propionaldehyde were undertaken. Use of (*S*)-phenylpropionaldehyde **28** gave two diastereomers **29** and **30** in a ratio of 2.5:1 and 68% yield. The major one is still the *anti*-Cram/Felkin compound **29**, which is quite unusual because **28** usually provides high selectivity in favor of the Cram/Felkin adduct **30** [28].

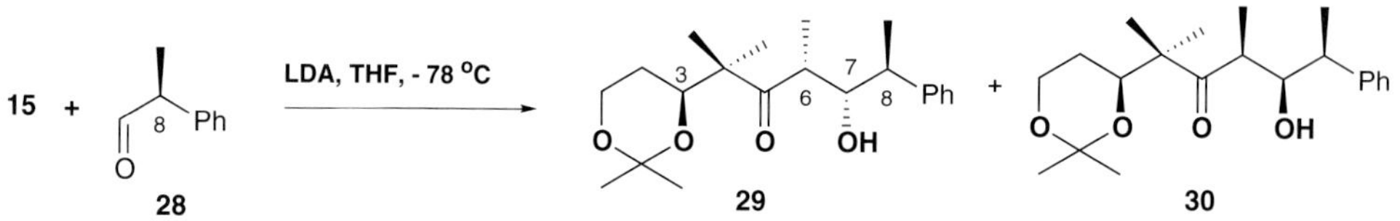

Scheme 7.9
Aldol reaction with (*S*)-phenylpropionaldehyde

This is a very good example where the inherent selectivity of the enolate overrides the inherent selectivity of the aldehyde with the 6(*R*),7(*S*)-configuration resulting from double stereodifferentiation using the aldehyde with the (*S*)-configuration. On the other hand, the same reaction with (*R*)-phenylpropionaldehyde **31** yielded two diastereomers **32** and **33** in a ratio of 40:1 and 77% yield as shown in Scheme 7.10 [26].

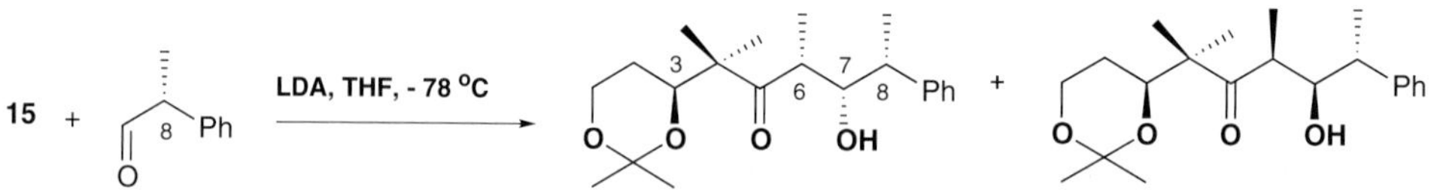

Scheme 7.10
Aldol reaction with (*R*)-phenylpropionaldehyde

Scheme 7.11
Double stereodifferentiating aldol reactions with (*R*)-nucleophile **34**

The major isomer was the Cram/Felkin product **32** with the 6(*R*),7(*S*),8(*R*)-configuration which clearly indicated that in the (*S*),(*R*)-combination of chirality both effects work in the same direction and produce a matching case with very high selectivity. The same holds for the use of the (*R*)-enolate system as shown in Scheme 7.11 [26].

In contrast, α-chiral aldehydes of type **7** show the opposite behavior, favoring the *anti*-Cram/Felkin product as the major isomer. In these cases, the matched case is the (*S*),(*S*)- or (*R*),(*R*)-combination of chirality benefiting of both chiral elements.

These correlation studies indicate a dominating influence of the chiral enolate **15** versus the chirality of the aldehyde. In the case of (*S*)-phenylpropionaldehyde **28** the chirality of enolate **15** overrides the directing effect of the aldehyde chirality, still producing a 2.5:1 ratio in favor of the 6(*R*),7(*S*),8(*R*)-stereochemical triad. The use of the (*R*)-phenylpropionaldehyde **31** forms exclusively the 6(*R*),7(*S*),8(*R*)-stereo chemical triad in a 40:1 ratio. In this case the matched pair is the (*S*)/(*R*)-combination of chirality. The (*R*)-enantiomer of the enolate reverses the stereochemical outcome.

In connection with a synthesis of 6-desmethyl-epothilones Schinzer used methyl ketone **38** instead of the ethyl ketone **15** to study the influence of the missing methyl group. As seen in Scheme 7.12, the selectivity dropped to 1.7:1 still favoring diastereomer **39** with the (*S*)-configuration at C7 [29]. The lower selectivity can be easily explained by the lack of 1,3-diaxial interactions in the chair-like Zimmerman-Traxler transition state model [30].

7.3
Model Study via Chelation Control in the Aldol Reaction by Kalesse

Kalesse *et al.* [31] described an aldol reaction where a silyloxy protected ethyl ketone **41** is coupled to an β-oxa-aldehyde **41** based on Roche acid [32] (Scheme

LDA, THF, - 78 °C

67%

38 + 7 → 39 + 40

Scheme 7.12
Double stereodifferentiating aldol reactions with methyl ketone **37**

7.13). This is a very useful example in connection with the model studies to approach the epothilones.

As a result of a chelation-controlled aldol reaction exclusive formation of compound **43** with the desired 6(*R*),7(*S*),8(*R*)-triad was observed. The structure has been confirmed by X-ray analysis after transformation to a crystalline six-membered lactone. In their macro-lactonizsation strategy to synthesize epothilone B Nicolaou *et al.* used the same ketone **41** in an aldol reaction with aldehyde **19**. The diastereoselectivity in the aldol addition was only moderate, giving a ratio of 3:1 in favor of the "natural" epothilone configuration [33].

OTBS OTBS O + OBzl → LDA, THF, - 78 °C → OTBS OTBS O OH OBzl

41 42 43

Scheme 7.13
Chelation-controlled aldol reaction by Kalesse *et al.*

7.3.1
Different Aldol Strategies: Mulzer's Total Syntheses of Epothilones B and D

A series of similar aldol reactions were carried out in Mulzer's lab. In his study towards the total synthesis of epothilone B and D Mulzer *et al.* used first an achiral ethyl ketone of type **44** in an aldol reaction with the known (*S*)-aldehyde **45**. In this particular case only the chiral element at **44** directed the addition giving a 4:1 mixture of diastereomers **46** and **47**. The major di-

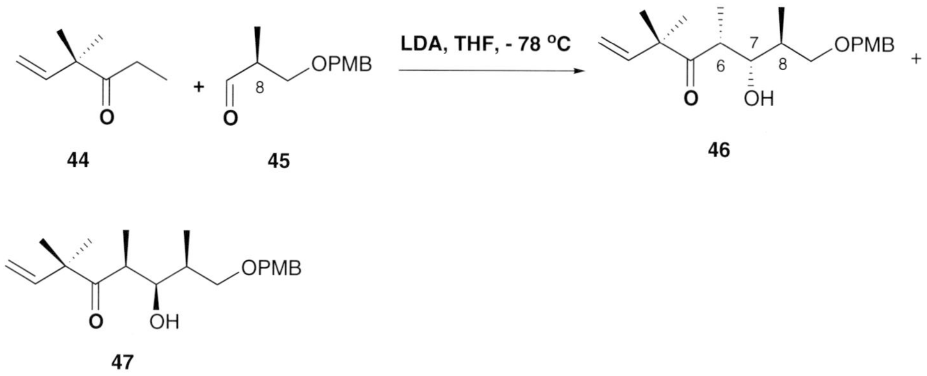

Scheme 7.14
Chelation-controlled aldol reaction by Mulzer *et al.*

astereomer with the (*R*)-configuration at C6 and the (*S*)-configuration at C7 is formed as a result of a chelated Cram-type aldol reaction.

In a later study Mulzer presented examples of double stereodifferentiating aldol reactions with (*S*)-C3 protected nucleophiles [35]. The same double TBSO-protected (*S*)-ethyl ketone **41** used before by Kalesse *et al.* gave 6:1 ratio in an aldol reaction with an α-(*S*)-chiral aldehyde **48** as a result of matching chirality (70% yield). Again, the major isomer **49** had the "natural" epothilone configuration at C6 and C7.

A further modification of the nucleophile **50** yielded the almost exclusive formation (19:1; 92% yield) of the "correct" diastereomer **52** in the presence of the sensitive epoxide function [36].

Both examples shown in Schemes 7.15 and 7.16 are aldol reactions with

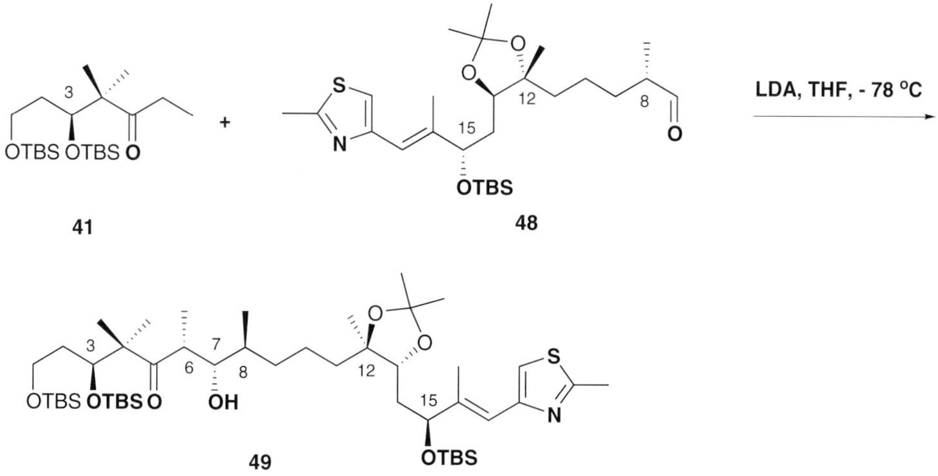

Scheme 7.15
Double stereodifferentiating aldol reactions by Mulzer *et al.*

Scheme 7.16
Double stereodifferentiating aldol reactions in the presence of the epoxide functionality by Mulzer *et al.*

(*S*)-chiral nucleophiles protected at C3. The stereochemical outcome can most likely be explained by the cyclic transition states discussed in Scheme 7.7.

All types of nucleophiles protected at C3 show a predominant effect in driving the aldol reaction in the direction of favoring the *anti*-Cram/Felkin product. This characteristic seems to be independent of the nature of the aldehyde. The best selectivities were obtained with nucleophile **50** and even superior to that, nucleophile **15** protected as an acetonide. The final part of this chapter will focus on a special type of aldehyde where a long range effect improves the stereochemical outcome of the aldol reaction.

7.4
Long-range Structural Effects on the Stereochemistry of Aldol Reactions

In a study connected with an improved synthesis of epothilone B and deoxyepothilone F, Danishefsky *et al.* conducted aldol reactions with functionalized ethyl ketones and an γ,δ-unsaturated aldehyde. Two major studies were carried out in Danishefsky's lab: One with an achiral system [37] and one with a chiral enolate bearing the correct oxidation level at C1 and a TBS-protected alcohol at C3 [38].

First, in a model study Danishefsky *et al.* investigated the reaction of a new type of enolate **53** with several aldehydes. The typical Cram/Felkin-directing (*S*)-phenyl-propionaldehyde **28** gave the expected all-*syn* orientated compound **54** (Scheme 7.17, **54**/**55** = 11:1).

t-BuO OTES O O 53 + O 8 Ph 28 LDA, THF, - 78 °C t-BuO O OTES O OH Ph 54 +

t-BuO O OTES O 6 7 8 OH Ph 55

Scheme 7.17
Model study from Danishefsky *et al.*

A quite interesting trend was observed with other α-chiral aldehydes. (*S*)-Methyl pentane aldehyde **56** (R = ethyl) behaved only slightly more selectively generating a diastereomeric mixture **57** and **58** of 1:1.3. Unsaturated groups improved the selectivity to about 1:2–1:5.5 (see Scheme 7.18). Danishefsky explained these results by a special long range effect connected to the distance between the formyl and the terminal olefin group via a non-bonding interaction of the carbonyl group of the enolate and the olefin group. The maximum effect can be obtained with a γ,δ-unsaturated aldehyde **56** (R = allyl; 1:5.5).

t-BuO O OTES O 53 + O 8 R 56 LDA, THF, - 78 °C t-BuO O OTES O OH R 57 +

t-BuO O OTES O 6 7 8 OH R 58

R = ethyl, Ph, allyl, dimethylvinyl, OBzl

Scheme 7.18
Aldol reactions of **53** with various aldehydes

In a further series of experiments Danishefsky employed chiral ketones of type **59**. Both enantiomers were available with high optical purity and could be involved in investigations in the double stereodifferentiating aldol reaction. However, the lithium anion of **59** (R = TBS) could not be effected in useful yield due to the sensitivity of the β-silyloxy system to elimination. The less basic titanium enolate of **59** gave mixtures of diastereomers in moderate yields. The stereochemical outcome of these reactions showed that the configuration at C3 rather than C8 had a larger effect on the newly

Scheme 7.19
Double stereodifferentiating aldol reaction by Danishefsky *et al.*

formed centers at C6 and C7. However, the selectivity of about 3.5:1 was quite low (even lower than the result shown in Scheme 7.18), indicating that neither of the enantiomers of **59** and **60** benefited from matching chirality.

The difficulty in the lithium series of these aldol reactions caused Danishefsky to investigate the lithium dianion **59** (R = H) as the nucleophile. Indeed, these reactions did not induce β-elimination and provided quite unexpected results with a high level of stereoselectivity [38].

When the lithium dianion of (*S*)-**59** (R = H) was treated with (*S*)-**60** a 2:3 mixture of diastereomers **61** and **62** was obtained in 53% yield, corresponding to the (*S*)-configuration at C6 and the (*R*)-configuration at C7 as the major diastereomer. Thus, the major isomer **62** corresponded to the "unnatural" stereochemistry of the epothilones. The minor isomer presented the required "natural" configuration of the epothilones. When the same reaction was then performed with the lithium dianion of (*R*)-**63** (R = H) and the (*S*)-configurated aldehyde **60** it gave rise to a single diastereomer **64** that contained the "natural" configuration of the epothilones at C6 and C7. This is a result of the matching chirality of the two chiral reaction partners, but represented a major drawback because the center at C3 proved to be (*R*)-configurated. Therefore, it could not be used as a precursor for epothilone syntheses without a troublesome inversion at C3.

Despite that, these undesired results from the dianion aldol reactions can be used to understand some principles of the double stereodifferentiating reactions. The C3 protected series, such as Schinzer-type aldol reactions described before required the (*S*)-configuration at C3 to establish the (*R*)-configuration at C6 and the (*S*)-configuration at C7 with a similar α-chiral (*S*)-aldehyde. The matching chirality in the protected series corresponds to the mismatched case in the unprotected series of these double stereodifferentiating aldol reactions. This disparity could be a result of different transition states.

Danishefsky proposed five factors governing such aldol reactions: (i) the chair-like transition state leading to a *syn*-aldol, (ii) chelation of the lithium counter ion by the β-oxygen at C3, (iii) a *syn*-relationship of the α-proton of the aldehyde and the methyl group at C6 of the enolate, (iv) *anti*-attack of

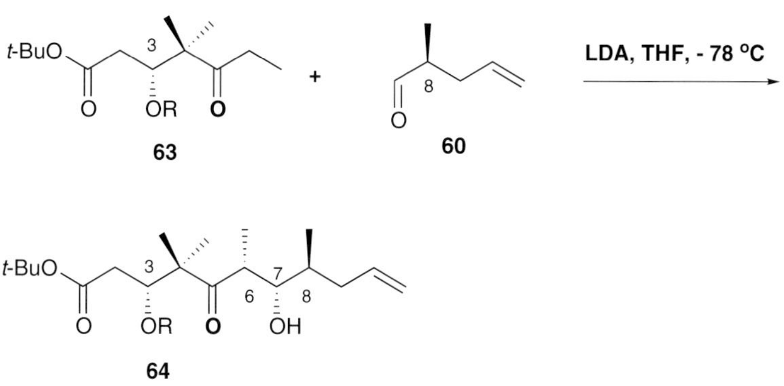

Scheme 7.20
Double stereodifferentiating aldol reaction by Danishefsky *et al.*

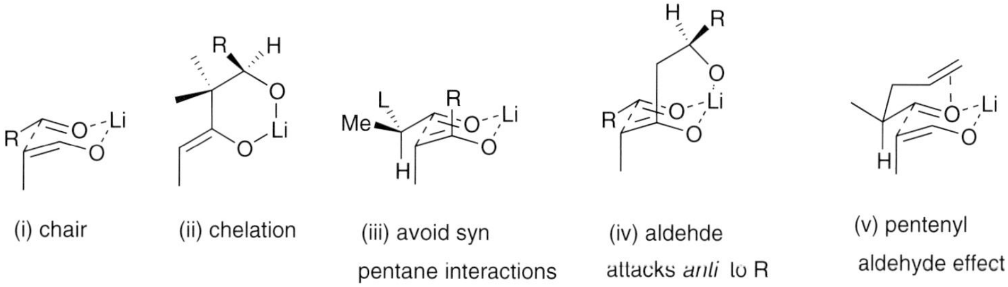

Scheme 7.21
Factors governing the aldol reaction

the aldehyde *versus* the large group of the enolate, and (v) the special γ,δ-unsaturated aldehyde effect [38].

As a result of all these energy-lowering factors, Danishefsky came up with the following transition state model leading to **64** [38].

A final experiment was designed to check the possibility of an internal kinetic resolution. For this purpose, a racemic mixture of the nucleophile

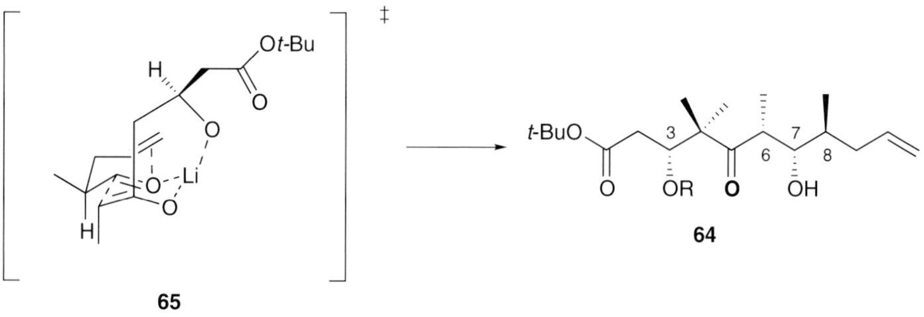

Scheme 7.22
Transition state of Danishefsky-type aldol reaction

59 (R = H) was reacted with homochiral (*S*)-**60**. The reaction generated only diastereomer **64** which supported the concept of matching and mismatching issues at the kinetic level [38].

7.5
Summary and Conclusion

This short chapter gave some insight in the troublesome and quite complicated aldol reaction in the context of epothilone chemistry. All the examples presented were related to the major stereochemical issue around this fascinating class of natural products that is the absolute control of the stereochemical triad C6–C7–C8. Many groups have contributed to that particular stereochemical problem because of the importance of this class of natural products an and optimal solution to this problem obviously is highly desirable. In the meantime, many analogues of the natural epothilones have been synthesized by academic groups and pharmaceutical companies – some of these are even more potent in biological systems than the natural compounds. This could lead to important developments for new *anti*-cancer drugs in the near future. To our knowledge, the simple acetonide-protected ethyl ketone is the backbone of many industry-based analog programs to achieve the central aldol coupling. This ketone is known as a simple and robust coupling partner yielding exclusively the desired stereochemical outcome with a large number of chiral aldehydes – even on multi kilogram scale.

Typical procedure [17]:

Synthesis of (4*R*,5*S*,6*S*,4′*S*)-2-(2,2-dimethyl-1,3-dioxan-4-yl)-5-hydroxy-2,4,6-trimethyl-10-undecen-3-one 16: A solution of ethyl ketone **15** (1.17 g, 5.45 mmol) in THF (1.0 mL) was added to a freshly prepared solution of LDA [*n*BuLi (3.34 mL, 1.6 M solution in hexanes, 5.35 mmol, 0.98 equiv) was added to a solution of diisopropylamine (749 μL, 5.35 mmol) in THF (4.0 mL) at 0 °C] dropwise at −78 °C. The solution was stirred for 1 h at −78 °C. Aldehyde **7** (688 mg, 5.45 mmol, 1.0 equiv) was added dropwise and stirring was continued for 45 min at −78 °C. The reaction mixture was quenched by dropwise addition of saturated aqueous NH_4Cl solution at −78 °C. The organic layer was separated and the aqueous layer was extracted with Et_2O. The combined extracts were dried over $MgSO_4$ and concentrated in vacuo. Flash chromatography (pentane/Et_2O = 10:1) of the residue afforded *anti*-Cram aldol product **16** (1.36 g, 73%) and Cram aldol product (57 mg, 3%) as colorless oils.

References

1 G. Höfle, N. Bedorf, H. Reichenbach, (GBF), DE-4138042, **1993** [Chem. Abstr. **1993**, 120, 52841].

2 K. Gerth, N. Bedorf, G. Höfle, H. Irschik, H. Reichenbach, *J. Antibiot.* **1996**, *49*, 560–563.

3 H. Reichenbach, G. Höfle, in: Biotechnical Advances, **1993**, *11*, 219–277.

4 G. Höfle, N. Bedorf, H. Steinmetz, D. Schomburg, K. Gerth, H. Reichenbach, *Angew. Chem.* **1996**, *108*, 1671–1673; *Angew. Chem. Int. Ed. Engl.* **1996**, *35*, 1567–1569.

5 a) D. M. Bollag, P. A. McQueney, J. Zhu, O. Jensens, L. Koupal, J. Liesch, M. E. Goetz, C. Lazarides, M. Woods, *Cancer Res.* **1995**, *55*, 2325–2333; b) R. J. Kowalski, P. Giannakakou, E. Hamel, *J. Biol. Chem.* **1997**, *272*, 2534–2541.

6 Moos, P. J.; Fitzpatrick, *F. A. Proc. Natl. Acad. Sci. USA* **1998**, *95*, 3896–3901.

7 Pratt, W. B.; Buddon, R. W.; Ensminger, W. D.; Maybaum, J. *The Anticancer Drugs*, Oxford University Press, Oxford, **1994**.

8 D. Schinzer, *Eur. Chem. Chron.* **1996**, *1*, 7–10; b) M. Kalesse, *Eur. Chem. Chron.* **1997**, *2*, 7–11; c) L. Wessjohann, *Angew. Chem.* **1997**, *109*, 739–742; *Angew. Chem. Int. Ed. Engl.* **1997**, *36*, 739–742; d) K. C. Nicolaou, F. Roschangar, D. Vourloumis, *Angew. Chem.* **1998**, *110*, 2120–2153; *Angew. Chem. Int. Ed. Engl.* **1998**, *37*, 2014–2045.

9 Wartmann, M.; Altmann, K.-H. *Curr. Med. Chem.* **2002**, *2*, 123; b) Harris, C. R.; Kuduk, S. D.; Danishefsky, S. J. *Chemistry for the 21st Century* **2001**, *8*.

10 A. Balog, D. Meng, T. Kamenecka, P. Bertinato, D.-S. Su, E. J. Sorensen, S. J. Danishefsky, *Angew. Chem.* **1996**, *108*, 2976–2978; *Angew. Chem. Int. Ed. Engl.* **1996**, *35*, 2801–2803.

11 Z. Yang, Y. He, D. Vourloumis, H. Vallberg, K. C. Nicolaou, *Angew. Chem.* **1997**, *109*, 170–172; *Angew. Chem. Int. Ed. Engl.* **1997**, *36*, 166–168.

12 D. Schinzer, A. Limberg, A. Bauer, O. M. Böhm, M. Cordes, *Angew. Chem.* **1997**, *109*, 543–544; *Angew. Chem. Int. Ed. Engl.* **1997**, *36*, 523–524.

13 N. Miyaura, A. Suzuki, *Chem. Rev.* **1995**, *95*, 2457.

14 K. C. Nicolaou, Y. He, D. Vourloumis, H. Vallberg, Z. Yang, *Angew. Chem.* **1996**, *108*, 2554–2556; *Angew. Chem. Int. Ed. Engl.* **1996**, *35*, 2399–2401.

15 K. C. Nicolaou, Y. He, D. Vourloumis, H. Vallberg, F. Roschangar, F. Sarabia, S. Ninkovic, Z. Zang, J. I. Trujillo, *J. Am. Chem. Soc.* **1997**, *119*, 7960–7973.

16 D. Schinzer, A. Limberg, O. M. Böhm, *Chem. Eur. J.* **1996**, *2*, 1477–1482.

17 D. Schinzer, A. Limberg, A. Bauer, O. M. Böhm, *Chem. Eur. J.* **1999**, *5*, 2483–2491.

18 F. Stuhlmann, unpublished results, University of Magdeburg.

19 A. Limberg, PhD thesis, Technical University of Braunschweig, **1998**.

20 a) D. Schinzer, A. Bauer, J. Schieber, *Chem. Eur. J.* **1999**, *9*, 2492–2500; b) D. Schinzer, A. Bauer, J. Schieber, *Synlett.* **1998**, 861–864.

21 a) R. E. Taylor, Y. Chen, *Org. Lett.* **2001**, *3*, 2221–2224; b) R. E. Taylor, G. M. Galvin, K. A. Hilfiker, Y. Chen, *J. Org.*

Chem. **1998**, *63*, 9580–9583; c) R. E. Taylor, J. D. Haley, *Tetrahedron Lett.* **1997**, *38*, 2061–2064; d) N. Yoshikawa, Y. M. A. Yamada, J. Das, H. Sasai, M. Shibasaki, *J. Am. Chem. Soc.* **1999**, *121*, 4168–4178; e) D. Sawada, M. Shibasaki, *Angew. Chem.* **2000**, *112*, 215–219; *Angew. Chem. Int. Ed. Engl.* **2000**, *39*, 209–213; f) D. Sawada, M. Kanai, M. Shibasaki, *J. Am. Chem. Soc.* **2000**, *122*, 10521–10532.

22 a) K.-H. Altmann, *Mini-Reviews in Medicinal Chemistry* **2003**, *3*, 149; b) U. Klar, W. Skuballa, B. Buchmann, W. Schwede, T. Bunte, J. Hoffmann, R. B. Lichtner, *ACS Symposium Series* **2001**, *796*, 131.

23 a) C. H. Heathcock, *Science*, **1981**, *214*, 395; b) C. H. Heathcock, C. T. White, *J. Am. Chem. Soc.* **1979**, *101*, 7076–7077; c) C. H. Heathcock, M. C. Pirrung, C. T. Buse, J. P. Hagen, D. S. Young, J. E. Sohn, *J. Am. Chem. Soc.* **1979**, *101*, 7077–7079; d) C. H. Heathcock, C. T. Buse, W. A. Kleschick, M. C. Pirrung, J. E. Sohn, J. Lampe, *J. Org. Chem.* **1980**, *45*, 1066–1081.

24 a) D. J. Cram, F. A. Abd Elhafez, *J. Am. Chem. Soc.* **1952**, *74*, 5828–5835; b) D. J. Cram, K. R. Kopecky, *J. Am. Chem. Soc.* **1959**, *81*, 2748–2755.

25 a) M. Cherest, H. Felkin, N. Prudent, *Tetrahedron Lett.* **1968**, 2199; b) N. T. Anh, O. Eisenstein, *Nouv. J. Chem.* **1977**, *1*, 61.

26 Muharram, planned PhD thesis, University of Magdeburg, 2004.

27 a) S. Masamune, W. Choy, F. A. J. Kerdesky, B. Imperiali, *J. Am. Chem. Soc.* **1981**, *103*, 1566–1568; b) C. H. Heathcock, C. T. White, *J. Am. Chem. Soc.* **1979**, *101*, 7076–7077; c) S. Masamune, W. Choy, J. S. Petersen, R. L. Sita, *Angew. Chem.* **1985**, *97*, 1–31.

28 C. H. Heathcock in *Asymmetric Synthesis, Vol 3* (Ed. J. D. Morrison), Academic Press, New York, **1984**, 111.

29 O. M. Böhm, University of Magdeburg, unpublished results.

30 H. E. Zimmerman, M. D. Traxler, *J. Am. Chem. Soc.* **1957**, *79*, 1920–1923.

31 E. Claus, A. Pahl, P. G. Jones, H. M. Meyer, M. Kalesse, *Tetrahedron Lett.* **1997**, *38*, 1359–1362.

32 K. Gerlach, M. Quitschalle, M. Kalesse, *Tetrahedron Lett.* **1999**, *40*, 3553.

33 K. C. Nicolaou, S. Nincovic, F. Sarabia, D. Vourloumis, Y. He, H. Vallberg, M. R. V. Finlay, Z. Yang, *J. Am. Chem. Soc.* **1997**, *119*, 7974–7991.

34 J. Mulzer, A. Mantoulidis, E. Öhler, *Tetrahedron Lett.* **1998**, *39*, 8633–8636.

35 J. Mulzer, G. Karig, P. Pojarliev, *Tetrahedron Lett.* **2000**, *41*, 7635–7638.

36 H. J. Martin, M. Drescher, J. Mulzer, *Angew. Chem.* **2000**, *112*, 591–593; *Angew. Chem. Int. Ed. Engl.* **2000**, *39*, 581–583.

37 A. Balog, C. Harris, K. Savin, X.-G. Zhang, T.-C. Chou, S. J. Danishefsky, *Angew. Chem.* **1998**, *110*, 2821–2824; *Angew. Chem. Int. Ed. Engl.* **1998**, *37*, 2675–2678.

38 C. B. Lee, Z. Wu, F. Zhang, M. D. Chappel, S. J. Stachel, T.-C. Chou, Y. Guan, S. J. Danishefsky, *J. Am. Chem. Soc.* **2001**, *123*, 5249–5259

Index

Numbers in front of the page numbers refer to Volumes 1 and 2: e.g., 2/250 refers to page 250 in volume 2

a

b

Modern Aldol Reactions. Vol. 1: Enolates, Organocatalysis, Biocatalysis and Natural Product Synthesis.
Edited by Rainer Mahrwald

ISBN: 3-527-30714-1